D1420891

Adhesive Bonding

Adhesive Bonding

TECHNIQUES AND APPLICATIONS

Charles V. Cagle

Epoxylite Corporation
South El Monte, California

McGraw-Hill Book Company

New York Toronto London Sydney

ADHESIVE BONDING

 Library of Congress Catalog Card Number 68-16167

09586

1234567890 MAMM 7543210698

To Glenda, Michael, and Valerie

Preface

This textbook was written because a definite need existed for a book geared to the basic fundamentals pertinent to adhesive bonding, and because of the encouragement of my very good friend Dr. Henry Lee. Having personally experienced the need for a book in this particular area, I accepted the challenge to write such a text. I am hopeful that the by-product of that decision will fill a gap which exists between the highly scientific publications available and the fine books that resulted from various adhesive symposiums, which do not have the continuity that is desirable for a good instruction text.

I have attempted to present factual information, in an honest and simple manner, which would be of interest to various adhesive activities. This book is geared primarily to the needs of the shop man, technician, or professional engineer who is not familiar with adhesive bonding.

Much emphasis has been placed on cleanliness, surface preparation, and training, which, in my opinion, hold the golden key to success in adhesive bonding. If I can enlighten but a small number, my two years of work will not have been in vain.

It would be impossible for a single individual to write a text of this nature without the cooperation and assistance of others. After visiting approximately 35 U. S. firms and posing rather embarrassing questions to their leadership, I received some rather frank and disturbing answers. The results of these visits and observations fully supported and confirmed my own thoughts on the subject which were formulated as a result of 14

Adhesive Bonding

CHAPTER 1

Background of Adhesives

An adhesive is defined as a substance capable of holding material together by surface attachment. This is a general definition and includes terms such as glue, mucilage, cement, and paste. Various descriptive adjectives are applied to the term "adhesive" to indicate certain characteristics. This may indicate physical form, e.g., liquid or tape adhesive. Chemical type is sometimes indicated, that is, silicate adhesive or resin adhesive. The material may be disclosed, e.g., metal-to-metal adhesive, paper adhesive, or label adhesive.

Adhesives are one of the oldest joining techniques, but adhesive technology progressed very little until the twentieth century. Records show that adhesives were used over three thousand years ago. Adhesives derived from bitumen and tar pits were used in early structures. They are known to have been used as mortar by the builders of the Tower of Babel. The Egyptians used glue formulated from tree resin and eggs. A Biblical example of the knowledge of adhesives is recorded in the twenty-second chapter of Ecclesiastes. It is indicated here that proper selection of an adhesive must be made to be compatible with the substrate being joined. Engineers are still struggling with problems similar to the one recorded in this Biblical verse: "He that teacheth a fool is like one that glueth potsherd together." (Potsherd was a fragment or broken piece of earthenware.)

The adhesive engineer would be foolish to use an acetate where the part was to be subjected to high humidity or a polyamide where design requirements specified extremely high temperatures. Care must be exercised when selecting an adhesive to do a specific job.

There was very little advancement in adhesive technology until the twentieth century. Great strides were made in metal bonding due to the military requirements of World War II. The early metal-bonding adhesives combined phenolic resins with neoprene and nitrile rubber, which produced a tough material with good peel and shear strengths. In 1950, experimentation began in earnest with the epoxy formulations, which offered equal strength properties. In general, they were preferred due to the low solvent content and 100 percent reactive ingredients.

The adhesive industry has grown at a rapid rate since the early fifties, but the major increases are reflected in the last five years. As an example, it is estimated that 7 million lb of adhesives were consumed in 1966. This is approximately a 15 percent increase over 1965, and a moderate estimate would be a 15 percent increase in 1967 as related to consumption in 1966. Many would challenge that figure as being conservative.

The field of structural adhesives will continue to grow. New formulations are in constant demand due to the fact that there is no universal adhesive. Design engineers are searching for the lightest possible design, with structures consistent with structural integrity.

The many types and classes have varying properties. New uses for adhesives are being discovered. Since there is no universal adhesive, the competition runs high and the rewards are rich to those engaged in advancing adhesive technology.

One of the smaller companies in the United States reports 36,000 problems a year confront their staff. A few of these problems are:

1. A North Carolina manufacturer wants a crystal-clear foam-rubber adhesive which dries rapidly leaving no depression tack and an invisible seam.

2. A New York contracting firm needs a waterproof coating to prevent rain and moisture from penetrating concrete and cinder blocks.

3. The United States needs a nontoxic cement for bonding contact lenses on sharks' eyes.

Today, approximately 175 manufacturers are listed as adhesive formulators. The brands and numbers run into the thousands. A single individual can only be acquainted with a small number of these varied formulations. The adhesive industry has become a highly specialized profession requiring full-time attention of several thousand individuals in the United States alone. The requirement for an adhesive engineer is much more than a degree in chemistry and a tube of cement. To predict accurately how an adhesive will react in a certain environmental condition without former experience in that area is impossible without first testing the material to that particular condition. Strong and permanent adhesive bonds are obtained only if a number of factors are favorable. Regardless of how reliable the adhesive appears under laboratory conditions, chances are remote that the same results will be produced in the shop. The designer may have recommended a system that is adequate for a particular assembly; all other conditions may be acceptable, but the shopman or technician has the final responsibility of producing a reliable assembly. A poor job at shop level dooms the assembly to failure. The shop function is surely not the largest problem in producing a good part –it is only *one* of the problems.

The adhesives industry has found its place in many industries and will surely spread to many other fields.

ADHESIVES IN THE BUILDING INDUSTRY

1. Ceramic tile adhesives; usually rubber-base mastics with a volatile solvent or with water as a vehicle.
2. Plastic tile adhesives; usually oil-resin, resin, or rubber-base with water vehicle.
3. Metal tile adhesives; same types of adhesives as for plastic tile.
4. Hardboard adhesives; rubber-based mastics with solvent vehicle, resin-based mastics, oil-resin-base mastics.
5. Plywood or wood-paneling adhesives; rubber-base mastics with solvent vehicle or synthetic rubber with solvent or water vehicle, known as "contact-bond adhesive."
6. Gypsum wallboard; casein or dextrine adhesives, rubber-base mastics, contact cements.
7. Cementatious material adhesives; repair of existing concrete surfaces, bonding new work to old.
8. Floor coverings; resilient floor tiles–mastics of various types.
9. Woodblock floors; rubber mastic-solvent vehicle, rubber mastic-water dispersion.
10. Ceramic floor tiles.
11. Sinks and counter tops.
12. Roof adhesives; rubber-base liquid adhesives with nonflammable solvents and fire-snuffing ingredient.
13. Ceiling tiles; resin-base mastics, rubber-base mastics, contact cements.
14. Insulation; rubber-base adhesives.
15. Preparation of sandwich panels for prefabrication-type structures. The mobile home industry uses a massive quantity of honeycomb panels fabricated with adhesives.

ADHESIVES IN THE ELECTRICAL INDUSTRY

Many of the demands placed upon adhesives in the electrical industry are comparable to those encountered in other structural adhesive applications. An added consideration is that electrical as well as mechanical failure may occur. The operating temperatures of modern electrical equipment often eliminate the use of practically all common thermoplastic resins as adhesives; thermosetting resin adhesives are commonly used.

For use in electrical equipment an adhesive must possess some or all of the following properties: good electrical qualities, such as high volume resistivity and low dielectric constant; chemical resistance; moisture resistance; low odor; good tracking resistance; radiation resistance.

Adhesives may be used in the fabrication of the following types of electrical equipment: transformers, switchgear components, capacitors, microwave devices, motors, generators, and insulators.

ADHESIVES IN THE AUTOMOBILE INDUSTRY

A number of different types of adhesives are currently used in the automotive industry. Thermosetting or structural adhesives are employed primarily for bonding brake linings and transmission bands. Thermoplastic adhesives are used primarily for bonding door weatherstrips and trim materials. A third type of material is sealants and sound deadeners.

Brake-shoe bonding was one of the earliest large-scale applications of thermosetting adhesives in the automobile industry. One manufacturer has bonded over 100 million brake shoes since 1949 without a single major brake failure attributable to the bonding process. The major advantages of bonded brake linings are the longer life gained by wearing the friction material down to the metal shoe rather than just to the rivet heads, the increased area obtained by eliminating rivet holes and chamfers, and reduction of the possibility of scoring drums by rivets. Rubber-resin and modified resin adhesives are used in bonding brake lining, the adhesive in widest use being nitrile rubber-phenolic resin.

Bonding of glass to metal in vent windows, movable side windows, and roll-type back lights requires adhesives with structural characteristics. The adhesive must be thermosetting for permanence, yet must allow sufficient flexibility to accommodate the different thermal expansion coefficients of glass and steel. The most successful adhesive for this purpose appears to be a nitrile rubber-epoxy resin cement.

Bonding of metal body sections is still the subject of study in Detroit but many plastic bodies, such as those used in the Chevrolet Corvette, have been adhesive bonded.

Themoplastic adhesives are used for the attachment of sponge-rubber weatherstripping as door seals, as trim attachment adhesives for fabrics, silencer pads, and floor mats.

The primary requirement for sealers and sound deadeners is that they adhere well and remain in place for the life of the car. Generally, the cohesive strength of such materials is low compared to that of adhesives.

Potential applications of adhesives in the automobile industry include the replacement of some spot welding, particularly in outer panels of deck lids, hoods, and doors or in the attachment of the roof to the roofside-rail. They also include bonding of roof bows to the car roof, bonding of transmission parts which cannot be easily welded or riveted, and for attachment of trim, body "soldering," and seam and body sealers.

ADHESIVES IN THE AIRCRAFT INDUSTRY

The advent of lightweight construction in aircraft design about 25 years ago, with its corresponding use of thin-gage metal and honeycomb construction, demanded the development of new concepts in methods of joining and fastening. In England the challenge was accepted and resulted in the development of the first structural adhesive, REDUX, used for joining load-bearing components in aircraft. Shortly thereafter the first commercial aircraft adhesive was produced in the United States. This adhesive was derived from a combination of a phenolic resin and a neoprene rubber. This type of adhesive is still in use today in modern jet aircraft such as the KC-135.

Not only did adhesive bonds succeed in efficiently transferring loads from one component to another, but they did this so efficiently that former problems associated with stress concentration in and around rivets, with resulting fatigue cracking, was dramatically reduced. The result was increased service life. The value of the stress distributing properties of a bonded joint was realized. Looking backward, the record is clear; there is no evidence of structural failure in aircraft due to failure of adhesive joints. The same cannot be said for riveted metal components.

In addition to their better strength and durability, adhesives also exhibit a cost advantage over riveting. A net fabrication savings of 13¢ per sq ft was realized for a typical external surface area by use of adhesive bonding instead of riveting. In addition, with adhesives 0.020-in. aluminum sheet could be used, whereas 0.051-in. aluminum sheet had to be used with rivets. This resulted in a further cost reduction and, of course, a great saving in weight.

fig. 1-1 *The U. S. Air Force C-5A transport. (Lockheed-Georgia Company, Marietta, Ga.)*

The supersonic transports and cargo carriers, such as the C-5A under development by the Lockheed-Georgia Company, will utilize a massive amount of adhesives (Fig. 1-1).

ADHESIVES IN THE AEROSPACE INDUSTRY

In the aerospace industry, adhesives really come into their own. It is safe to say that none of our present missiles and vehicles could exist

fig. 1-2 *The Surveyor spacecraft. (Hughes Aircraft Company, Culver City, Calif.)*

without the use of adhesives. For example, the attachment of an ablative heat shield, made of plastic material, to a metallic substructure can only be accomplished with adhesives, since welding is not possible and conventional methods, such as riveting or bolting, would result in these metallic fasteners being burned away during the fiery heat of reentry.

Adhesives have been found which can withstand the extremes of heat and cold encountered during a space voyage and still have sufficient strength to stand up to the stresses placed upon them during voyage. The Surveyor spacecraft utilizes many various types of adhesives combined with complex joint design (Fig. 1-2). The use of adhesives resulted in weight reduction and a large payload of instrumentation. To meet design requirements in weight reduction, fiberglass and plastics joined to metallics were used in many areas; thus adhesives were the best possible approach.

To accomplish a soft-landing on the moon, a landing gear was designed utilizing a honeycomb bonded crushable structure that would absorb the impact on landing (Fig. 1-3).

Hughes Aircraft Company has developed two missiles that are prime examples of the utilization of adhesives in the missile industry. The AIM-47 missile is a plastic structure designed for use with the Lockheed YF-12A interceptor, and the AIM-54A Phoenix is for the Navy's F-111B (Fig. 1-4). The Phoenix is a metallic structure utilizing a cork composite, adhesive-bonded ablative outer surface.

A masterpiece pertinent to adhesive bonding is shown in Fig. 1-5. Adhesives are used extensively in the fabrication of detector panels on

fig. 1-3 *Landing mechanism of Surveyor spacecraft. (Hughes Aircraft Company, Culver City, Calif.)*

fig. 1-4 *The AIM-47 missile (left) and AIM-54A missile (right). (Hughes Aircraft Company, Culver City, Calif.)*

fig. 1-5 *Detector panels for MMC micrometeoroid detection satellite. (Fairchild-Hiller Corporation, Hagerstown, Md.)*

the 90-ft wing of the MMC micrometeroid detection satellite, developed by the Fairchild-Hiller Corporation.

THE NEED TO KNOW

Many companies now realize the importance of training in the area of adhesive structural bonding. Some companies still have individuals in high positions that tend to overlook this problem. Many companies spend thousands of dollars to certify people in the welding parameter. It seems there exists no real problem in training people to drive rivets and install bolts, but just mention adhesives—no, no, no, it isn't necessary. Yet the very same companies scrap out several thousand dollars' worth of assemblies a month. If a rivet is not properly installed, the majority of installation errors can be corrected by the removal and replacement method. What about an adhesive-bonded honeycomb panel? Scrap.

The greatest asset in the adhesive-bonding class is motivation among the people involved. When people are instructed beyond the surface, and interest is shown in their work, interest on their part is sure to follow. The reason is evident. People do not train for the adhesive profession, they are usually drafted. Many are unhappy; they prefer some other area. They must be shown how important their job really is.

The United States, as well as the rest of the world, needs a professional organization for adhesive engineers. This would accomplish much in obtaining the recognition they deserve. This would help spread the knowledge accumulated by the individual engineer or the company he represents. An organization of this type would also aid in obtaining certification for people who shoulder the responsibility of building a sound structural assembly. A large percent of shop problems can be traced directly to processes.

Many factors influence the choice of fastening mode. Adhesives offer certain advantages over conventional methods of joining. The advantages are as follows:

1. *Allows fabrication of smoother parts.* They do not break through or deform the surface of an assembly. This is important aerodynamically for exterior air-frame, missile, or space structures. The reduction of drag reduces the injecture flight profile. This is also advantageous for such items as electronic modules where close or sliding fits are required. Where naturally smooth surfaces are a requirement, adhesives will eliminate any grinding and filling operations.

2. *Permits use of lighter weight materials.* Bonded thin sheets do not fail in bearing; on the contrary, their full strength can be realized. Because the entire area is attached, adhesive bonding minimizes the stress concentrations that commonly occur with screws, bolts, rivets, and

spotwelds. Uniform stress distribution provides greater strength and rigidity in the assembly, which may make it possible to reduce the thickness of the material or eliminate overall size.

One aircraft manufacturer reports 450 lb saved on the engine cowlings alone by changing to a honeycomb configuration.

3. *Serves as a vibration dampener.* Better stress distribution also means better fatigue resistance under vibration loads. Structural adhesives are capable of transferring, distributing, and absorbing stresses so that metals often fatigue before the adhesives. They may often be used in conjunction with rivets and bolts.

4. *Joins dissimilar substrates.* Many times, corrosion problems caused by their electromotive-series relationship occur when dissimilar metals are joined by conventional methods. The adhesive layer isolates the two materials, eliminating this problem. Adhesive bonding also provides a satisfactory method for joining metal to plastics. Solid propellants are successfully bonded to steel cases. This is the type of area in which adhesive bonding is really feasible (Fig. 1-6).

5. *Permits easier fabrication of unique contours.* Many complex and unique contoured surfaces may be adhesive bonded satisfactorily that would be very difficult to join with conventional joining methods.

6. *Acts as a seal.* Since the adhesive bond is continuous, it can be a seal against liquids and also against many gases.

7. *May be utilized as an insulator or conductor.* With the proper adhesive, the bond becomes an effective electrical insulator. Many adhesives are specifically formulated to act as conductors. Thus, a choice of dielectric or conductive properties is available across the substrate.

fig. 1-6 *(a) Structural adhesive uses; (b) machinery adhesive uses. (Loctite Corporation, Newington, Conn.)*

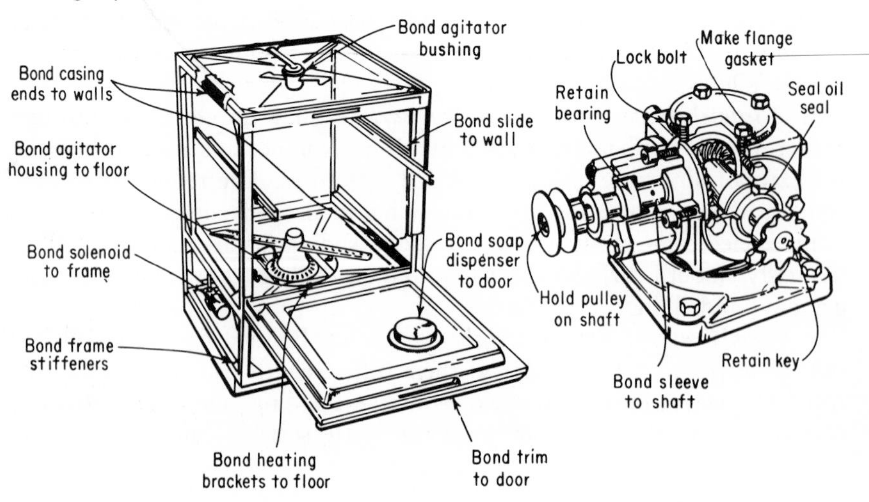

8. *Heavy-gage materials bond to thin substrates.* The assembly of heavy-gage metal to thin sheet metal by welding is difficult. Metals cannot be welded satisfactorily to non-metals. Adhesive bonding can be used to join such abstract combinations.

9. *May reduce cost.* This is a question that is hard to evaluate. It usually depends on the tooling required and the number of parts to be fabricated. Cost is not considered the primary reason for the utilization of adhesive bonding, but savings can be realized on large scale production, especially metal-to-metal assemblies (Fig. 1-7). It is significant that a large surface area can be bonded in approximately the same man-hours as required for one-half the area, whereas a riveted or spotwelded surface requires man hours proportionate to the size of the surface area.

Looking at the other side of the coin, it is evident that adhesives have many disadvantages and limitations. Some of these factors are listed as follows:

1. *Cure cycles.* Where high strengths are required at elevated temperatures, the adhesives are frequently cured at elevated temperatures—most conveniently in an oven, press, or autoclave. This may not be practical or desirable with certain materials, part sizes, or configurations.

2. *Critical processes.* To produce joints of optimum expectations, the processes become critical. For example, surface preparation is one of the more critical areas and requires the use of experienced personnel, expert handling techniques, and constant scrutiny by quality control.

3. *Environmental exposures.* Although moderate exposures to common fluids (such as water, lubricants, oil, and solvents) have little effect upon bond strengths, this varies with the type of adhesive and the length of exposure. Adhesives are usually considered to be weakened by exposure to sunlight and heat. Care must be taken to select a bonding process based on knowledge of the effects of environmental exposure in the joints. This is an area in which tests must be conducted to ensure a wise adhesive selection pertinent to a bonded assembly.

4. *Joint design.* Care must be taken to design the joints to assure that stresses will be in the direction where the adhesive bond is the

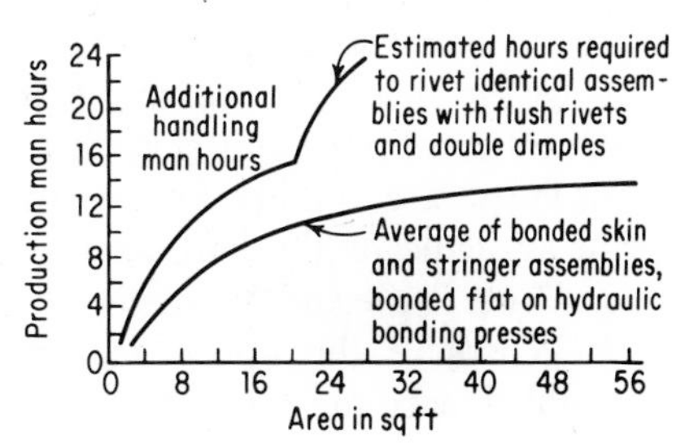

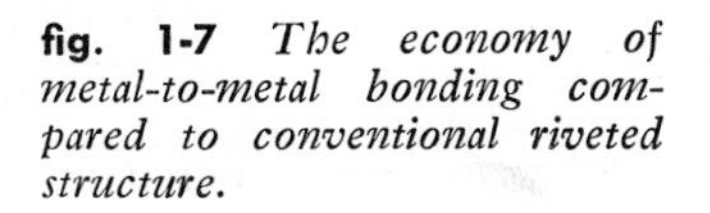
fig. 1-7 *The economy of metal-to-metal bonding compared to conventional riveted structure.*

strongest and to minimize the stresses in its weak direction. The primary idea here is that the design of the next assemblies may be influenced and must therefore be considered in relation to the design of the entire structure. Subsequent bonding operations must be considered, as many adhesives would deteriorate under static heat age conditions. It would be unwise to bond a subassembly with an adhesive that would deteriorate when subject to more cure cycles in later major assembly bonding operations.

5. *Service life expectation.* The life of the part must be considered. This includes environment and life expectation. Service life of many adhesives may have a shorter expected duration than is desirable for the end product.

6. *Solvent solution.* Solvents that are present in a great number of adhesives may tend to deteriorate the adherends. This can be controlled with careful selections based on knowledge of the substrate to be joined, the use of primers and conversion coatings, and a careful analysis of the adhesive-system candidates.

7. *Quality control and quality assurance more difficult.* The quality analysis of a bonded part is more difficult than one joined with conventional fasteners. The process controls must be rigidly adhered to and the assembly inspected at various stages during fabrication. The use of nondestructive techniques has increased the chances of satisfactory determination of quality.

8. *Trained personnel.* One of the major problems associated with adhesive bonding is the lack of trained personnel in the field. In the early adhesive movement, no emphasis was placed on training. The idea seemed to be, "just stick it together." The companies that neglected training were doomed to failure. As structures became more complex and the technology increased, the need for training became apparent. Today, although a time-consuming and expensive item, it is a universal consensus that well-trained personnel is a basic requirement for success in adhesives.

Before entering into the discussion relative to basic adhesive bonding, the fact is evident that careful consideration must be given to a part to be bonded. The following questions should be answered before the design concept originates.

1. What materials are to be joined? What particular alloys? What hardness conditions and surface finishes?

2. What are the thicknesses of the adherends? Is there a sketch or print available or the part or assembly?

3. What is the end use of the part? Is it a subassembly?

4. What are the upper and lower temperatures which the joint or assembled part must be able to withstand? Will these temperatures be

constant or intermittent? For what periods of time will these temperatures be seen during the service life of the joint?

5. Will the joint be exposed to a humid environment? If so, how humid, how long, and at what temperature?

6. To what qualification tests will the part be subjected? What military specifications govern the part, if any?

7. What chemical solvents, oils, and other fluids will come in contact with the bond? What temperature and what type of exposure?

8. Will the part need electrical continuity? If so, how much? Would the part be an insulator? What are the requirements as to volume resistivity, dielectric strength, etc.?

9. What strengths in the joint are required? What tensile, shear, peel, compression, impact, vibration, or other stresses must the joint withstand? What minimum bond-strength values are permissable for the upper and lower limits?

10. Is there any reason to prefer a tape adhesive over a paste system? Why? Is this true only for some portion or for the entire assembly?

11. What cure conditions of temperature and pressure can be tolerated by the bonded part or assembly?

12. Are there any conditions present that might limit the type of bonding tools or fixtures used?

13. What will be the bonding sequence and the recure conditions?

14. What type of quality control will be effective? This is determined by the complexity of the design concept, the size and mobility of the part, and the materials to be joined.

When these questions are adequately answered, four basic general factors must be evaluated. These are: (1) the materials involved, (2) the job requirement, (3) the joint design, and (4) the cost. The joint design will be discussed in a later chapter.

Cost is analyzed as follows: Cost involves adhesive preparation for use in the operation, waste and probable rejected parts, personnel needed for inspection and production, and also the adaptability of present bonding equipment. The equipment and means available to produce bonds of desired strength and performance levels is a most important cost factor. The method of adhesive application and time involved must be considered.

FUTURE EXPECTATION

The future advancement and consumption are practically beyond mental comprehension. Even today, as one's surroundings are observed, the use of adhesives, sealants, and coatings are associated with almost every product that is marketed. The average individual consumed approximately 35 lb of adhesives in 1966, if the total consumption in the United States is divided by the total population. The phenomenal growth in the adhesive industry must be prepared for and met by management of the manufacturing firms related to the field.

The fabrication of aircraft and the space vehicles will unveil adhesive systems with better performance characteristics. This text will not necessarily be limited to aerospace applications, but will be oriented to applications in that area. The aerospace industry has provided a large amount of information pertinent to adhesives because of the amount of research that was necessary to design and fabricate hardware with good performance and high reliability.

In the future, one can expect high-temperature adhesives with stability up to 800°F and possibly 1200°F. The indications are that adhesives that cure at lower temperature and less pressure will be available with even better properties than those in current use. The low temperature (–400°F) will be improved, especially in the area of handling and processing. The trend will be toward the development of adhesives that have a wider temperature operating range and that are convenient for handling more film types, and one-part adhesives will be marketed. Studies and research will be directed toward adhesives with better moisture and fluid resistance. It could also be assumed that the prices of the newer systems will decrease as volume increases and competition becomes keener, thus making adhesives more attractive to new fields of manufacturing and production. The newer systems, increased production volume, and virgin areas of utilization will also bring about more rigid process controls and the necessity for more, better trained, and better informed personnel.

An important point to ponder is what qualifications well-trained adhesive personnel must possess. At the present time, the profession is not as specialized as many others are. For example, design is a problem area because many engineers who were accustomed to designing with conventional fasteners find themselves designing with adhesives. If they are not fully aware of other aspects of the entire operation and the general design concepts of adhesive-joint stresses, problems are sure to follow.

The materials and process engineers must play an important role, even in design. They must consider tooling engineering as they are faced with difficult tolerances and often with critical pressure and heat requirements. Quality control faces problems that are not always visual but must be controlled by in-process inspection and nondestructive testing. This necessitates better trained floor inspectors equipped with well-documented procedures and knowledge of nondestructive testing techniques, many of which are complex.

The adhesive field requires that the worker have knowledge of many fields to be proficient and effective. At present, many companies are limited to a small number of people who are competent in the adhesives industry. These individuals must have some knowledge of chemistry to be fully aware of the adhesive's general properties and reactions. They must have a general idea concerning joint design and loading. They should have knowledge of tooling and mechanical testing equipment, which necessitates a knowledge of mathematics, physics, and mechanical engineering. They must have knowledge of metallurgy with the many surface treatments, conversion coatings, and exotic metals now in use. A knowledge of electronics and ultrasonics is a tremendous asset, due to the expanding field of nondestructive testing. The well-trained adhesives man is usually an instructor; he must be able to impress upon people the importance of their mission. He should be a salesman capable of selling the skeptics on adhesives and promoting confidence in their use.

In summary, the adhesive field will expand enormously. More complex systems will be available, coupled with more exotic metals and substrates to be joined. Assemblies will become larger, new equipment will present itself, and more complex test methods will be imposed. Better ultrasonic test methods as well as other nondestructive testing methods, which are in their embryo stage, will be in contention. The advances that lie ahead will certainly dwarf present technology. The positive approach to success will be better-trained personnel in all areas of this rapidly expanding field.

REFERENCES

Acquisition and Education, *Adhesives Age* (Editorial), November, 1966.

Austin, J. E. and L. C. Jackson: Management: Teach Your Engineers to Design Better With Adhesives, *SAE Journal,* October, 1961.

"Basic Adhesive Bonding Course," AVCO Corporation, Lowell, Massachusetts.

Been, J. L.: Adhesive Bonding, *Product Engineering,* May, 1955.

Cagle, C. V.: "Adhesive Bonding Training Manual," Marshall Space Flight Center, Huntsville, Ala., January, 1965.

DeLollis, N. J.: "The Use of Adhesive and Sealants in Electronics," Scandia Corporation, December, 1965.

Hoskins, C.: Formula to Prosper By: Diversification + Expansion + New Products, *Production,* May, 1966.

Kuno, J. K.: "Comparison of Adhesive Classes for Structural Bonding at Ultra-high and Cryogenic Temperature Extremes," Whittaker Corporation, February, 1964.

Losoncy, Jr., W. A., R. E. Greenlee, and C. Williams: "Structural Adhesives for Metal to Metal Bonding," U.S. Department of Commerce.

Merriam, J. C.: Adhesive Bonding, *Materials and Design Engineering*, September, 1959.

"The Place of Metal Bonding in Modern Aircraft Structures," Bulletin 158, Bonded Structures Limited, Cambridge, England, 1960.

Riley, M. W.: Adhesive Bonding: State-of-the-Art, *Materials and Design Engineering*, December, 1961.

Twiss, S. B.: Adhesives of Tomorrow, *Adhesives Age*, January, 1966.

"Wanted, Adhesive Problems," Adhesive Products Corporation, 1965.

Will They Replace Other Forms of Joining?, *Welding Engineer*, September, 1964.

Winter, D. E.: Adhesive Bonding, *Aircraft Production*, vol. 21, no. 4, 1958.

Wright, J. P.: Adhesive Bonding Methods, *American Machinist*, June, 1958.

Yates, H. R. M.: "Structural Adhesives for Metals and Plastics," presented at the University of Wisconsin, Madison, January, 1962.

CHAPTER 2

Adhesive Properties and General Characteristics

An adhesive or formulation is generally a mixture of several materials. The extent of mixture and the ratio usually depend upon the properties desired in the final bonded joint. The basic materials may be defined as those substances which provide the necessary adhesive and binding properties.

Solvents are employed in many systems to provide vehicle and viscosity control. In some cases, low-molecular-weight resins of high fluidity are added to the basic resin to help control viscosity.

Fillers such as metallic oxides, mineral powders and various fibers are sometimes used to control reinforcement, decrease shrinkage, lower the coefficient of thermal expansion, control temperature operating ranges, and—in some instances—provide a more satisfactory system for a special environmental condition. Fillers are also used to control viscosity, especially if a thixotropic paste is desired. The most common of these are the ultrafine mesh silicas such as Cab-O-Sil. Most fillers also lower the cost of the system. They also prevent waste by virtue of improving the handling properties. They are often referred to as "extenders."

Catalysts and hardeners are employed to activate the resin systems, especially where thermosetting resins are concerned, in order to speed up hardening and make the adhesive system practical. Acids, bases, salts, alcohols, sulfur compounds, and peroxides are a few of the basic catalyst materials. The selection must be based upon a knowledge of the mechanics of polymerization reactions which account for the curing or hardening of adhesives. Catalysts are very important in forming the final joint. The amount of catalysis is critical. Overcatalyzing may result in a poor joint, and the same holds true for undercatalyzing.

There are several classes, types, and groups of adhesives. These have been classified as to use, chemical composition, mode of application, setting factors, vehicle, etc. The first general classifications to be considered are structural and nonstructural adhesives. These classifications are sometimes difficult to clarify. A structural adhesive would normally be defined as one which can be employed where joints or load-carrying members associated with primary design are required. This type of adhesive will be subjected to large stress loads. The term "structural bonded joints" equates "structural" with the importance of its mission. In this concept, a further definition may be required where "primary structural" means loss of the aircraft or vehicle through joint failure and "secondary structural" means severe damage and impairment of the mission. The criteria in many cases have been defined on the basis of bond strength using the arbitrary top value strength of 1000 psi.

This is considered by many as a very poor definition, and since many people have disagreed on these terms, this discussion is included only to raise the question and allow the individual concerned to draw his own conclusion. It must be considered because a large portion of companies have specifications placed in these general categories. The major problem is that they all differ in context.

Nonstructural adhesives are not capable of supporting appreciable loads and are generally required to locate parts in an assembly. They will be employed many times where only a temporary bond is required. Their failure would not usually result in the loss of a vehicle. Adhesives, sealants, and coatings usually fall into this category, but could be responsible for the full accomplishment of the mission.

The type of adhesive material is easier to define and usually falls into three categories:

a. Thermosetting
b. Thermoplastic
c. Elastomeric

Thermosetting resins are synthetic organic substances which can be converted by chemical reaction into a permanently hard, practically infusible, and insoluble solid. These resins are high-molecular-weight polymers which react by polymerization to form hard substances, usually rigid and possessing high strength properties. Thermosetting resins usually have a high modulus of elasticity, do not support combustion, and resist the action of most chemicals. When reacted, the thermosetting system will not be liquified by heat but will deteriorate or decompose under heat ranges beyond its limitations. We might compare this to the baking of bread. Once it is catalyzed (baking powder), further baking will only burn it.

The thermoplastic resins are often employed in metal and plastic bonding and usually adhere well to both. They do not lend themselves to use as good load-bearing adhesives, especially if they would be subjected to elevated temperatures. They will soften when heated and harden when cooled. An example here would be butter, placed in a molten liquid state by heating and becoming solid upon cooling.

The more common thermoplastic resins include the polyvinyls, acrylics, polystyrenes, cellulosics, and polyamides. They are sometimes used effectively with thermosetting resins for specific formulations.

Elastomeric resins are used widely for modification of the thermosetting systems. They generally fall into a distinct class; e.g., natural and synthetic rubber. A true elastomer is usually defined as a material that will stretch twice its original length without inherent loss of elastic properties. When used as a modifying agent for other resins, they usually induce flexibility and increase peel strength of the systems. They are often used alone or in slightly modified form for sealants, but lack the strength to be used alone for structural applications. Examples of this class are the butyls, nitriles, polysulfides, and neoprenes.

The remainder of this chapter will be devoted to chemical types. This enables the chemist to define adhesives a step farther and place them into chemical classes.

1. Epoxies

A thermosetting system, 100 percent reactive when in a pure state, the epoxies are very desirable and more widely used than any other chemical type. Epoxy is one of the newer types and has penetrated more fields of manufacturing operations in a shorter space of time than any of its predecessors. The epoxies have been formulated from more materials than any other class. They are very versatile and can be formulated to do any job, limited only by heat. Some formulations will withstand 800°F for short periods. The heat ranges of the epoxies are usually determined by the catalysts utilized to harden the system. The many catalysts used with epoxies produce systems of variable properties. The most common are the aromatic amines and cyclic anhydrides. The amines produce the low-temperature cure cycles and limited heat range, while the anhydrides usually require higher cure cycles and withstand higher operational temperatures. Table 2.1 shows the general properties of a basic epoxy resin hardened by various catalysts with 301 stainless steel adherends.

Epoxies are available in liquid, paste, and film forms (supported and

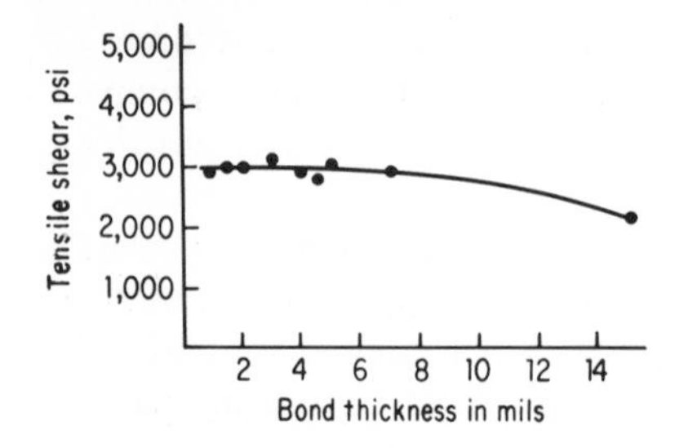

fig. 2-1 *Effect of bond-line thickness on an epoxy.*

unsupported). The two-component systems are more widely used because of the extended shelf life. They may be stored for long periods and naturally will not activate until mixed. A few of the early epoxies were one part, in a stick form that was heated prior to application, which proved to be impractical from an application standpoint. Epoxy adhesives are not widely used in the film form unless they are modified, that is, alloyed with another adhesive system.

The epoxies are not affected by bond-line thickness as compared to many other structural adhesives (Fig. 2–1). This is important for application and processing, because the epoxies require very little pressure, and become very fluid when heated prior to the gel or "B" stage. The thin bond line is preferred, but is sometimes difficult to control with a paste or liquid. If bond-line control is essential, it may be accomplished by utilizing glass beads of the desired size in the resin, which do not adversely affect the mechanical strength unless used excessively. The bond-line control is one of the prime advantages of the film-type adhesives, especially if a carrier is utilized. When utilizing the epoxy with a carrier, care must be exercised in pressure application, because bond-line starvation will occur, due to the very fluid state of the resin under heat and pressure.

The epoxies have low peel and impact strength as compared to many other structural adhesives because of their brittle nature after cure. To improve the undesirable properties, they are alloyed with various other adhesive systems to produce a system to meet the demand of design requirements.

The field of epoxies is so large and varied, it is impractical to cover in this document. There are many volumes written on epoxies alone. "Handbook of Epoxy Resins," written by Henry Lee and Kris Neville and published by McGraw-Hill Book Company, is recommended for further information.

In summary, these factors should be remembered pertinent to epoxy resins:

1. *Adhesion.* The expoxies have high specific adhesion to metals, glass, plastics, ceramics, paper, concrete, wood, and various other substrates.

Because of their brittle nature, epoxies are not recommended for bonding the rubbers and elastomeric adherends, although they will adhere to these types of materials. The epoxies can be formulated to create mixtures of low viscosity and improved wetting, spreading, and penetrating action. If the substrate to be joined is cleaned and processed properly, adhesion presents very few problems.

2. *Cohesion.* When properly cured, the cohesive properties are considered very good, but are usually the limiting strength factor. The adhesive properties are superior to the cohesive properties in most formulations, thus cohesive failures will be experienced during testing from room temperature to the maximum operating limits of the system.

3. *100 percent solids.* The epoxies in the unmodified state cure without releasing water or other condensation byproducts. This makes them desirable where contact pressures are necessary for manufacturing. They are also convenient for bonding such materials as glass or thermoplastics, where high heat and pressures would be unsatisfactory. This characteristic also makes them desirable as potting compounds, since the possibility of air bubbles or inclusions is reduced. The addition of silver, carbon, or other conductors has proven very successful in varying the electrical properties of epoxies without the problems of discontinuities in the bond line and also without adversely affecting the mechanical properties of the system.

4. *Low shrinkage.* The epoxies cure with only a fraction of the shrinkage of vinyl-type adhesives such as polyesters and acrylics; consequently less strain is built into the glue line, and the bond is stronger. The shrinkage can be reduced to a fraction of 1 percent by incorporation of silica, aluminum oxide, or other organic fillers. A shrinkage factor of 3 percent would be considered extremely high for epoxies.

5. *Low creep.* They maintain their shape under prolonged stress better than thermoplastics and many thermosetting systems. This is an important asset in favor of the use of epoxies, because creep is considered a major problem in structural adhesive bonding, and an area of prime concern by designers. Creep, in all probability, has hampered the use of adhesives and plastics in the building industry more than any other single factor.

6. *Resistance to moisture and solvents.* The epoxies are resistant to moisture. Moisture does not effect an epoxy in the least but will migrate through the joint and deteriorate the substrate. When epoxy bonded joints are subjected to moisture or water immersion, the failures usually occur at the interface. This indicates the importance of proper surface preparation of the adherends. Their resistance to solvents is considered outstanding and accounts for their rapid advancement in the coating field. Because fluids do migrate through an epoxy with little or no effect to

the system the substrate problem does exist, which makes other systems more desirable for use in long-term exposure to such fluids as fuels, although when modified with an elastomeric system, for example, they may possess very desirable properties in these areas.

7. *Versatility applicable to modification.* The properties of the epoxy may be changed by:

a. Varying of the base resin and curing agents.
b. Varying cure cycles, both temperature and cure time.
c. Alloying the compound with another resin.
d. Compounding the various fillers. This may affect the cost factor, but the economics of epoxies are governed more by the type of catalyst utilized.

They are effective barriers to heat and electric current, yet at the same time may be modified easily for conduction of electricity. They are versatile in applying due to their wide range of modification, and may be applied manually, semiautomatically, or automatically.

2. Phenolic adhesives

The phenolics or phenol-formaldehyde resins are formed by the condensation reaction of phenol and formaldehyde. This material was discovered in 1872. The phenolics are very rigid, strong, and have excellent resistance to fungi. They have moderate to good resistance to moisture, and very good high-temperature properties. The phenolic resins have been used extensively in the lamination of plywood and in filament-wound structures. They enjoy a wider range in the structural-adhesive category when alloyed with other materials.

There are two basic classes of phenolic resins: resoles and novalacs, and both begin as phenol alcohols. They are catalyzed with either an acid or an alkali. Regardless of the formulation of phenolic resins, they are considered to have high resistance to deteriorating influences encountered in service. They would not be considered excellent in resistance to stresses caused by thermal expansion, and extenders should not be used in attempts to correct this weakness. When combined or alloyed with other adhesive systems, they become excellent structural adhesives and are widely used in this manner throughout the aerospace industry.

3. Nitrile adhesives

The nitrile rubbers are elastomers and copolymers of unsaturated nitriles and dienes. The nitriles are not used as structural adhesives in this form, but yield many one-part adhesives that are used for bonding small nonstructural parts, especially in the electronics and plastics industries. The nitrile rubbers, when prepared for use as a cement, are milled on tight cold-mill rolls, broken down, and rendered soluble in some type of solvent. The most widely used nitrile rubber adhesives are cured by the solvent escape drying method, but they may be catalyzed by the utilization of sulfur compounds and cured at room or elevated temperatures. The nitriles are available from the manufacturers in a variety of formulations, but the important role of this rubber system for structural adhesives comes as a result of being alloyed or mixed with another resin. The nitriles, like the phenolic resins, do not have the desired properties for structural bonding when used alone, but, for example, if the nitriles and phenolics are combined their mechanical properties change to a system with excellent properties for structural use. The nitriles give the rigid resins flexibility that produces high peel strengths and better than average shear strengths.

4. Vinyl adhesives

The vinyl polymers do not stand alone as a structural adhesive, but hundreds of adhesives are formulated by the use of this class of polymer. Vinyl is the univalent radical $CH_2{:}CH^-$, derived from ethylene, a compound which undergoes polymerization to form high-molecular-weight resins. More generally, the term "vinyl polymer" has been used to include a variety of resins, plastic films, and elastomers obtained by polymerizing monomers having one or more unsaturated double or triple bonds, including diolefins, such as butadiene, vinyldienes such as vinyldiene chloride or methyl methacaylate, and unsaturated compounds such as maleic anhydride.

The vinyls are important to adhesive bonding not only from the adhesive standpoint, but because the films derived from these substances are widely used as vacuum bags, slip sheets, etc. The more widely used ones are polyvinyl chloride, polyvinyl alcohol, and polyvinyl fluoride.

5. Neoprene

Neoprene was the first synthetic elastomer developed that possessed properties comparable to natural rubber. It is defined as an oil-resistant synthetic rubber obtained by polymerizing chloroprene. The neoprenes were limited in use due to the cost factor until the shortage of natural rubber in World War II. At that time, neoprene was the only synthetic rubber available for use in adhesives and as a result, formulators began to experiment with it. They found that neoprene adhesives were just as good, if not better in many cases, than those based on natural rubber. The neoprenes are used in three general capacities in the adhesive industry. They are used structurally when alloyed with another resin, as a rubber cement, and as a noncuring "tacking" paste. The neoprene cements are usually dispersed in solvents such as toluene, which is one of the more widely used. It may be dissolved in mixtures of aromatic and aliphatic hydrocarbons.

The neoprene cements may be cured at room temperature or by the use of heat, depending upon the accelerator used. Magnesium and zinc oxides are two of the more common accelerators which effect a slow, room-temperature cure.

The maximum operating temperature does not usually exceed 170°F and would show signs of degradation if used at that temperature for long periods of time. As neoprene ages, traces of hydrochloric acid are formed by decomposition of the chlorine-containing molecules; this acid tends to deteriorate most fabrics such as cotton, rayon, linen, etc.

Neoprenes are used extensively in the shoe industry and moderately in the automobile industry for bonding weatherstripping. They are used in the aerospace industry for bonding rubbers and plastics where high strength is not required.

6. Polyurethanes

A wide variety of polyurethanes can be formed by crosslinking highly reactive isocyanates with various polyols. This elastomeric material provides a bond which resists not only the shear and tensile stresses satisfactorily but has very high impact resistance and excellent cryogenic properties. This has brought them into widespread use in space applications, especially for insulation problems. They are also widely used as sprayable coatings for aircraft wing assemblies, and for bonding solid propellants. They are utilized for bonding metal to metal, elastometers, foam, plastics, nylon, glass, ceramic, and the fluorocarbons. Due to the flow characteristics they are not considered a good material for honeycomb construction.

The cohesive strength is usually better than the adhesive strength, but good cohesive failures are obtained by careful processing with a majority of formulations. They yield from 3000 to 5000 psi in shear at room temperature, but shear strength varies with cure conditions and pressure. There is a correlation between the bond-line thickness and shear strength, the ideal bond-line thickness being in the range of 2 to 6 mils. They will yield up to 8000 psi in shear at −423°F, but are limited to approximately 250°F at elevated temperatures (Fig. 2-2).

The polyurethanes are not considered ideal. They pose processing problems due to their reaction with water and their gaseous nature. The systems that are MOCA catalyzed require hot mixing to diffuse the catalyst into the resin. The ratio of MOCA to resin has varying effects on the final joint and should be carefully controlled. They may be degassed before application, but the amount of degassing affects the pot

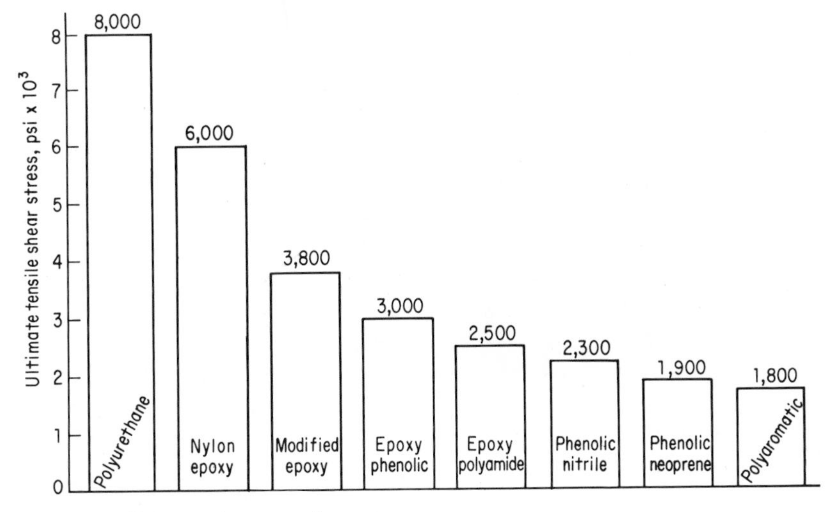

fig. 2-2 *Effects of cryogenics on various adhesives.*

life. When applied before gelling starts, they are very fluid and sometimes tend to cause starved bond lines. This has been controlled in special applications by the addition of 6 to 12 percent of nylon fibers.

Another problem associated with the polyurethane system applies to storage. Storage must be maintained that will inhibit fractional crystallization; in fact, it should inhibit any crystallization and water accrual in the raw material.

In summary, the polyurethanes are excellent cryogenic materials, exhibit excellent shock properties, are more difficult to process than many systems, suffer from excessive creep at room temperature, and show changes in properties on aging, some of which are undesirable. They still hold the answer to cryogenic application, but have poor elevated-temperature strength.

7. Silicones

Silicones are semi-inorganic polymers made up of a skeleton structure of alternate silicone and oxygen atoms with various organic groups attached, and are thermosetting-type resins. A large variety of the RTV (room-temperature vulcanizing) compounds are formulated utilizing the silicone resins. They do not possess the mechanical properties to be used as structural adhesives, but are widely used as sealants and potting compounds. A prime example of the use of silicone rubber adhesives is shown in Fig. 2-3 in a flexible heat exchanger for NASA's Apollo spacecraft (by North American Aviation, Space Division), which consists

fig. 2-3 *Flexible heat exchanger.*

of a series of light copper tubes bonded together into a continuous parallel series with silicone rubber sealant systems. The unit was designed to pass heat from the electronic equipment to the outside radiation system.

The silicones vary in curing temperatures from room temperature to 250°F, depending on the formulation and vulcanizing agent. The majority of these systems require only contact pressures during cure.

The silicones have many very desirable characteristics as listed:

1. Good high-temperature properties. They have good thermal and oxidative stability at temperatures up to 600°F and will withstand short exposures up to 800°F.
2. Silicones are good thermal insulators, which accounts for their utilization as thermal insulation and heat sinks.
3. They have good low-temperature properties when compared to many other systems. The methyl silicones have brittlepoints at –100°F, but the methyl-phenyl silicones may be used to –175°F.
4. They maintain good electrical properties over a wide temperature range.
5. They have adequate resistance to aging and weathering and remain stable when exposed to ozone, corona, and sunlight.
6. They have fair resistance to water and moisture.
7. Generally, all silicones will withstand radiation; however, the most effective group is the silicone resins, followed closely by the silicone rubbers. In all probability, the most outstanding characteristic is their ability to resist combined heat and radiation.
8. The ease of handling and low-temperature cures brand the silicones for future growth.

The silicones are handicapped by low shear strength and many do not possess the adhesion or tack quality level desired. Adhesion may be promoted by the use of primers. They deteriorate under constant contact with fuels, which limits their usage in fuel areas.

8. Polyesters

The reaction of organic acids and alcohols produces a class of materials called "esters." When the acids are polybasic and the alcohols are polyhydric, they can react to form very complex esters. They are usually called "alkyds" and have long been useful as surface coatings and glass-reinforced plastics. This same principle, utilized with various modifications, brings the polyesters into the adhesive field. The polyester adhesive systems cure rigid, and have a temperature operating range up to 500°F. They reveal poor adhesion to metals, especially aluminum.

The polyesters are attacked by most solvents and have a high shrinkage rate when compared to other adhesives. Shrinkage rates may run as high as 4 percent. Attempts are made to combat the high shrink rate by the utilization of fillers such as calcium carbonate and aluminum silicate.

Recently a new polyester has been developed that is flexible. The evaluation of this system is incomplete but indications are that it adheres better to metals than many of the earlier polyester systems.

9. Acrylics

The acrylics are a group of thermoplastic resins formed by polymerizing the esters or amides of acrylic acid. They are usually transparent, low viscosity, polymerizable liquids and were developed primarily for use as liquid locknuts. They are now used as adhesives, but are more important pertinent to structures as transparent sheets (plexiglass and Lucite).

The acrylic adhesives have indefinite shelf life when stored at ambient temperature with access to oxygen. When oxygen is excluded by applying the material in a thin film between two mating surfaces, gelation occurs at room temperature in a matter of minutes. To prevent gelation before application, the liquid is packaged in a low-density polyethylene container permeable to oxygen. Curing may be accelerated by elevating the temperature using an oven, heat lamp, or press. The acrylics have been cured successfully in a vapor degreaser when small details are being joined. The heat causes gelation to occur before the solvent extracts the adhesive from the joint. Perchloroethylene, with a boiling point of 250°F results in a more rapid cure with less leaching than trichloroethylene with a 190°F boiling point. Also, the vapor degreaser removes the thin film of liquid that is kept from curing by contact with the air.

If the liquids are applied to sensitive electromechanical devices, be sure the uncured surface liquid is removed. Outgassing and condensation of volatiles in sealed systems may cause problems in service or storage.

Certain metals, such as zinc, cadmium, and gold, do not promote cure of these materials. For these metals an organometallic activator is supplied in solution in a chlorinated solvent, which is applied directly to the metal and allowed to dry. Zinc- and cadmium-plated surfaces can also be activated by a chromic rinse prior to sealing the surface.

Strength is far too low for the acrylics to be used as structural adhesives in lap joints, but resistance to torque shear is outstanding for joining cams, sleeves, pulleys, and gears to shafts in lieu of conventional fasteners. In this type of application, they have better properties than the epoxies and are easier to use, but are limited in service to temperatures below 200°F. They have been used successfully for bonding films, such as Tedlar, to metals, where high temperatures are not required.

10. Rosin (sometimes called colophony)

Spirit-soluble thermoplastic materials are available in two forms: gum rosin, the more popular form of which is obtained by distillation of the exudation fom pine trees, and wood rosin, which is prepared from pine trees.

Rosin adhesives are used for metal-container labeling either as hot melts or in solvent solution, often with added plasticizers. It is also used in the modification of other resins. Rosin is used in the powder form for bonding wood components in aircraft, but is not usually termed a structural adhesive. It has very good strength, and good water and moisture resistance, but poor resistance to fuels and solvents.

11. Polysulfide rubber adhesives

The polysulfide adhesives are synthetic polymers obtained by the reaction of sodium polysulfide with organic dichlorides such as dichlorodiethyl formal, alone or mixed with ethylene dichloride.

The polysulfides are used as adhesives where high strengths are not required, but are used more often as sealants. They are sometimes used as binders for solid propellants. This system offers good resistance to light, oxygen, oils, and solvents, and impermeability to various gases. They adhere well to almost any adherend, but have poor tensile properties. They exhibit poor properties when subjected to high humidity conditions and have an operating temperature range of —67 to +250°F.

The polysulfides are usually procured in two-component paste or liquid systems, have good shelf life, and require no special storage facilities. They may be catalyzed, mixed, and frozen for several days to eliminate production handling problems. The polysulfides are considered a wise choice (if the service requirements do not exceed their capabilities) because of the economy involved.

12. Ceramic adhesives

A typical formulation of ceramic adhesives may contain silica, sodium nitrate, boric acid, and ferric oxide. These materials are heated above 2000°F, blended, and then crystallized. A hard frit is formed, dried, milled, and then passed through a screen of the desired size to ensure uniform grain size. Oxides and water are then added and the results are ceramic adhesives. The viscosity may be controlled by the amount of water added to the mixture.

In recent years, investigations have been carried out to adapt the ceramic adhesives to structural bonding. They possess high shear strengths up to 1500°F (1800 psi in shear) and may reach 5000 psi in shear at room temperature.

The prime disadvantages are low-peel and flexural strengths coupled with very-high-temperature cure cycles. Much attention has been diverted from the ceramic systems since the newer polyaromatics became a reality. At present, the ceramics are not practical but research is currently in progress to improve the unfavorable properties they possess. The ceramics must be improved to become a sound structural adhesive but show great promise as an encapsulation material for high-temperature rocket nozzles and nose cones.

13. Cyanoacrylate adhesives

A special-purpose proprietary cyanoacrylate adhesive (Eastman 910) is a one-part, clear, watery liquid. It is free of solvents and cures at room temperature in contact with many surfaces without the addition of a catalyst or hardener. The system sets by an anionic polymerization mechanism which is catalyzed by weak bases such as traces of moisture on most surfaces in contact with the atmosphere. Surfaces such as phenolic, polyester, polyethylene, and polystyrene plastics tend to inhibit the curing rate, and may be pretreated with a diluted solution of an activator, phenylethylethanolamine (910 surface activator). However, most common metals, glass, wood, and rubber surfaces bond very rapidly and cure in a matter of minutes.

Shear strengths up to 4000 psi can be obtained, but peel and impact strengths are poor. When exposed to elevated temperatures the mechanical properties are poor, and aging as low as 160°F reflects degradation. When exposed to temperatures above this, the adhesives turn yellow and decompose. The prime advantage is the fast cure, and relatively little effort is required for well-mated joints. It is widely used for bonding small electrical components and as a tacking adhesive; however, it is expensive.

Problems have been encountered on production lines, as the adhesive presents an operator hazard because of its strong and rapid adhesion to the skin. Curing may be slow when the humidity is low, but this problem may easily be solved by placing an open container of water near the parts being bonded, but this material should never be placed in an oven for cure.

14. Polyaromatic adhesives

The polyaromatics are the first commercially available adhesives designed for temperature operating ranges above 600°F. The polyaromatic systems are not ideal, as fabrication problems still exist; however, they show great promise for high-temperature performance.

The two systems that are currently attracting more attention are the polybenzimidazoles (PBI) and polyimides (PI). The polybenzimidazole adhesives are based on polyaromatic compounds which are reaction products of a tetra-amine and bisphenylester. Polyamide adhesives are based on resinous reaction products of a diamine and a dianhydride. These systems are available in both liquid and supported film form, the supported film being more practical as a structural adhesive. The polyaromatic adhesives provide tensile shear strengths comparable to the epoxy-phenolics at temperatures up to 700°F, but have much better heat stability than the epoxy-phenolics and provide better strength retention after long-time exposure to elevated temperatures (Fig. 2-4). These systems open up a new era for structural adhesives for joining beryllium, stainless steel, and titanium, and will be used on future supersonic aircraft, missiles, and space vehicles.

The polyaromatics are utilized for honeycomb structures, metal-to-metal, and metal-to-plastic bonding. They are also used as prepregs and foams. When used in honeycomb structures, perforated honeycomb is essential due to the gas liberation of these systems during cure.

The primary objection to the use of the present polyaromatic systems is the high temperatures required for curing (450 to 700°F) and the high pressures required for a satisfactory bonded joint (50 to 200 psi). In many cases the high-temperature performance is improved by post curing in the

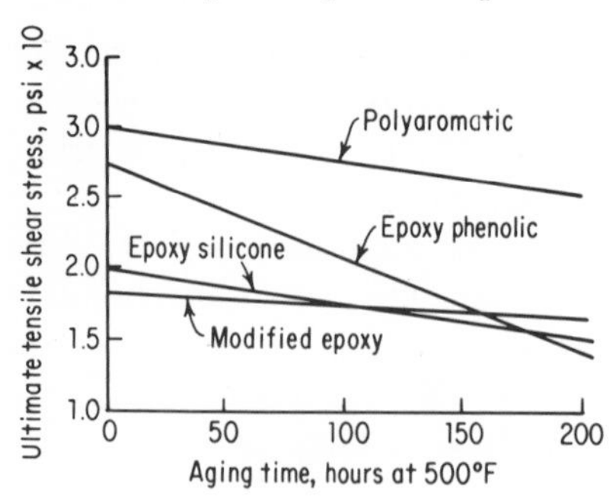

fig. 2-4 *Aging at 500° with various types of adhesives.*

range of 500 to 700°F; this is highly dependent on the substrate being bonded.

The polyaromatics are normally utilized with a primer, which is also a polyaromatic. Long-time aging characteristics are improved by incorporating an inorganic arsenic compound into the system. For this reason, contact with the skin and inhalation of vapors should be avoided. Bonding should be accomplished in enclosed autoclaves or hooded areas so the toxic vapors are removed from the bonding area.

Further data on these adhesives is contained in the book "New Linear Polymers" by Henry Lee and Kris Neville, published by the McGraw-Hill Book Company.

15. Vinyl phenolic adhesives

The vinyl-phenolics were the first adhesives used on a large-scale production operation for bonding sandwich structures. These systems are composed of phenolic resin combined with polyvinyl formal or polyvinyl butral. They are available in the form of emulsions, tape form, or the liquid and powder system, the latter being one of the first used extensively.

The vinyl-phenolics produce high shear strengths (5000 psi) when utilized as a metal-to-metal bonding agent. They produce joints with T-peel strength in the range of 45 to 60 lb. When utilized in sandwich structures, climbing-drum peel tests produce values from 45 to 55 lb, and flatwise tensile strengths of up to 700 psi. They have an operating temperature from -67° to +250°F. They have good impact strength and creep resistance, fair weathering characteristics, and are considered very good from the economic standpoint.

Some of the vinyl-phenolic liquid systems require rigid adherence to prescribed time-temperature-pressure cycles. This often means heating to a given temperature at contact-pressure, then "breathing" the bond line by complete removal of pressure, followed by a steady rate at full bonding pressure until completion of cure. Parts are often coated, open-face dried to boil off solvents, mated, and cured under prescribed pressure.

The vinyl-phenolics are still used in large amounts, but the lower-temperature systems are replacing them very rapidly. Studies show the lower-temperature cured systems offer basically the same properties and offer a great advantage pertinent to the effect on substrates, especially aluminum, as thermal problems are reduced.

16. Neoprene phenolic adhesives

The neoprene phenolics were one of the early structural adhesives developed in the United States. They are used primarily in metal-to-metal applications and produce lap shear values in the 4000 to 6000 psi range. They have good peel properties, excellent resistance to fatigue, good low-temperature characteristics, and good resistance to salt spray, hydraulic oils, moisture, and most solvents.

They are normally used with a primer to increase adhesion and peel strength. Cure cycles are usually in the 300 to 400°F range and pressures from 25 to 200 psi. They do not have good flow characteristics, and thus are not considered good systems for sandwich construction.

17. Epoxy-silicone adhesives

The epoxy-silicones offer good high-temperature characteristics and have proved useful in applications up to 500°F. One of the more commonly used systems is the asbestos fabric supported tape.

The primary disadvantages of this system are low shear strengths (1500 to 2000 psi), poor peel strength (8 to 16 lb), high cure cycles (500 to 600°F), high cure pressures (50 to 100 psi), and that many times, post cures are necessary to produce bonds of desired quality. The epoxy-silicones are not widely used, as better systems are now available.

18. Epoxy-polysulfide adhesives

The polysulfide elastomers were one of the first elastomer modifiers for epoxies used to improve strength, elasticity, and peel strength of the epoxies. This combines the toughness and elastic properties of the polysulfide with the strength of the epoxies. This system will produce shear strengths to 3000 psi and peel strengths in the range of 30 to 60 lb, depending on the percent of polysulfides. They have good resistance to oils, fuels, and solvents, and fair resistance to moisture. They have an operating range of –100 to +300°F. These systems will cure at room or elevated temperatures; the elevated-temperature-cured systems usually have better chemical resistance. They will produce good joints with contact pressure, but may be slightly pressurized during cure.

The epoxy-polysulfides adhere well to most substrates and are used extensively as adhesives and sealants. They are not considered a structural adhesive but are used for details where high strengths are not required.

19. Epoxy-nylon adhesives

The epoxy-nylons are one type of the more widely used structural adhesives. They are available as two-component liquids and as films, both supported and unsupported; the unsupported film being used in larger quantities. They are used for metal-to-metal honeycomb sandwich structures and for various types of plastic adherends. The epoxy-nylons are usually cured at 325 to 350°F with low pressures (10 to 30 psi). They will produce shear strengths up to 7000 psi and peel strengths to 80 lb, and are considered one of the stronger structural adhesives available at room temperature. They are limited to a maximum service temperature of 200°F, but possess good low-temperature and cryogenic properties (Fig. 2-2). They have fair resistance to solvents, excellent impact and shock properties, but deteriorate very rapidly when exposed to moisture or high-humidity conditions. Creep resistance is fair at room temperature but becomes poor when exposed to slightly elevated temperatures.

The epoxy-nylons have good flow properties and wetting characteristics but are affected by bond-line thickness (Fig. 2-5). When properly applied and cured, they exhibit better adhesive than cohesive properties. These systems have produced problems in production that are difficult to detect. In the event of a tooling-pressure problem or dimensional mismatch, the adhesive will actually swell, adhere to the substrates, and fill a gap up to 0.020 in. when utilizing a film 0.010 in. thick. The result is naturally a poor joint, and one subject to much faster degradation when exposed to environmental conditions than the identical joint with a preferred thin glue line (0.0005 to 0.003 in.).

A primer may be utilized with the epoxy-nylons but is not usually considered necessary. In many cases, the primer will become the weak link in the joint. If primer is applied to protect the substrates from cleaning to cure, then it would be considered a worthy venture. The use of a primer will not improve the system at elevated temperatures or improve its moisture resistance or aging characteristics because the problems do not occur at the interface but in the bond line and are of a cohesive nature.

20. Epoxy-phenolic adhesives

The epoxy-phenolic adhesives are used for metal-to-metal sandwich structures, high-temperature plastics, ceramics, and in other areas where the structures will be subjected to elevated temperatures. They retain good shear properties up to 500°F and will withstand short-term exposures up to 1000°F. Long-term heat aging affects mechanical properties, but

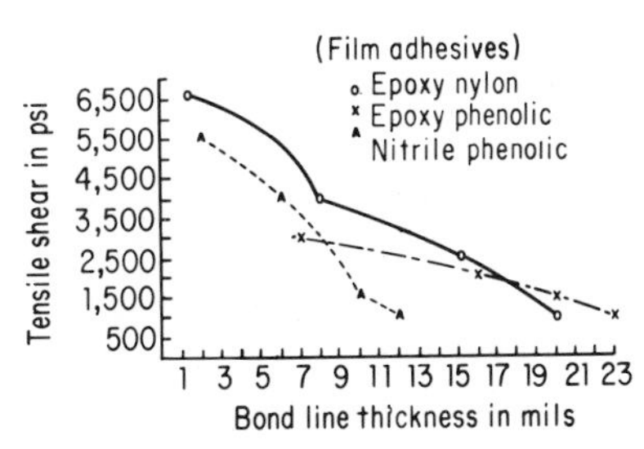

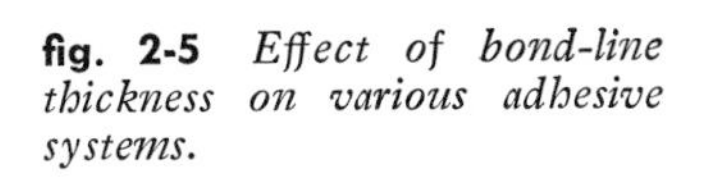

fig. 2-5 *Effect of bond-line thickness on various adhesive systems.*

they are still superior to most structural adhesive systems.

The epoxy-phenolics are available in film form and two-component liquids, the glass-supported film being more popular. They may be used with or without a primer, but are more frequently used without primers because they have better adhesive than cohesive properties. The liquid type has better elevated-temperature properties because thinner bond lines are attained, and with the glass-supported films the glass becomes a factor.

The epoxy-phenolics will produce shear strengths of approximately 3000 psi with aluminum, but have poor peel properties (10 to 20 lb). Impact resistance is poor but creep resistance is excellent. They have good resistance to moisture, solvents, and chemicals.

One of the major production problems associated with these systems is the liberation of gases and moisture during cure. They are cured at temperatures from 260 to 350°F, with 10 to 40 lb of pressure. They become very fluid during cure and starved bond lines are not uncommon with these adhesives. They are not as dependent on bond-line thickness for good strength properties as are many adhesives systems (Fig. 2-5). They are considered adequate for cryogenic temperature application but do not compare in mechanical properties to many other systems at low temperatures. In all probability, their greatest asset and reason for widespread usage is the wide temperature range in which they will perform satisfactorily.

21. Nitrile-phenolic adhesives

The nitrile-phenolics are available in the liquid form, supported, and unsupported films. The unsupported-film forms are more widely used today because the liquid forms have been replaced by better systems applicable to the use they were designed for, and the supported film offers few advantages over the unsupported-film types.

This system is used primarily as a metal-to-metal, metal-to-plastic, or plastic-to-plastic adhesive. Long-term aging properties below 300°F are considered excellent. The operating range is approximately –67 to +350°F.

The nitrile-phenolics have a cure range of 260 to 350°F and pressure requirements from 50 to 200 psi. The pressure naturally controls bond-line thickness, and shear strengths vary with bond-line (Fig. 2-5). They have better cohesive than adhesive properties and are normally utilized with a primer system. They have excellent resistance to oils, solvents, fuels, and moisture. They produce shear strengths up to 5500 psi and peel strengths (metal-to-metal) in the area of 60 to 70 lb. The nitrile-phenolics are not recommended for sandwich structures because of the high pressure requirements and poor flow characteristics.

TABLE 2.1 Tensile Shear Strength of Epoxy Cured with Various Hardeners

Hardener	Cure conditions	Tensile shear strength, psi at 77°F Unaged	77°F Aged*	180°F Unaged	180°F Aged	250°F Unaged	250°F Aged	500°F Unaged	500°F Aged
DETA	60 min 200°F	2000	195	1625	35	690			
DEAPA	90 min 200°F	1315	895	1450	890	1440			
Versamid 115	90 min 200°F	2510	340	1285	50	290	20		
BF_3 Complex	2 hr 320°F	1655	320	520	542	165	460	200	90
Phthalic Anhydride	3 hr 320°F	1875	355	1500	210	1635	420	130	325
HHPA	3 hr 320°F	1875	845	2925	1435	585	1710	245	130
1, 8 DAPM	3 hr 300°F	1845	180	2145	170	2235	—	125	235
4, 4′ MDA	2 hr 320°F	1775	230	1740	245	1750	900	225	80
Dicyandiamid "Dicy"	2 hr 350°F	2060	145	3025	225	1600	200	175	280
PMDA	1 hr 350°F	2250	1755	2050	1680	1700	1580	950	645

* Specimen aged 200 hr at 500°F.

22. Modified epoxy intermediate curing films

The intermediate-curing epoxy films are one of the newer structural adhesives to be available commercially. They develop full strength with a 250°F cure and possess excellent properties from −100 to +200°F. They have good fatigue resistance, excellent shock properties, and good resistance to oils, fuels, solvents, and most chemicals.

These systems are used with a variety of substrates and perform well in both metal-to-metal and honeycomb structures. They do not require high pressures during cure, are normally utilized with a primer, and have good handling characteristics. They require storage at 0°F or preferably below, and degradation occurs very rapidly when they are allowed to remain at room-temperature conditions.

These films become very fluid during cure and perform better when vacuum bagged and autoclave cured.

Only a small portion of the hundreds of adhesive formulations have been scanned in this chapter. The chemistry of many adhesives is very complex, but if the chemical type is known, individuals familiar with adhesive systems have a general idea of their inherent properties. They can only be modified within a limited range and still remain in a particular chemical category. Chapter 8 presents more information on the properties of various adhesive formulations.

REFERENCES

"Aerospace Engineers Reference to Silicone Materials," General Electric, Waterford, N.Y., January, 1966.

Blaick, Jr., C. F.: Adhesion of Urethane Elastomers, *Adhesives Age,* September, 1964.

Bloomingdale Rubber Company, "Properties of Various Structural Adhesives," presented at AVCO Corporation, Nashville, Tenn., April, 1962.

Bodner, M. J., and W. H. Schrader: "Investigation of Eastman 910 Adhesive," Department of Army, Project 583-22-005, October, 1957.

"Bonded Structures," A Selection of Literature, Bonded Structures Ltd., Duxford, Cambridge, England, March, 1964.

Bonding Tedlar to Steel with Acrylic Adhesive, *Adhesives Age,* October, 1964.

Carmichael, E. P., and W. F. Gross: "Phenolic Based Adhesives," Narmco Inc.

Colovos, C. C., T. R. McClellan, and K. W. Raush: Polyurethane Sealants . . . Some Improved Systems, *Adhesives Age,* November, 1966.

Davis, M. L., and J. M. McClellan: Urethane Elastomeric Sealants: Present and Potential Markets, *Adhesives Age,* May, 1964.

Elam, D. W., "Epoxy-Resin Base Adhesives," Shell Chemical Company.

"Epon Adhesives Manual," SC 60–112 Revision, Shell Chemical Company.

Epoxy Tape Adhesive Bonds Jet Fuselage without Heat, *Adhesives Age,* March, 1963.

"FM-34 Adhesive," American Cyanamid Company, Havre de Grace, Md., April, 1966.

Forlana, Krumwiede, Thorton, Benet, Nelson, "Research on Elevated Temperature Resistant Ceramic Adhesives," WADC-TR-55-491, University of Illinois, Urbana, October, 1962.

George, D. A., F. Roth, and P. Stone: Polysulfide Sealants: Parts 1 and 2, *Adhesives Age,* March and April, 1965.

Heine, E.: Strengthening Adhesives with Nylon, *Adhesives Age,* June, 1963.

House, C. I.: Structural Adhesives . . . Some Types Designed for Metal to Metal Bonding, *Welding Engineer,* September, 1964.

House, P. A.: Microorganism Resistant Coatings for Aircraft Fuel Tanks, *Adhesives Age,* April, 1965.

Hudson, G. A.: Formulating Two-component Polyurethane Sealants, *Adhesives Age,* November, 1965.

Kausen, Robert C.: Adhesives for High and Low Temperatures, *Materials in Design Engineering,* September, 1964.

Kempf, J. N.: Adhesive Bonding, *Product Engineering,* August, 1961.

Kraus, Herbert S.: Epoxy Bonding, *Adhesives Age,* April, 1959.

Krieger, R. B., Jr.: "Two Types of Structural Adhesives: Polyamide Epoxies and Duplex Films," National Metal Congress, November, 1962.

———, and R. E. Politi: "High Temperature Structural Adhesives," American Cyanamid Company, Havre de Grace, Md.

Lee, Henry, and Kris Neville: "Epoxy Resins, Their Applications and Technology," McGraw-Hill Book Company, New York, 1957; also, "Handbook of Epoxy Resins," 1967.

———, "Use of Dianhydrides in High-temperature Epoxy Resin Formulations," presented at the National Symposium on Adhesives and Elastomers for Environmental Extremes, Society of Aerospace Material and Process Engineers, Los Angeles, May, 1964.

LeFave, G. M., F. Y. Hayashi, and R. Gamers: One-part Polysulfide Joint Sealants, *Adhesives Age,* October, 1962.

Levine, H. H.: "Polybenzimidazole Resins for High Temperature Reinforced Plastics and Adhesives," presented at symposium sponsored by Materials Central, Dayton, Ohio, December, 1962.

Long, R. P.: "Cryogenic Adhesive Application," presented at Structural Bonding Symposium, Marshall Space Flight Center, Huntsville, Ala., March, 1966.

"Material-Adhesive-Strength at −65°F-AF-31 Lap Shear-Effect of Increased Line Strength," General Dynamics Report No. FTDM-2464, Fort Worth, Tex., September, 1962.

Merriam, J. C.: Adhesive Bonding, *Materials in Design Engineering,* September, 1959.

Noble, M. G.: New Developments in Silicone Rubber Compounds for Extruded Electrical Insulation, *Wire and Products,* February, 1957.

Pattee, H. E., G. E. Faulkner, and P. J. Rieppel: "Adhesive Bonding of Titanium," Battelle Memorial Institute, Columbus, Ohio, June, 1958.

Poet, R.: "The Nature and Application of Heterocylic, Polyaromatic Linear Polymers," Narmco Materials Division, Costa Mesa, Calif.

Pratt, D. S., J. E. Shoffner, and H. C. Turner: "Development and Evaluation of High Temperature Ceramic Adhesives," Report 8926-109, General Dynamics/Convair, March, 1960.

"Proceedings from the Seventeenth Annual National Forum," Washington, D.C., May, 1961.

Scales, M.: Epoxy Based Structural Adhesives, *Adhesives Age,* November, 1964.

Schwider, A. M., I. M. Zelman, and A. J. Tuckerman: "Recent Aerospace Applications for Structural Adhesives," The American Society of Mechanical Engineers, March, 1965.

Sterman, S., and J. B. Toogood: How to Promote Adhesion with Silicones and Silanes, *Adhesives Age,* July, 1965.

Twiss, Sumner B.: Structural Adhesive Bonding, Parts 1 and 2, *Adhesives Age,* December, 1964, January, 1965.

Zakim, J., and M. Shihadeh: Sealants and Caulking Compounds, *Adhesives Age,* August, 1965.

CHAPTER 3

Applying Adhesives

The selection of the correct method of application of an adhesive may be as important to the success of the bond as the choice of the adhesive itself. Each different type of adhesive may require its own special type of application. In the aerospace industry, the most common method of applying adhesives is manual, or by hand. The manual application may utilize a dip method, brush application, roller application, or spreading with a knife, trowel, spatula, or special applicator of some type (Fig. 3-1). Edge sealants and potting compounds are usually applied manually (Fig. 3-2). Many times, a unique application may present itself when conventional methods or standard equipment may not be adequate; therefore, a special technique or device must be devised to apply the system. The assembly, the sequence of operation, or the adhesive system may present the problem. The special application problem must be

fig. 3-1 *Manual adhesive application.*

fig. 3-2 *Sealing edge of honeycomb panel. (Furane Plastics.)*

solved by cognizant personnel, or a vendor specialist in the field of adhesive application may be consulted.

The semiautomatic application includes spray guns, caulking guns (Fig. 3-3), or other air-pressure devices (Fig. 3-4). The latter device utilizes an old technique, the syringe, which, coupled with air pressure, performs well on small parts.

The automatic applicators are used for high-speed production applications (Fig. 3-5). The automated devices are usually geared for a particular part or assembly that is to be mass produced; thus it is impractical when parts are to be joined in a limited number.

Before discussing the application of adhesives, the layup or application area must be discussed because this is of utmost importance. Regardless of the caliber of equipment, the knowledge of the technicians, the cleanliness of parts, the assembly area must be adequate.

Humidity and temperature must be controlled, the control depending

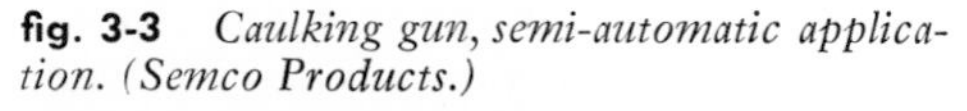

fig. 3-3 *Caulking gun, semi-automatic application. (Semco Products.)*

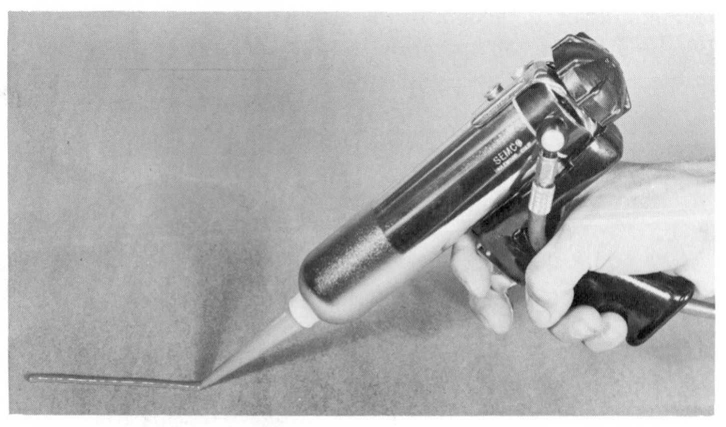

on the type of adhesive being utilized and the criticality of the operation. The layup room must be kept clean, free of dust and contaminants. Tooling should be cleaned before being transported into the clean room and there should be a special area for tool cleanup. Tool cleanup is not only important from the contamination standpoint, but also to ensure smooth parts from the next operation, this being especially true in the fabrication of honeycomb-bonded assemblies.

The mold releases for tooling must be carefully selected. The standard waxes are very poor and the silicones should never be used. There are water-soluble mold releases designed for bonding that are relatively safe, but cleanup and handling must be accomplished with caution.

There are many insignificant and hard-to-detect ways of contaminating parts during layup. For instance, an individual may have rubbed his hand over his hair which is loaded with that "greasy kid stuff." Someone may clean his eyeglasses with a silicone treated paper or fail to wash his hands thoroughly after devouring a greasy pork chop for lunch. The only way to control these things is complete education pertinent to these problems and the conscientiousness of the employee. Mixing and application tools should be kept clean and free of contaminants. Personal cleanliness is imperative, not only from the standpoint of adhesive contamination but also to prevent the individual from getting dermatitis from various resins and additives. Personnel should wash with soap and water as soon as possible after coming in contact with adhesive materials.

When removing liquid adhesives from the container, care should be exercised so the next individual to use the same adhesive does not find

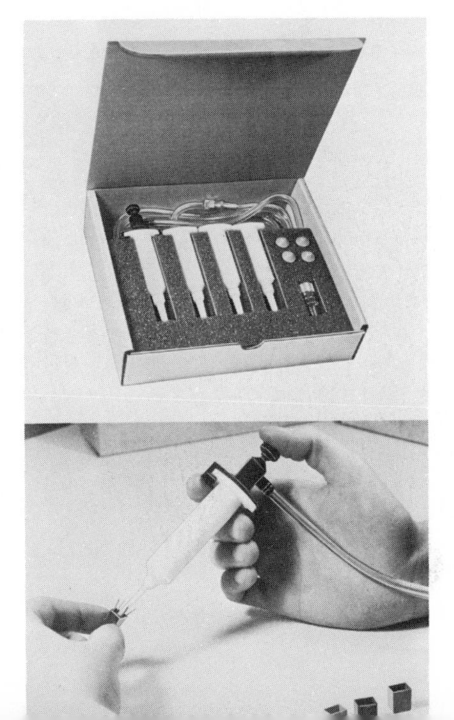

fig. 3-4 *Air-pressure syringe, semi-automatic application. (Armstrong Products.)*

the container lid "bonded" tight. Adhesives may be removed from a container by the use of two "tongue depressors" to avoid contact with the container top. These items may seem of relatively little importance, but a messy group of containers in a storage cabinet indicates carelessness and poor workmanship.

When inspecting or evaluating a controlled layup room, here are some considerations:

1. What type of waxes or solvents is being used?
2. Are the gloves being worn by personnel clean? White gloves are recommended for this reason; the dirty ones may be spotted.
3. Are parts being assembled first in, first out, after cleaning?
4. Is an inspector watching each operation?
5. Do the employees wash after handling the adhesives?
6. Are all the tools clean?
7. Are there any open containers of adhesives?
8. If films are being used, are they first in, first out of the freezer?
9. Are the humidity and temperature of the layup area within the required range?
10. What type of trim is being used in the case of films? Is it being trimmed so that no scratches are apparent in the substrate? Has the film been qualified and do records indicate this?
11. If a mixer is being used, is the adhesive being mixed for the proper length of time? Even so, does it show uniform color during the application?
12. Have the details been prefitted before assembly? Are there records indicating this?
13. Do the parts respond favorably to the water-break-free test

fig. 3-5 *Automatic adhesive applicator.*

pertinent to cleanliness? Are they being assembled within the required lead time from cleaning to layup? Do records indicate this?

14. Are layup tables and floors clean?

15. Are scales and weighing devices clean? Have they been checked for accuracy and labeled within a specified time limit?

16. How is the clean room located in regard to the priming area and the curing area?

17. If an autoclave operation is in order, when is the part bagged and where? Could this be improved?

18. Are the layup area and bonding area well isolated from such operations as the machine shop, where oils and lubricants are being used?

All of the above are important but there are probably many more questions, depending on the particular operation.

The location of the clean room is important. A clean room exists in the bonding shop of a major firm in the United States with rest rooms located in such a manner that hundreds of people a day pass through the layup room daily. This company needs new plant engineers, or a complete education of the ones it has.

One-part liquid adhesive

The one-part liquid adhesives are not widely used as standard adhesives, but with structural bonding they fall more into the primer class. They are often used for secondary bonding sealants, and many of the rubber-base, solvent-evaporation cure adhesives fall into this category. The one-part adhesives may be applied in almost any manner.

Since it is obtained in the proper proportion, the only problem prior to application of the one-part adhesive is assurance that the mixture is homogenous, that is, that none of the components have settled to the bottom of the container during storage. They should be stirred or agitated before use and it may be desirable to degas the system prior to use. When applying an adhesive of this type, an excessive amount should be applied to both surfaces and pressed together to remove the entrapped air bubbles in the joint.

Primers fall into this category, and again a controlled room is necessary. Many of the newer primers are very sensitive to humidity and temperature. Primers may be brushed, dipped, or sprayed, but spraying is most common. Primers require thinning in many cases; thus, the use of the proper thinner in the proper amounts is important. Control of the primer thickness poses frequent problems. There are various devices commercially available to measure the adhesive-film thickness. It may be necessary to fabricate standards or "color chips" and to control the thickness visually by color comparison. Listed below are some of the problems that may pose difficulties during a spray operation:

1. Shutdown due to clogged nozzles
2. Waste, that is, inability to confine spray to designated area
3. Variations in coating weight and thickness
4. Slow speeds, because several passes are required
5. Increased fire and health hazards due to high solvent content

Various modifications of the basic spray system have been designed to overcome some of its shortcomings. Such systems are now air operated or hydraulically operated. One method introduces a transfer roll, and the adhesives are sprayed onto the roll rather than onto the substrate. This will yield a more uniform coating.

The automatic roller and metering device is utilized on one-part adhesives but is seldom used in the aerospace industry. This type of equipment is expensive and would be justified only if many parts are to be fabricated which must be geared to a specific operation. Dip coating is seldom used in structural application, but is used extensively in other industries such as book binding.

The multiphase adhesive

The desirability of many two-component adhesives which include epoxies, polyurethanes, and polyesters poses many problems. Control of costs and improved manufacturing efficiency require selection of the proper application technique. Although automated equipment is available, a large portion of multicomponent adhesives are applied manually.

Preparation of the multicomponent adhesives begins with adequate

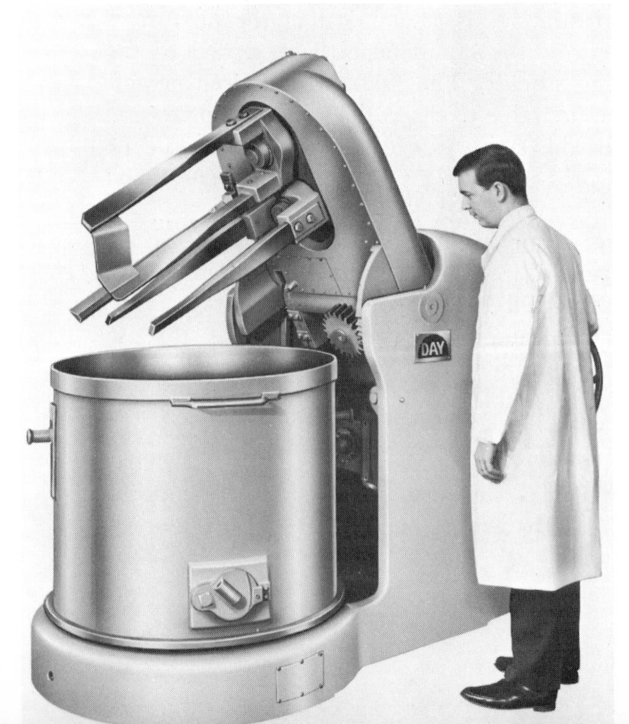

fig. 3-6 *Floor-type automatic mixer. (J. H. Day Company.)*

fig. 3-7 *Large mixer (52 Cu Ft capacity.) (J. H. Day Company.)*

mixing. If filled resin is used, it will be advantageous to place the unopened container on a shaker and agitate the contents for a period of time. For base resin filler or hardener filler combinations blended by the user, test data rather than theory should be used to determine the mode of mixing. If simple stirring is adequate, then use it. Where mechanical mixing is not satisfactory or practical, or when a filler such as chopped glass fibers is extremely hard to wet, a mill may be required.

There are a variety of mechanical mixers available from the small hand mixer, and the floor blender (Fig. 3-6), to the giant blenders for large-scale production (Fig. 3-7). A study of mechanically mixed versus automatically mixed adhesives indicates that mechanical mixing is generally satisfactory for most compounding operations. Since mechanical mixing is usually more controllable, less costly, and more versatile it should be employed whenever possible. If small batches are to be mixed, a spatula or tongue depressor used by hand will perform a satisfactory job.

If small amounts are to be mixed, care should be taken to ensure that all components are mixed individually before they are blended together. An example is Epoxylite 5302, a liquid and powder epoxy system. The solid content tends to settle to the bottom of the container during storage. The powders contain a blend of various densities and will also settle out. Mistakes have been made on a system of this type by stirring the liquid, but failing to agitate the powders. Both components must be agitated before blending together.

Weighing is also an important factor in mixing adhesives and extreme care should be exercised in this operation. It has been stated earlier in this text that undercatalysis can produce a poor joint; overcatalysis may create the same problem. Weighing scales should be kept in excellent condition, clean at all times, and checked each time before weighing begins. A defective set of scales or an error on the operator's part has resulted in massive losses in labor and materials.

The most common method of spreading multicomponent adhesives in the aerospace industry is by hand, using a flat paddle or trowel. Sometimes solvents are incorporated to make spreading easier, but they often lead to inferior bonds if the gases are not allowed to escape prior to final joining. Sometimes a viscous adhesive may be warmed in order to facilitate spreading. This technique is suitable for adhesives of sufficiently long pot life, but care should be exercised so as not to heat the adhesive excessively. When the adhesive is warmed, the pot life or working life is usually reduced. The warming may be accomplished in an oven at approximately 14 to 150°F for a few minutes or with the hair-dryer type of heater.

In utilization of one-part or multicomponent adhesives, the bond-line thickness is usually important. Listed below are some of the reasons:

1. It requires greater force to deform a thin bond line.
2. Probability of flow or creep is greater as thickness increases.
3. Internal or "locked in" stresses at the adhesive interfaces and thermal stresses due to differential expansion are reasonably proportional to the thickness of the film.
4. If the adhesive is hard or rigid a thin film is more resistant to cracking when the joint is flexed.
5. The greater the volume of adhesive in a joint, the greater the probability that it will contain an air bubble, an unbonded solid particle, or other source of weakness.

It is generally agreed that for maximum strength and rigidity, an adhesive layer in a joint should be as thin as possible without starving the joint. To prevent starvation, the amount of adhesive must be sufficient to fill the gap. The bond-line thickness may be controlled by the use of glass beads mixed with the adhesive.

Film tape form

The tape form of adhesive is the most common in structural bonding applications. Such form is convenient for application to flat or slightly curved surfaces, but difficult to apply to intricately shaped parts. This form is convenient to handle, is usually stored under refrigeration, and has a limited storage life when compared to the multicomponent systems. The tape adhesive may include a fabric carrier, such as woven nylon or glass cloth, or it may be a film calendered from the basic adhesive system. In utilization of this form of adhesive, the amount of adhesive applied has been determined largely by the manufacturer. A popular system is usually available in several thicknesses and weights. The thickness may vary, but resin content remains the same, depending on the carrier; or the resin content may vary and the glass remain constant. Practically all adhesive tapes are received from the manufacturers with a polyethylene protective backing sheet on each side.

It may be desirable for special applications to fabricate a film or impregnate a cloth to meet the requirements of an application. This may be necessary because the proper weight cannot be purchased or the adhesive is not available on the proper type or weight of cloth, or a basic resin has been modified that is not available commercially. If a cloth is to be impregnated, the catalyzed adhesive is impregnated into a glass cloth carrier (Fig. 3-8). The cloth is pulled through a reservoir of adhesive and a doctor blade controls the adhesive thickness and weight. A vinyl separator film is used on the underside to prevent the adhesive from sliding on the table top and permits ease of handling. Impregnation of cloth may be accomplished, especially in small pieces, by simply laying the cloth on a slip sheet and brushing on the adhesive, or spreading with a trowel, large putty knife, or a piece of Teflon material of the desired size and configuration. The commercial film is considered more practical, with the exception of a few isolated cases.

The films should be removed from the freezer (if frozen) and allowed to thaw to room temperature before the wrapping is removed. This reduces moisture condensation on the film and allows easier handling.

fig. 3-8 *Impregnating glass carrier cloth. (North American Aviation, Columbus, Ohio.)*

Parts should be prefit before cleaning, to ensure that the faying surfaces have the proper clearance. This clearance is naturally dependent on the type of adhesive that is to be utilized. If severe gaps are apparent, they can be corrected by shimming or adhesive buildup in that area. All parts should be indexed to allow the parts to be reassembled in the identical manner as the prefit.

After the parts are clean and ready for film application, the film should be cut slightly oversized for the part. The polyethylene film should be removed from one side only and placed to a mating surface. If this is a sandwich construction, it would be placed on one of the face sheets. After it is in place, the wrinkles are worked out and all air bubbles removed. If air bubbles are present in the film, small holes may be pricked or punched in the film to allow the air to escape before mating. A majority of the film-type adhesives will adhere to a cleaned or primed surface, but occasionally, it is necessary to use a tacking paste to hold the film in place and to prevent slippage or movement during assembly. When using a tacking paste, use only enough to hold the film in place. There are case histories of a tacking paste that was spread as if primers were being used. After the film is smooth and free of air bubbles on one mating surface, it must then be trimmed to the desired size. The flow properties of the adhesive and the nature or configuration of the part determines the trim. It may be practical to trim the adhesive 0.030 to 0.060 in. short of the edge of the part to eliminate excessive cleanup. One of the errors made frequently is trimming the adhesive without a backup material under the film to prevent scratching or damaging the adherend. Many failures result from fatigue when the substrate has been cut or scratched during the trim operation. A piece of clean Teflon sheet stock will serve as an excellent backup material, but if it is practical, the use of a pair of scissors eliminates any doubt.

If the assembly is large and more than one piece of film is necessary, butt slices should be used and gaps not to exceed 40 to 50 mils. Whenever possible, splices should not be located near joints, metal splices, or core splices.

Many types of film are available from the high-pressure tapes for metal-to-metal bonding to the composites especially designed for honeycomb assemblies. The composites utilize the two-adhesive systems, usually an epoxy and phenolic composition. This gives good node filling for the core and a bond with better mechanical properties to the face sheet. The major problem here is the tendency for technicians to reverse the film, which produces a poor assembly. If this occurs it can usually be detected by observation of the flash. This will be discussed in the chapter on quality control.

Curing

Curing should take place with closely controlled temperature and cure-cycle time. The selection of cure time, rate of heat rise, and cool-down rate are dependent on the adhesive formulation, the type of joint, and the expected service condition of the bonded joint. Thermocouples should be installed in the bond line during heat-cure cycles to permit a close check of temperature conditions. The iron-constantan wire is adequate, and should be embedded in the adhesive joint or in the flash adjacent to the edge of the part. If the part is large, at least one thermocouple should be used for every 10 sq ft, and the thermocouples should be checked for continuity before the operation begins.

Pressure control

Pressure should be maintained over the entire bonded area during the cure cycle. Pressure must be applied in such a manner as to be uniform and to prevent movement of the assembled parts or displacement of the adhesive. Bonding pressure is important to:

1. Obtain uniformly thin adhesive bond layer.
2. Overcome viscosity of the adhesive film at the curing temperature.
3. Overcome internal pressure exerted by the release of adhesive solvents and water vapors.
4. Overcome the surface imperfections and dimensional mismatch between the mating surfaces during cure. The pressure required for various systems may vary, depending on the task the final assembly may be required to perform. Pressure is a function of the size of bonded parts, the perfection of the mating surfaces, viscosity of the adhesive, and the gas liberation of the adhesive system.

The nitrile-phenolics may be utilized, for example, bonded from 30 to 200 psi bonding pressure. When bonded using 30 psi they will develop approximately 2800 psi in shear, but when bonded at 200 psi they will develop 6000 psi. The bond-line thickness and psi in tensile may vary in cure cycle from 250 to 350°F. This also affects their properties as the lower cure cycles improve peel strength but result in a loss in shear, and vice-versa for higher cure cycles. The chemical resistance of many adhesive systems is improved by the elevated-temperature cure cycles. Therefore, the bonding pressure and cure cycles must be established for each project as the project is designed and scheduled. It is the responsibility of all concerned with the program to see that specified requirements are carried out properly.

Out-of-station bonding

It is apparent that all bonding operations cannot be performed in a controlled area. In final assembly, attachment of clips or electronic details is a prime example; this would also apply to in-field repairs. This is normally labeled out-of-station bonding and may be applied to any bonding operation which uses equipment such as heat lamps, electric heater strips, portable ovens, etc. Portable pressure devices may also be used, that is, mechanical vacuum tools, vacuum bags, and spring-loaded devices. Bonding operations of this nature may utilize room-temperature curing systems and be held in place by tape, screws, etc.

It is advisable to avoid out-of-station bonding whenever possible because the bonding cycle must be performed under the same rigid controls as in-station bonding, without the luxury of all the fine controls available in a well-equipped layup room. Production costs are considerably higher with out-of-station bonding, not only because parts are added to the assembly in small quantities or even individually, but also because it is a slower, nonroutine operation, and may adversely affect subsequent activity. If the out-of-station bonding is critical, it may be feasible to construct a portable clear-plastic enclosure. This is used frequently for these types of operations and with portable heaters, pumps, and filters, the humidity and temperature can be controlled to some extent.

If a part must be attached out-of-station and it is known and planned for during the cleaning operation, it may be primed for protection, thus preventing the necessity of an unfavorable cleaning operation after many other parts have been incorporated into the assembly. When a cleaning of the attachment or bonding area is necessary, a thixotropic mixture would be in order. Care should be taken that the area is masked off to prevent the cleaning agent from penetrating an area where it might not be removed; this presents a possible corrosion problem. (See Chap. 5.)

Repairs of adhesive bonded structures and components

Repair procedures are developed with the objective of attaining a strength as nearly as possible equal to the strength of the original part with a minimum of increase in weight, aerodynamic characteristics, and electrical properties where applicable. This can only be accomplished by replacing damaged material with identical material or an approved substitute. In order to eliminate dangerous stress concentrations, abrupt changes in cross-sectional areas must be avoided whenever possible or practical by tapering joints, by making small patches which are round or oval shaped in lieu of rectangular, and by rounding corners of all large repairs. Smoothness and appearance of outside surfaces is sometimes necessary and always desired, and consequently, patches that project above the original surface may not be desirable. The extent of the defect may be of a nature that would only necessitate a small repair or adjustment. Small inclusions or air pockets in small units, metal-to-metal, or plastic assemblies may only require a small amount of filling with a room-temperature system by manual methods, possibly utilizing a spatula or a hypodermic needle. If a film was used to bond an original part, a good liquid substitute is usually available.

For the repairs of small dents and abrasions on metal skins, a metallic filled epoxy may be used. The "ding" is filled, cured, and sanded smooth to the contour of the assembly. If a heat cure is used, the original adhesive must be taken into consideration. Many adhesive systems would be affected, or aged, by a second or third cure cycle. A crack or split in either metal-to-metal joints or honeycomb assemblies may be repaired by drilling stop holes at each end of the rupture and adding an exterior bonded patch. The stop hole will prevent propagation of the break or crack, and the skin patch aids in load distribution (Fig. 3-9). This type of repair is used frequently, especially if the assembly is not a prime load-carrying structural member. Sandwich structures usually create more problems when repairs are necessary. The most common problem occurs in the production of honeycomb assemblies and sandwich structures as a result of careless handling, especially with thin face sheets. Many times, when not under close observation, a shopman will attempt to repair a

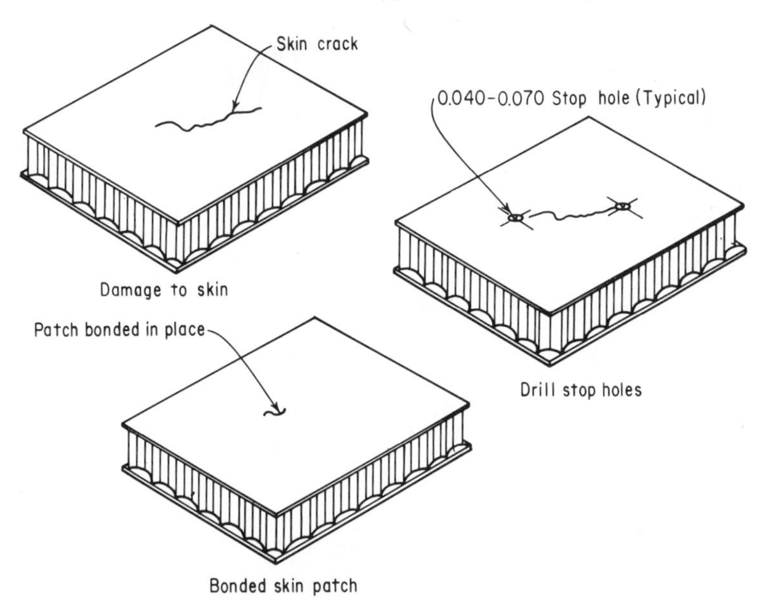

fig. 3-9 *Stop drill crack and add exterior patch.*

"ding" or abrasion or fill a gap in the adhesive without consulting the cognizant personnel, and this usually results in a "coverup," not a repair. Personnel with little or no experience with materials and techniques involved in making repairs will have difficulty in making acceptable repairs on the first attempt. Personnel should be instructed and trained in making repairs before they are allowed to make them on actual parts.

Scratches on the facings of honeycomb structures should be sanded smooth to prevent stress concentration. The area that is sanded should be treated (if metal) to prevent corrosion. Scratches which exceed a specified depth or length would be treated as a hole in the face sheet.

Dents in the facings will sometimes damage the core and cause the skin to be unstable over a large area. Dents in which the core is damaged are repaired by pulling the facing to meet the contour of the part (Fig. 3-10). The damaged core is then replaced with a potting compound. The potting compound will stabilize the face sheet and prevent buckling (Fig. 3-11).

Holes in the face sheets are repaired by removing the damaged area. If the hole is small, approximately 1 in. in diameter or smaller, it may be repaired as shown in Fig. 3-12.

If the damaged area or hole is larger than 1 in., but not over 3 in. in diameter, it may be repaired using a flush or external patch as shown in

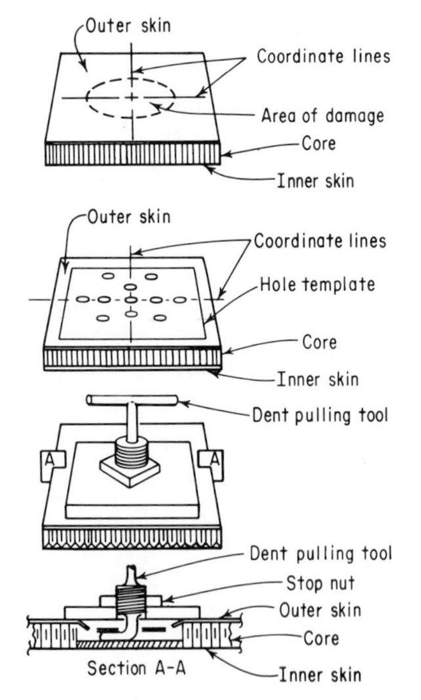

fig. 3-10 *Dent repair, pulling.*

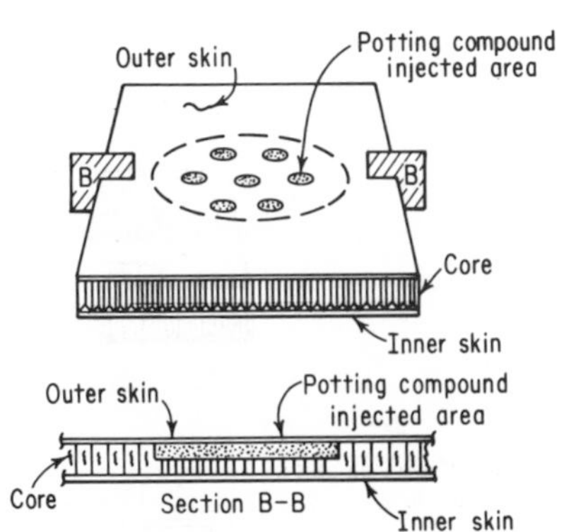

fig. 3-11 *Potting a dent repair.*

Figs. 3-13 and 3-14. If a hole is over 3 in., but not over 6 in. in diameter, it may be repaired by using a fiberglass laminate as shown in Figs. 3-15, 3-16, and 3-17. After the layup, as shown in Fig. 3-17, pressure may be applied by the use of a vacuum bag or a mechanical device, and heat may be applied with heat lamps or an electrical heating pad. The repair area should be covered with a bleeder to allow the gases to escape during cure if a vacuum is used.

Safety:

During mixing and curing, the individual must keep safety factors in mind. Material handling of adhesive systems may be carried out success-

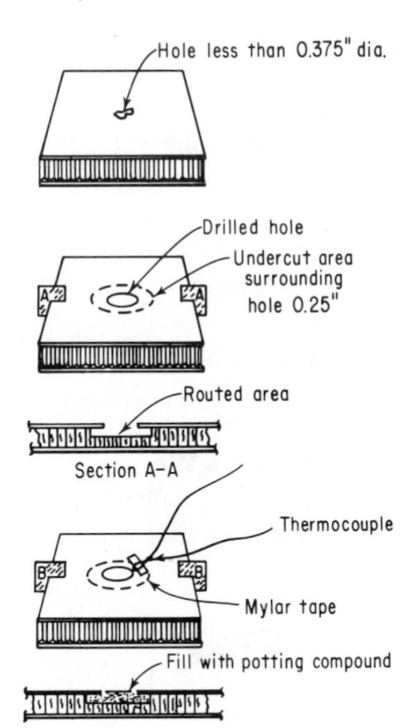

fig. 3-12 *Repair of small diameter holes.*

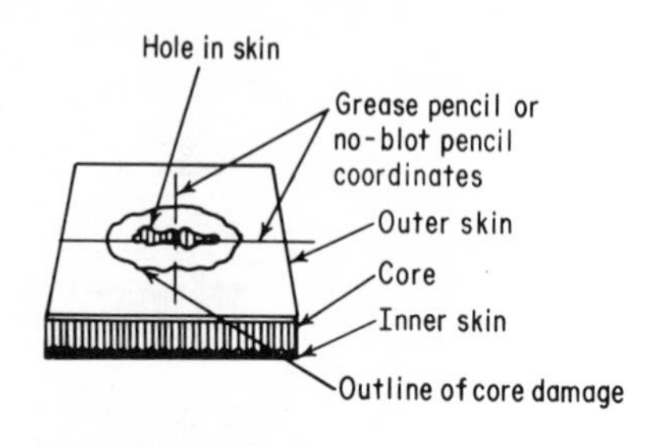

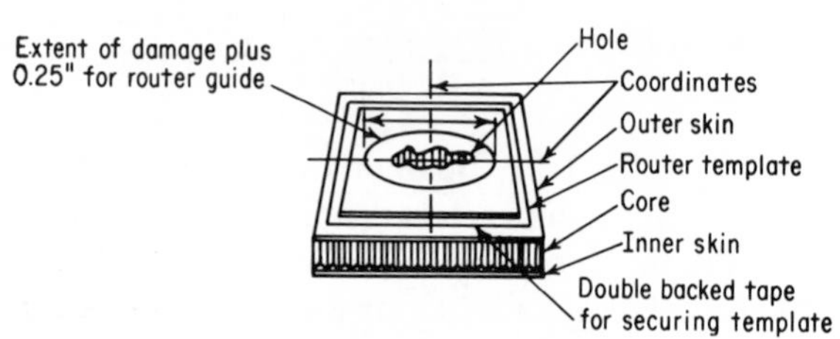

fig. 3-13 *Patch-plate repair.*

fully if precautions are taken to prevent exposure to substances or situations that are harmful. Listed below are some points to consider:

1. Individuals who have histories of allergies should not work with adhesives.
2. Operators and supervisors should always realize the importance of cleanliness.
3. Work areas should be provided to allow good hygienic protection to be followed.
4. Have washing facilities, including mild soaps, good skin creams, and clean towels available.

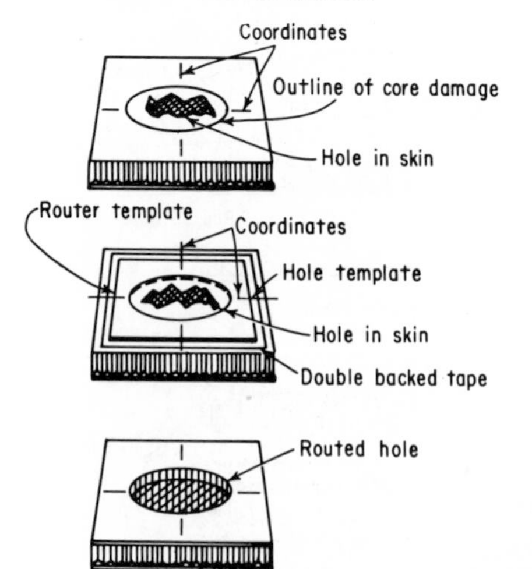

fig. 3-14 *Patch-plate repair.*

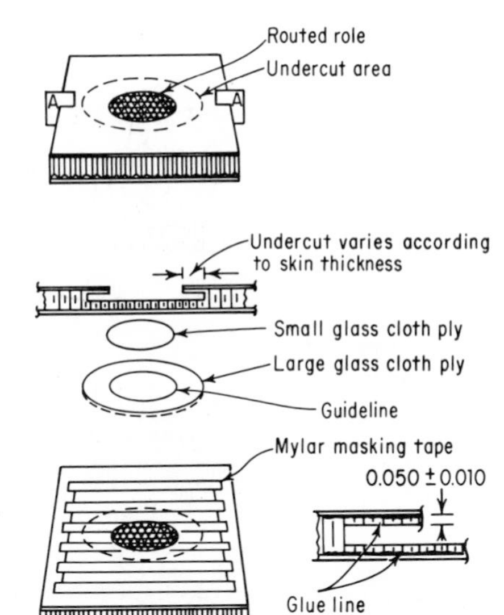

fig. 3-15 *Repair of damage with solid glass cloth laminate.*

5. Set up a systematic procedure to inspect work areas and operating procedures and follow through.

A person who becomes sensitized is never able to work in contact with adhesives again. It is costly to remove a technician from the bonding area

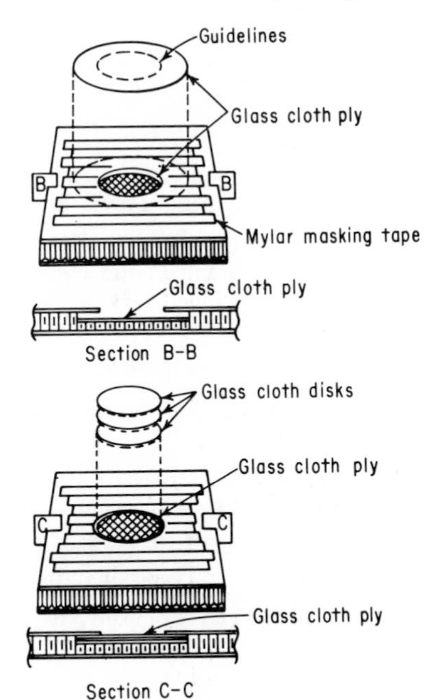

fig. 3-16 *Repair of damage with solid glass cloth laminate.*

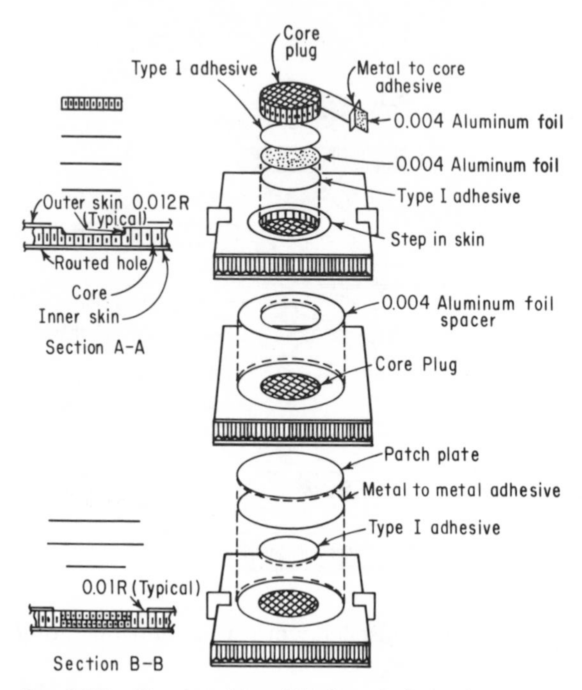

fig. 3-17 *Repair with solid glass cloth laminate.*

who is responsible and well trained. There is evidence that many of the solvents utilized in the bonding shop contribute to the problem.

REFERENCES

Heebink, B. G., and A. A. Mohaupt: "Effects of Defects on Strength of Aircraft-type Sandwich Panels," Forest Products Laboratory, March, 1956.

———, F. Werren, and A. A. Mohaupt: "Effect of Certain Fabricating Variables on Plastic Laminates and Plastic Honeycomb Sandwich Construction," Forest Products Laboratory, November, 1953.

Panek, E., and B. G. Heebink: "Repair of Aircraft Sandwich Structures," Forest Products Laboratory, March, 1956.

"Specialized Equipment through Research and Engineering," Semco Sales and Service, Inc., 1966.

Specially Equipped Spreaders Save Time and Labor in Timber Lamination, *Adhesives Age*, March, 1966.

Stiles, W.S.: "Secondary Bonding," North American Aviation, presented at Marshall Space Flight Center, Huntsville, Ala., March, 1966.

Brochure, John Day Company, 1966.

CHAPTER 4

Design Criteria and Joint Design

The objective of this chapter is to familiarize the individual with various joints, how they are stress loaded, and the reason they are utilized, but not to expand on complex bonded-structure designs. The average shop technician should be able to identify the various types of joints and have a basic knowledge of their function and the primary reason for a particular joint design selection. This is important even in the fabrication of specimens for laboratory test purposes.

The design of adhesive-bonded joints involves selecting the proper geometry, consideration of the adhesive and substrates to be employed, the size and dimensions of the joint, and the ease by which it can be tooled for, fabricated, and mass produced. The design engineer must consider the type of structure, service requirements, mechanical strength factors, service environment, and fabrication cost. These considerations are usually borne out by testing, either in the laboratory or assembly test performance, or both. This may include environmental tests, service tests, nondestructive tests, and destructive tests.

The materials and process engineers should be consulted during the design phase for assurance that the material and process requirements are practical. Many problems erupt in production because an adhesive system or joint design has not been properly researched and documented. It is essential that the design function consult tooling engineering to be sure that practical tooling can be fabricated at a reasonable cost. The coordinated effort of the above engineering function is very important when a new design is in the embryo stage. It may not always be practical, but if possible, the materials and process engineers should be standing by the drawing board as a new design concept unfolds. This is a practice with many companies at present, and the reason is obvious; changes can be made easily at this stage without an interruption in schedule.

The theme of this text has been and will be, "Adhesive bonding requires a coordinated, streamlined effort from design concept to the finished product." To illustrate this point, Company X, a few years ago, deemed it necessary to fabricate assemblies that would be joined via the adhesive-bonding technique, if they were to be competitive in their particular field of endeavors. It all seemed very simple, yet management realized they were far behind the other competitors in adhesive technology. The idea was simple: just stick it together; this is not a science, but an art; personnel can be easily trained. The first thought, which was logical, utilized the sheet-metal fabricators in the shop. They can easily be converted to bonding technicians as they have the metal fabrication know-how. An on-the-job training program proved disastrous. The people were not motivated, no job classifications existed that provided incentive, the union opposed such classifications, and the result was utter chaos. Company X must now implement a training program that is efficient, motivational, informative, and practical. The shop personnel accepted the challenge and were soon engulfed in a new and challenging venture. Conflicts then arose between workers and supervisors. Technicians were told by their supervisors to do things that were not in harmony with the training they had received. Company X must now make another adjustment, and train the supervisors in the art of adhesive bonding. The supervisors' attitudes were that this is a small scale operation that is destined for failure; it will not survive for very long. Bonding airplanes and missiles together was thought to be a ridiculous brainstorm from "upstairs." This problem was met by the same minority of forward and progressive engineers who realized the advantages of adhesive bondings, and supervisors were trained in this area. Next came the conflict between manufacturing and quality control. Quality-control workers must also be trained if they are to perform satisfactorily. Then came a program for planners, shop coordinators, and manufacturing support engineers. Materials and process engineering and the shop had gained considerable experience, but tooling had to be improved; the "C" clamps and spring devices were obsolete and had to be replaced, and modern tooling concepts employed. The pressure was then on tooling engineers, and they met the challenge. Plant engineering must provide clean rooms and adequate cleaning facilities. After two years elapsed, Company X had units in the field for system testing, and parts were failing right and

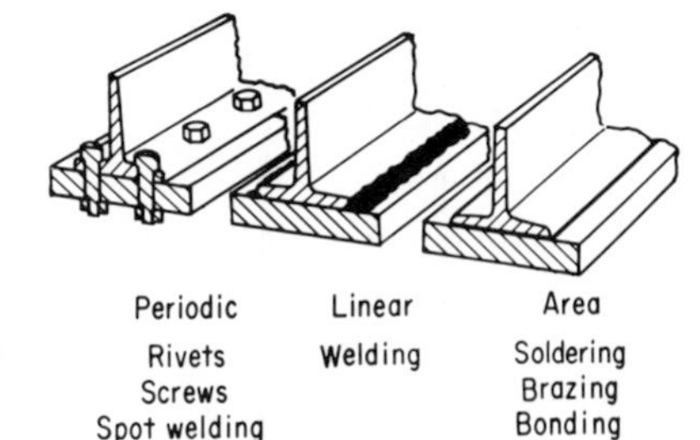

fig. 4-1 *Basic fastening methods. (Loctite Corporation, Newington, Conn.)*

left. Then the pressure was on from the customer for delivery. A complete analysis of the problem stunned the management of Company X: 80 percent of the problems were in the realm of joint design. Design engineers were still designing for rivets and bolts. Company X had to take it once again "from the top."

This is a typical example of the problems encountered in adhesive bonding until a company is oriented in the associated problems. When the company personnel are well trained and aware of the pitfalls, problems still occur. And the probability of success becomes higher with experiment but is always dependent on a sound, well-developed design.

The primary reasons for the utilization of adhesive-bonded structures were discussed in Chap. 1. The substrates or adherends carry considerable weight in the selection of fastening modes. The basic fastening methods for a particular type of joint are shown in Fig. 4-1. The joint configuration would naturally be altered to obtain maximum strength, regardless of the method selected. As an example, if adhesives were used to join this type of configuration, more tape would be proper in the flange area.

When joining thin-gauge substrates or plastics, adhesives show exceptional performances. This is illustrated in Fig. 4-2, which again indicates the importance of the substrate that is to be joined. Figure 4-3 indicates the strength of adhesive versus mechanical fastening when loaded in shear, and the substrate thickness factor is evident. Stress concentrations are considerably higher near a mechanical fastener as the material is

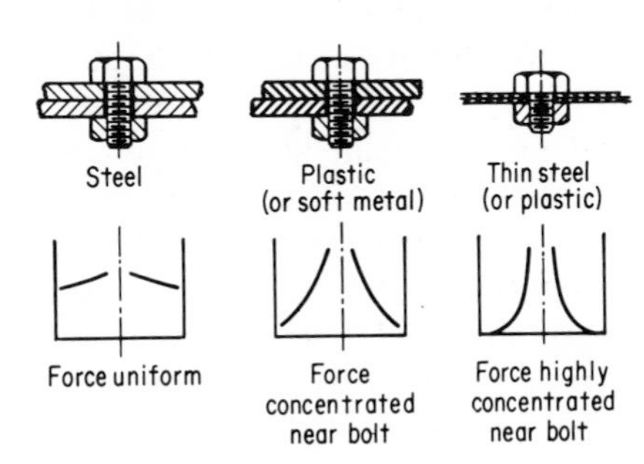

fig. 4-2 *Distribution of bolt clamping forces. (Loctite Corporation, Newington, Conn.)*

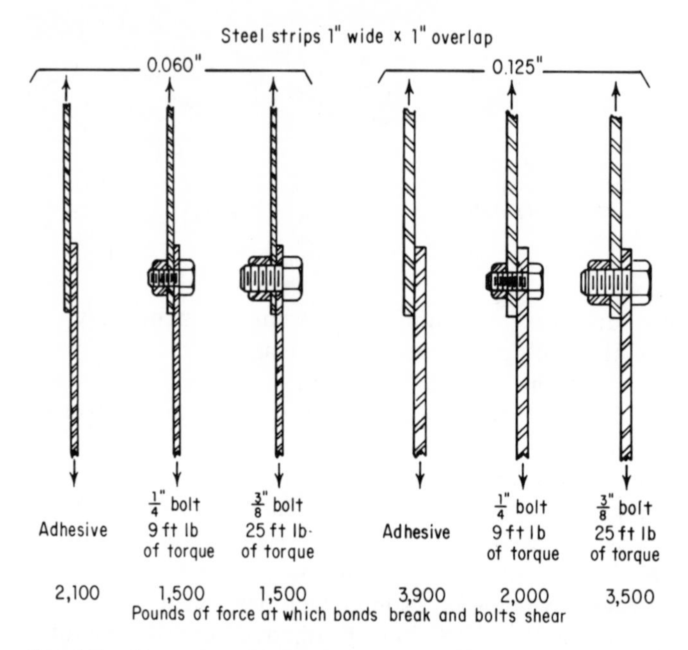

fig. 4-3 *Shear strength of bolts versus adhesive on varying thickness of substrates. (Loctite Corporation, Newington, Conn.)*

reduced in thickness. A simple experiment will illustrate this point, and is performed by cutting strips of paper, 1 in. wide and 5 in. long, from a standard writing tablet. Join some with any paper adhesive or mucilage (one-half inch overlap or more) and the others with staples, which represent the mechanical fasteners, using a few or as many as desired. Hold the specimens at the extreme outboard ends pertinent to the joints and slowly apply pressure in shear. As the pressure increases, observe the stress concentrations near the staples and note the mode of failure that is typical of both types of specimens.

Figure 4-4 depicts two types of joints that require the form of adhesive to be considered and the loading modes. The joint at the left would employ a liquid adhesive, possibly room-temperature curing, and would not require the strength that would be required for the common lap joint in the right of the figure. The lap type joint would possibly employ a film, with much higher strength. As an example, if the joint at the left were utilized on machinery, it might become necessary to remove the cylinder; thus the adhesive would be selected with that factor in mind. The structural joint at the right may require more permanence; therefore a stronger and entirely different type and form of adhesive system must be considered.

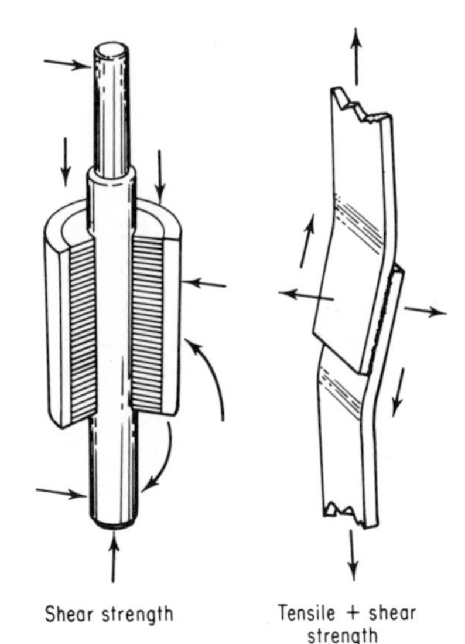

fig. 4-4 *Two types of structural joining. (Loctite Corporation, Newington, Conn.)*

METAL-TO-METAL BONDING

This term is applied to mated faying surfaces and usually refers to the lap type joints. This can be misleading as honeycomb structures may be metal to metal, but fall into a category of entirely different joint design considerations. The factors influencing the joining of two flat surfaces will be discussed in this section and shall be referred to as

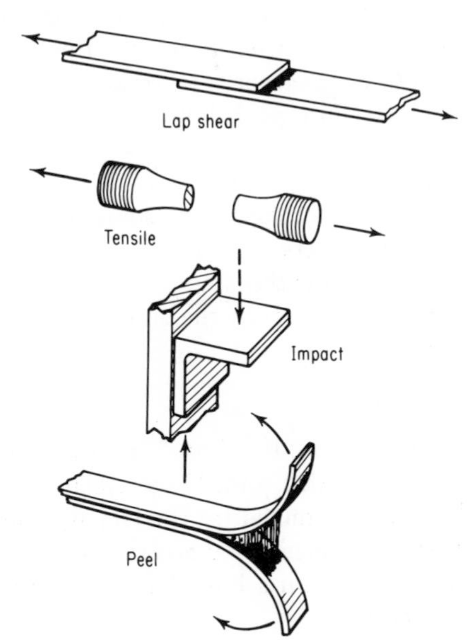

fig. 4-5 *Stress loading factors applicable to continuous faying surfaces. (Loctite Corporation, Newington, Conn.)*

metal-to-metal joints, although the term also applies to plastics and other adherends having a continuous faying surface. Figure 4-5 shows four basic loading modes applicable to continuous faying bonded surfaces, but it must be realized that other factors may influence these joints. For example, the joint labelled impact may be loaded in cleavage, or the peel factor may be present, especially if the material is thin gauge. Attempts are always made to design the cleavage and peel factors out of the design, but this is not always the case in the finished product. A joint that is designed for a shear load may be loaded in that manner, unless a rupture or break should occur near the edge of a flexible member. It may then become loaded in peel or cleavage. In combating a situation of this nature, a mechanical fastener may be installed near the edge of the member to hold it "in line" and prevent propagation of the break if the member has sufficient thickness to merit a stability fastener installation. However, it is not implied that a mechanical fastener will strengthen a bonded assembly, because it will not. The adhesive will carry the load until the bond ruptures, and the mechanical fasteners would not carry the load imposed, especially on thin-gauged materials; if they were installed for instance, the chances are they would not be properly spaced. If they were properly spaced, the extra weight would be added and the adhesive would be an absolute waste of effort. In simple terms, to bond an assembly and then rivet it actually weakens the structure and displays lack of confidence in adhesive bonding. There are exceptions because the adhesive may not be utilized for mechanical strength properties, but for their dielectric, shock, or vibration characteristics. A prime example of this is the utilization of a thin film of adhesive in a riveted aircraft structure to absorb vibration, reduce impact loading, or elevate the fatigue-failure probability factor. This is not uncommon, but the important point here is that the adhesive bond is not expected to carry heavy structural loads. Adhesives that are fully capable of carrying the expected service load may sometimes be used in heavy mechanically fastened members to isolate dissimilar metals, thus preventing galvanic corrosion; or the reverse may be in order, i.e., to establish electrical continuity between the substrates.

LAP JOINTS

The term "lap joint" is used to describe the continuous faying surface bonded joint. The selection of the type of joint configuration is largely determined by the end use, but there is always plenty of latitude within mechanical or artistic limits for design to achieve optimum bond strength within a boundary of feasibility when equated to service limits and cost.

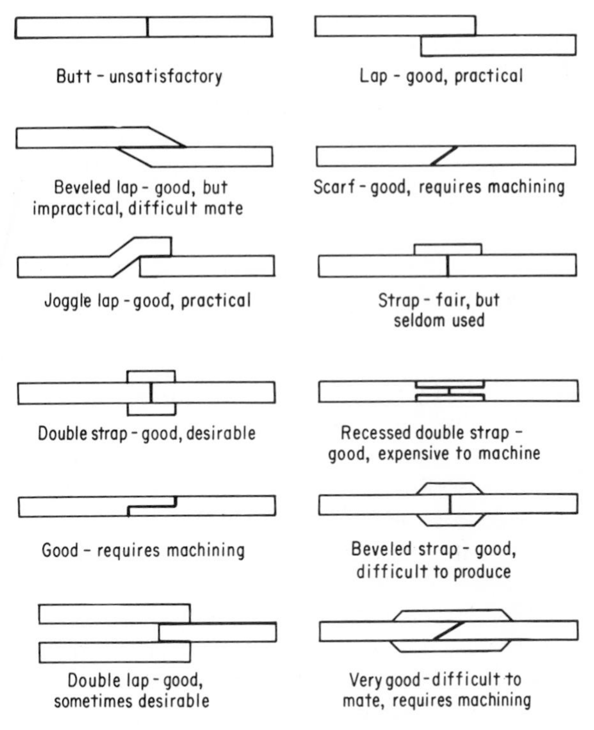

fig. 4-6 *Joints for metal to metal, plastic to plastic, and combinations.*

The lap joints are relatively simple in design, as shown in Fig. 4-6. In reference to metal-to-metal joints, the requirements can be summed up as follows. The joint should be designed to:

1. make the bonded area as large as possible
2. make the maximum proportion of bonded area contribute to strength
3. stress the adhesive in the direction of its maximum strength
4. minimize stress in the direction of which the adhesive is weakest

Generally, good designs avoid the types of loads and joints which concentrate stresses in a small area or on an edge. Because adhesives are generally stronger in shear, joints which allow the adhesive to be loaded in shear are preferable. A joint should never be loaded in the peel direction.

Two of the major factors influencing the design of lap joints are the magnitude and direction of the load the joint will have to bear. Most of the adhesives used for bonding flat surfaces are relatively rigid, strong in shear, and not so strong in peel or cleavage. Thus, by designing the

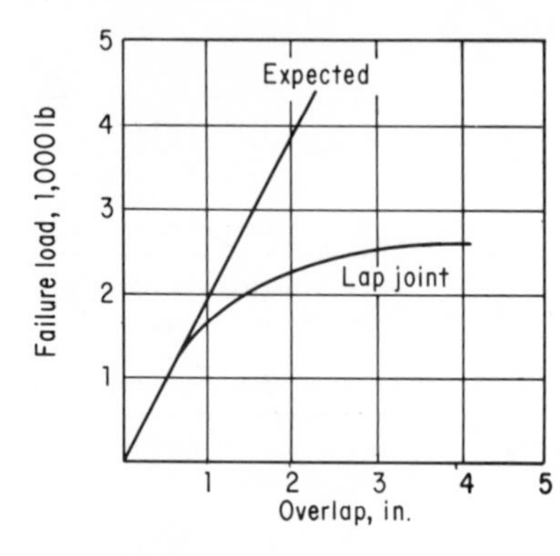

fig. 4-7 *Relationship between failure load and overlap.*

joint so that the adhesive is in shear, the effect of peel and cleavage stress is minimized.

The shear strength of a lap joint is directly proportional to the width of the joint. For example, by increasing the width from 1 to 2 in. (with the depth of the lap and gauge held constant), the strength is doubled. Strength may also be increased by increasing the depth of overlap. In this case, however, the relationship is not linear. Since the edges of the lap carry a relatively higher proportion of the load than the interior portion does, the unit increase in strength lowers as the length of overlap becomes greater. A relationship between failure load and overlap is realized by viewing Fig. 4-7.

The strength of a lap joint can also be increased by increasing the thickness of the substrate. In fact, unless the metal is of sufficient gauge, the yield strength may be lower than the shear strength of the adhesive bond. A curve showing this relationship for both aluminum and stainless steel is given in Fig. 4-8.

It should be noted here that if joints are beveled or utilize backing plates, joggle members, or other reinforcements, the mechanical strength may be increased, even doubled, especially in shear-loading capabilities.

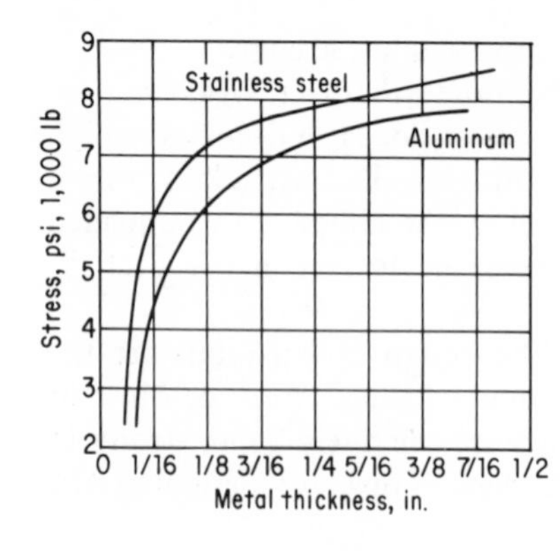

fig. 4-8 *Strength of lap joints when the adhesive is strong. The characteristics of the substrate may control the bond strength.*

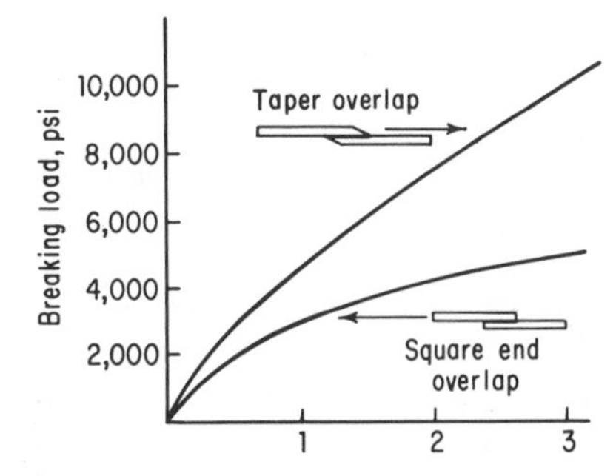

fig. 4-9 *Comparison of breaking strength of tapered joints with conventional lap joints.*

Figure 4-9 shows the comparison of tapered joints and simple lap joints pertinent to shear strength. Note that to a given point, the simple lap gives adequate results in comparison, but beyond this limit, the tapered joint is stronger. The scarf joint would be even stronger than the beveled joint because of the wider distribution of stress concentrations. When the scarf is employed with beveled straps, the optimum is realized because the stresses are distributed over a much larger area. The design consideration here is cost compared to the design requirements.

Design of joints for rigid plastics follows the same practice as for metals; some reinforced plastics of phenolic, polyester, and epoxy resins have strengths comparable to metal. However, they are often anisotropic; that is, their strength properties are directional and direction of load application becomes important. Anisotropic sheets would be oriented to obtain maximum strength properties.

In glass-filled laminates, there is usually a vast difference in shear and compressive strengths, depending on the orientation of the glass. These factors must be considered when selecting a joint configuration for joining these materials utilizing lap shear designs.

In assembling thin-gauge sheets to metal, particularly when bonded behind the metal, the sheet should be overlapped to place the adhesive in shear. The overlap should be adequate to place minimum unit stress on the adhesive.

As was stated earlier in this chapter, there is a lack of correlation in lap joints, as the length is increased or decreased, due to a non-linear relationship between the total bond strength and the bond area, or thickness of adherents, which is termed the T over L ratio. This condition is caused by deformation of the adherends under load, which causes stress concentration in the joint, and is geared directly to the yield strength of the substrate. This behavior is avoided in the standard lap shear test by using specimens of constant overlap and thickness (MMM-A-132) but is a significant factor in the lap-shear mechanical test of actual adhesive-bonded assemblies. For example, starting with a

0.2 in. overlap in length in a lap shear tensile specimen, as the lap increases in length or adherend thickness, the psi decreases up to a definite point. Using this theory a quality curve may be established by fabricating specimens of varying overlap and using this formula:

$$\frac{T}{L} \times 100 = \text{ratio}$$

where T = thickness of material
L = length of overlap

The quality curve may be generated by plotting the actual shear strength in psi versus T over L ratios (Fig. 4-10). The 100 percent curve may be developed by starting at 0.2 in. overlap and increasing overlap strength until the psi remains constant (which usually occurs at approximately 2.0 in.). The 90 percent curve may be developed by blocking out 10 percent of the joint in the direction of shear (using 0.002-in.-thick Mylar sheet stock). The 80 percent is obtained by blocking out 20 percent of the bond line, etc.

The yield line Y may be developed for the adherend in question (Fig. 4-10). This line can be found by expressing the shear function as a function of the T over L ratio, with the yield strength of the adherend as the parameter (Fig. 4-11). The effect of yield strength on the adherend is calculated as follows:

$$\bar{\tau} = \frac{P\,\text{max}}{b \cdot L} \quad \text{in adhesive}$$

$$\sigma = \frac{P\,\text{max}}{b \cdot t} \quad \text{in adherend}$$

$$\frac{\bar{\tau}}{\sigma} = \frac{\dfrac{P\,\text{max}/}{/(b \cdot L)}}{\dfrac{P\,\text{max}/}{/(b \cdot t)}}$$

$$\bar{\tau} = \sigma_Y \frac{t}{L}$$

The yield line is found as follows: $\bar{\tau} = \sigma_Y \dfrac{(T}{L)}$

where $\bar{\tau}$ = average shear stress, psi
σ_Y = yield strength of adherend (0.2 percent offset), psi
T = thickness of adherend, in.
L = length of overlap, in.

In the area to the right of the yield line Y (Fig. 4-10) the yield strength of the adherend is not exceeded. Therefore, the shear is directly proportional to the bond strength of quality. In the area to the left of the yield line, the failure is influenced significantly by the nonportional deformation of the substrate. Therefore, quality lines may

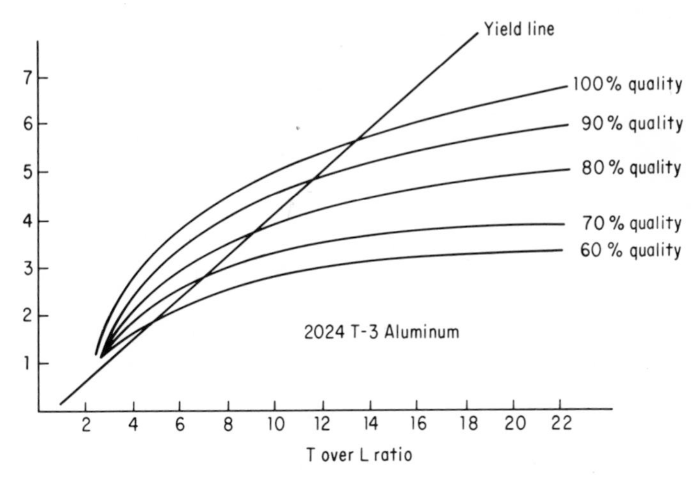

fig. 4-10 *T over L ratio curve.*

be determined by preparing a specimen with variable bond areas as illustrated in Fig. 4-12. This is calculated as follows:

$$\text{Quality percent} = \frac{B_1 + b_2}{b} \times 100$$

The yield strength of the adherend is very important in the design of a lap joint. The yield strength of the adherend could be one-half of the ultimate strength, yet when lattice slippage occurs underneath a bonded joint, the joint is limited and at this point is near failure. Many joint failures can be traced directly to this problem area. The T over L curves should be consulted to prevent this problem.

When stiffening members are attached to thin sheets, the sheets may deflect in service and load the adhesive in peel. If the flanges on the stiffening section can deflect with the sheet, the peel problem may be

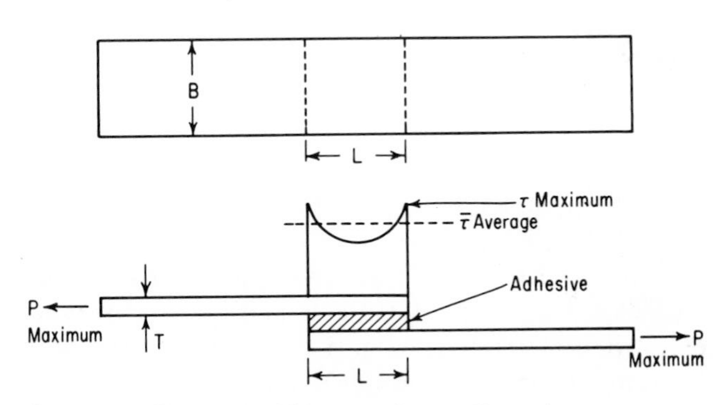

fig. 4-11 *Effect of yield strength on adherend.*

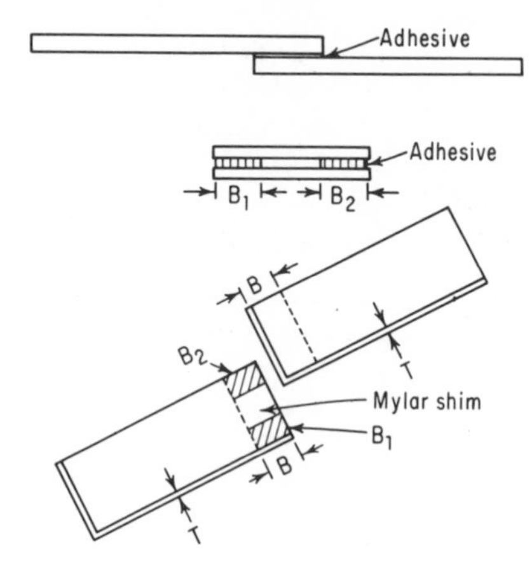

fig. 4-12 *Variable bond area.*

minimized, and reducing the stiffness of the flange on the stiffening section would result in an improved attachment. Figure 4–13 shows simple examples of the elimination of peel stresses.

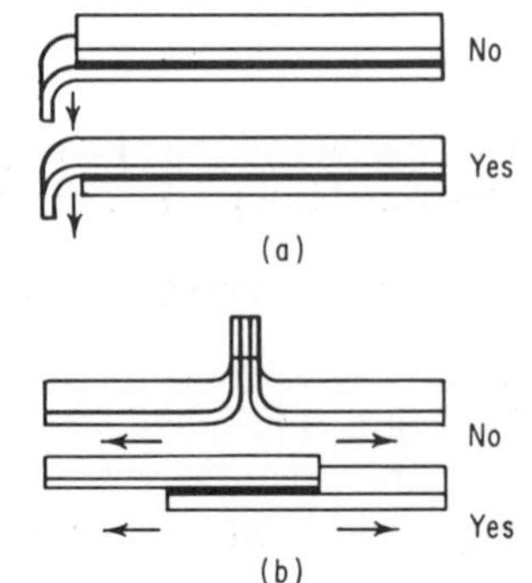

fig. 4-13 *Eliminating peel stresses.*

CORNER JOINTS

Corner joints and close-out areas create many problems in metal-to-metal fabrication. It would be considered very foolish to design out peel and cleavage, have a basically sound assembly, and fail to devote the necessary time to the development of good corner joints. This could very easily be the weak link in the assembly because it is often forsaken for the simplicity of tooling. A simple, thin corner close-out addition may be the difference between success or failure of the assembly (Fig. 4–14).

ANGLE JOINTS

Angle joints are not usually preferred, but are sometimes necessary, especially in attaching clips and attachment angles to accommodate other assemblies. Care must be exercised here to prevent peel loading. Consequently, rigid members aid in elimination of this problem. The bonding area should be made as large as possible and designed so the maximum proportion of the bonded area contributes to the strength and minimizes stress in the direction the adhesive is weakest (Fig. 4-15). If, in sample *A*, the load is in the direction of the arrow, high strength would be realized, but if the stress load is not exactly at right angles to the plane of the joint, as shown in sample *B*, a cleavage force would result; this is considered poor design. In sample *C*, the joint is shown with increased area and considered much improved as compared to *A* and *B*. Sample *D* has the same increased area and a beveled flange which reduces the stiffness at the edge of the joint. This spreads the stress concentrations over a large area and would be considered optimum strength design.

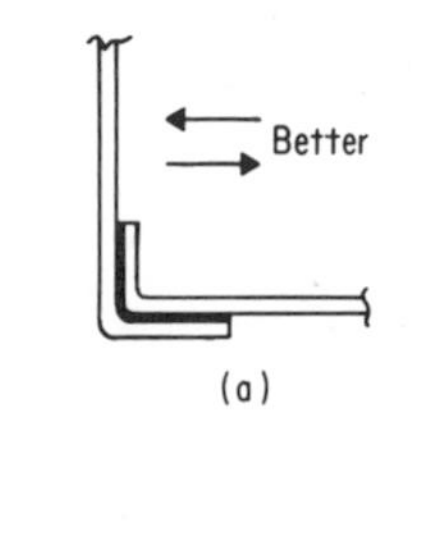

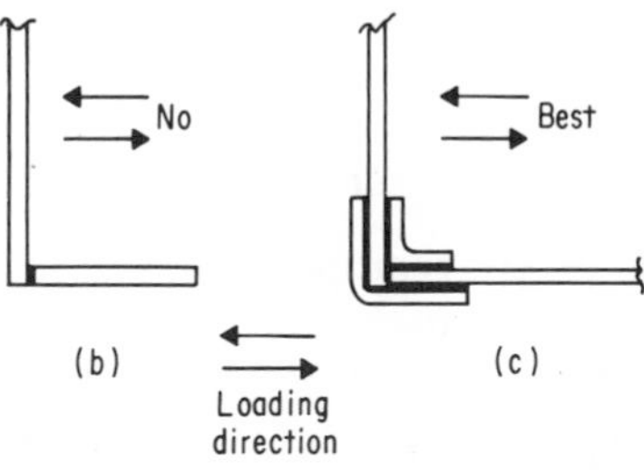

fig. 4-14 *Close-out corners.*

FLANGE JOINTS

The flange joint is widely used in aircraft construction as it adds rigidity and strength to the assembly. This joint does not usually pose a major problem in design. The flat flange section can be improved by tapering the bonded area, which reduces the flange stiffness. The flange stiffness may also be reduced by small grooves, if substrate thickness permits, on the top side of the bonded area (Fig. 4–16). In sample *A*, the squared flange is shown, which in many cases may be considered adequate. Sample *B* shows a similar flange with tapered bonding surfaces that reduces stiffness and would be considered an improvement when compared to sample *A* because the maximum proportion of the bonded area contributes to the strength, and stress is minimized in the direction in which the adhesive is weakest. Sample *C* shows the flange stiffness reduced with grooves which would be comparable in strength to sample *B* but are more expensive to produce. However, this type of flange capitalizes on the advantage of the ultimate strength properties of the adhesive.

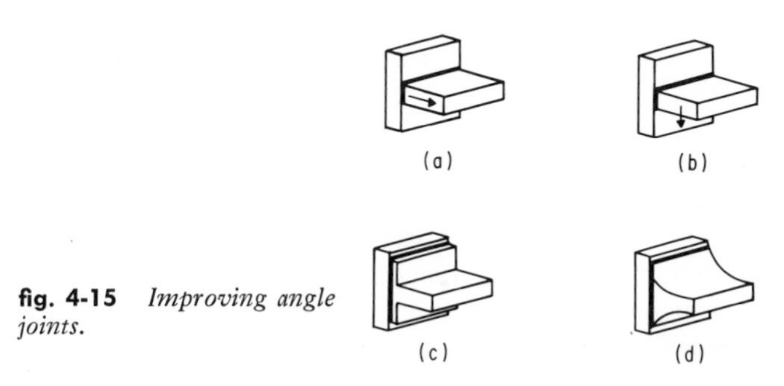

fig. 4-15 *Improving angle joints.*

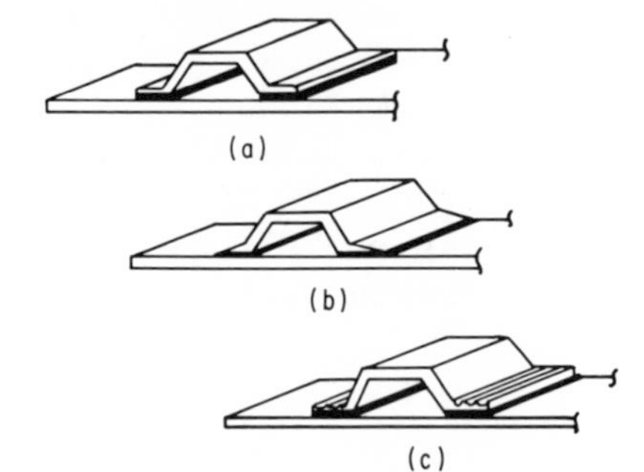

fig. 4-16 *Improving flange joints.*

GUIDELINE, SPECIFICATIONS

A convenient way to provide information as to the expected strength of structural adhesives is to observe the classification and requirements of MMM-A-132 "Adhesives, Heat Resistant, Airframe Structural, Metal to Metal." They are classified as follows:

Type I: For long time exposures (192 hr) to temperatures from −67 to +180°F
- *Class 1:* High T-peel and blister detection
- *Class 2:* Normal T-peel and blister detection
- *Class 3:* No T-peel or blister detection required

Type II: For long time exposures (192 hr) to temperatures from −67 to +300°F

Type III: For long time exposures (192 hr) to temperatures from −67 to +300°F and short time exposures (10 min) from 300 to 500°F

Type IV: For long time exposures (192 hr) from −67 to +500°F

The mechanical shear properties are in Table 4.1, and are used as guidelines by designers, especially in military hardware. Where strengths in excess of the military specification requirements are important to the design of an assembly, or if data are not given in the military specifications that must be considered, tests must be performed to determine these criteria. MMM-A-132 has an adhesive qualified product list table (4.2) that has been qualified to meet the requirements of this specification.

HONEYCOMB AND SANDWICH CONSTRUCTION

Sandwich structure is the largest area of bonding in the aerospace industry. Millions of square feet are bonded each month. It is difficult to define design requirements because of the many complex and changing designs, plus changing service requirements. In the design of an adhesive-bonded sandwich construction, it is important to remember that the nature of stress that the panel or assembly will be subjected to depends on the configuration of the structure and the magnitude and orientation of imposed loads. The loading modes should be known pertinent to the structure in question or tests must be performed to obtain these data. The simple flat panel is almost something out of the past. It is usually more difficult to produce a perfectly flat panel than one with compound curvatures, especially when utilizing high-temperature adhesives and elevated cures in presses or autoclaves.

There are some basic requirements that must be considered in sandwich design:

1. The facings must have sufficient thickness and strength to withstand the tensile, compressive, and shear stresses induced by the design load.

2. The core must have sufficient strength to withstand the shear stresses induced by the service loading.

3. The core must be thick enough and have sufficient shear modulus to prevent overall buckling under the design load.

4. Young's modulus of the core and the compressive strength of the facings must be sufficient to prevent wrinkling of the face sheets under the service requirements.

5. The core cells must be of proper dimension to prevent intracell buckling under the design load.

6. The core must have sufficient compressive strength to resist crushing by design loads acting normal to the panel facings or by compressive stresses induced through flexure.

7. The structure must have adequate strength in the flexural and shear rigidity parameter to prevent deflections under design load.

8. Ease of carving, forming, and tooling must be considered.

9. The type and weight of the adhesive must be considered. In extremely lightweight structures (face sheets 0.010 in. or less) the weight of the adhesive becomes an important factor. The operating temperature of the assembly must be considered, because adhesives have a tendency to creep at elevated temperatures.

10. The operating environment must be considered, as well as the temperature factor, moisture, vibration, and expected length of service, just to mention a few.

The following loading modes should be kept in mind pertinent to parts involving sandwich-type construction:

Creep. Resistance to creep loading in honeycomb sandwich structures is primarily dependent on the adhesive bond. Creep, in this case, is the amount of movement, strain, or deformation in excess of the natural amount, which results from the elastic qualities of the structure, metal, or adhesive system. The service environment of many structures may allow for a certain amount of creep, while with others it is a major problem to overcome. For instance, in wing structures of aircraft, creep deformation could alter the aerodynamic performance of the wing. Creep deformation of a bonded structure is dependent upon:

a. Temperature of the service environment.
b. Panel configuration pertinent to thickness and size.
c. Face material, type of alloy or substrate.
d. Stress in the face sheets.
e. Core material, alloy, metal, plastic, etc.
f. Density of core.
g. Cell size of core.
h. Stress in the core system, which includes the adhesive load.
i. Orientation of the core ribbon.
j. Type of adhesive; this also includes the cure pertinent to temperature elevation, and type of tooling used, and universal heating during cure is important.

Rapid or extreme temperature changes pose a tremendous problem. As an example, a missile loaded on the launching pad with LOX would subject the structure to a cryogenic condition and minutes later, when fired, these same components would be subjected to elevated temperatures.

Fatigue: Mechanical or sonic fatigue does not tend to be a major problem in sandwich structures. Adhesives tend to absorb this type of load. Failures due to repeated fatigue loading usually occur in the face sheets of a sandwich panel. This indicates that the adhesive adheres to the face sheet and core and performs its function during the entire fatigue life of the panel. This could become a problem as aircraft are designed for much faster speeds and utilize the newer and more exotic metals. The attachments and inserts are the more probable danger points applicable to fatigue life of the assembly.

Impact: If impact loads of magnitude are expected, the number one thought would be to increase the thickness of the skins and distribute the load over a wider area, thus reducing the stress. Impact load becomes a major problem at cryogenic temperatures and the choice under these conditions would be an adhesive that retains flexible properties at below-freezing temperatures. If it was anticipated that a sandwich panel would be subjected to impact loads, other provisions might be in order, such as elastomeric coatings, modified attachments, or foam-filled core. Probably the greatest impact loads on honeycomb panels occur in the shop during handling or placing subassemblies into the final part. The "hammer" mechanics can be very dangerous.

Peel: The peel factor should be eliminated if possible, but many times, this is impossible. A panel may be primarily loaded in shear until a break occurs or propagation starts; then the peel factor is present. The greatest insurance against the peel problem is the use of primers. Primer systems can improve resistance to peeling and bending stresses.

Compression: A panel loaded in compression is prone to buckle inward or outward. The core is usually the limiting factor in compression failure. If the buckle occurred in the outward direction, the flatwise tensile strength of the adhesive would be the prime factor; if the buckle occurred inward, the core would be the prime factor.

In the bonded sandwich structure the most important factor is the bond between the face skin and the core, regardless of the attachments, reinforcements, etc. The face-to-core bond must sustain approximately the same loads as the core. It should resist failure due to rupture or creep over the entire range of service and should satisfy all the loading requirements. In addition to shear strength, the core-to-skin bond must have sufficient flatwise tensile strength to prevent wrinkling of the facings under load. This type of defect tends to occur under edgewise loading. The shear strength of the core depends upon the nature of the core material, and its density, and thickness, to some extent upon the facing materials and their thickness, and upon the operating temperature. Data from the materials suppliers are available on these parameters, such as density, modulus, core shear, strength, node strength, etc.

The strength requirements and conservative design values for adhesives used in bonding facings to metal cores are found in MIL-A-25463, "Adhesive, Metallic Structural Sandwich Construction." The adhesives are classified as follows:

1. Type I For long time exposures to temperatures from −67 to +180°F
2. Type II For long time exposures to temperatures from −67 to +300°F
3. Type III For long-term exposures from −67 to +300°F and short time exposures from 300 to 500°F
4. Type IV For long-term exposures from −67 to 500°F
   a. Class 1—For bonding metal facing to metal cores only
   b. Class 2—For bonding metal facings to metal cores, inserts, edge attachments and other components of completed sandwich structures

A sandwich panel is analogus to an I-beam with facings and core corresponding to the flanges and web, respectively. The facings carry axial compressive and tensile stresses. The core sustains shear stresses and prevents wrinkling or buckling of the facings under axial compressives loading. Typical construction of a simple, basic honeycomb sandwich panel is shown in Fig. 4-17.

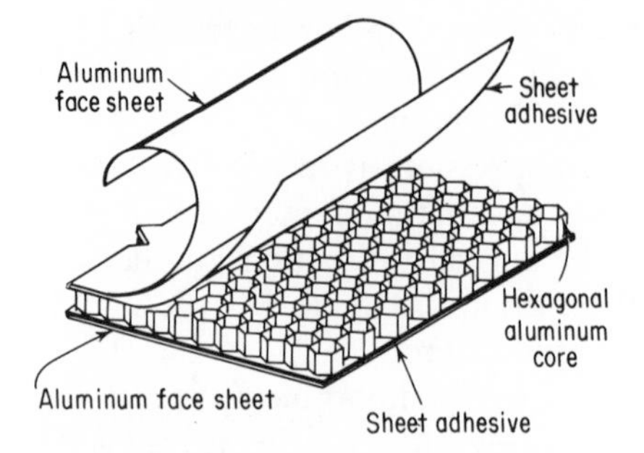

fig. 4-17 *Honeycomb sandwich panel.*

The following symbols will be used in the formulas of this section:

P = total load, lb
A = distance between support and load point, in.
L = distance between support, in.
h = panel thickness, in.
b = panel width, in.
C = core thickness, in. = h-$2F$
F = facing thickness, in.
F_c = core shear stress, psi
F_f = facing stress, psi
M = bending moment, in./lb
y = distance from the neutral axis to the fiber under construction, in.
i = rectangular moment of inertia with respect to the neutral axis, in.
E_f = modulus of elasticity for the facings, psi
$K = 1 - s^2$
s = Poisson's ratio of facing material

$$U = \frac{G_c bh(h+c)^2}{2c}$$

$$D = \frac{E_f bF\ (h+c)^2}{8k}$$

G_c = shear modulus of the core, psi
d = deflection, in.
P/D = slope of load deflection curve used to determine G_c, lb/in.

For facing selection, consider a sandwich panel with symmetrical facings loaded as a simple supported beam, and refer to the analogy of an I-beam. The average stress in the facings is then determined from the following basic beam formula:

$$F_f = \frac{MY}{i}$$

where $M = \dfrac{PA}{2}$

$$y = \frac{(h+c)}{4}$$

$$i = \frac{bf(h+c)^2}{8}$$

For facing stress then, facing stress is given by the formula:

$$F_f = \frac{PA}{bf(h+c)}$$

By letting $A = L/2$ this formula may be applied to a single-point loading (Fig. 4-18).

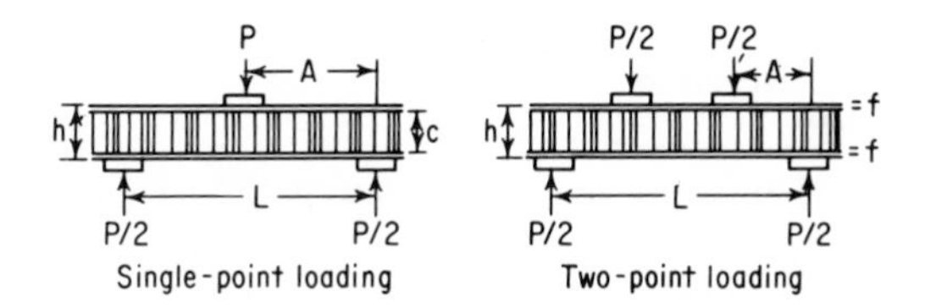

fig. 4-18 *Flexural loading applicable to honeycomb panels.*

In selecting the core, it must be remembered that the compressive and shear strengths of honeycomb and its shear modulus are nearly proportional to its density. Where necessary, local high-compressive stresses can be carried by densification of the core by an insert of metal, plastic laminate, or a casting resin. In some applications, such as flooring, core damage can be eliminated by using a thicker face sheet on the top side to aid in the impact-load distribution.

The shear stress in a sandwich core is estimated by the following formula:

$$F_c = \frac{P}{b(h+c)}$$

The allowable core shear stress is naturally influenced by span, facing thickness, specimen width, and core thickness, as well as adhesive variables when using the simplified formula.

The deflection of a simply supported sandwich beam loaded at midspan is given in the following formula:

$$d = \frac{PL^3}{48D} + \frac{PL}{4U}$$

The second term of this expression considers the deflection due to the shear deformation of the core and is normally minute. For this formula, one must determine core shear modulus. If the deflection of a centrally loaded sandwich beam is measured at midspan, this may be computed by the following formula:

$$G_c = \frac{\frac{P}{d}\,Lc}{2hb(h+c)\left[i - \frac{(P/d)}{(L^3/48D)}\right]}$$

This formula is not reliable when

$$\frac{PL^3}{d48D} \geq 0.6$$

When possible, the designer should take advantage of the fact that honeycomb is stronger in shear and has a higher shear modulus when the L, or ribbon, direction is parallel to the span.

Cell size must be emphasized again as an important variable in core selection. Smaller cells provide increased bond strengths and improved wrinkling stabilization to facings loaded in compression. In aluminum-faced sandwich, a maximum ratio of ten to one of the cell size of the facing thickness will ensure exceeding the compressive yield strength of the facings.

The bond between core and facings must sustain approximately the same shear stress as the core, which has been mentioned before, but this is an important factor in the adhesive selection. The adhesive joint must resist failure in rupture and creep over the entire range of service temperature and must be consistent with all other bonding and service requirements. The adhesive selected should have high peel strength if possible, although this is not a prime factor in panel design. Adhesives having average peel strengths, but high shear, fatigue, and creep strengths are used successfully in aerospace sandwich construction.

The use of adhesives in the form of a supported film or tape is desirable in several ways. It ensures uniform distribution of the adhesive over the area to be bonded and facilitates close control of adhesive thickness and weight. The presence of a carrier fabric also improves the peel characteristics of the adhesive joint, although it may be the limiting factor at elevated temperatures. It is important that a fillet of bonding material be developed between the face sheet and the honeycomb cell walls, thus the mode filling characteristics of the chosen system must be considered. Tests must be conducted with simulated panels similar to the panel to be designed. The satisfactory performance of a bonding system in a normal lap joint is not necessarily an indication that it will perform well in sandwich construction, although many adhesives perform well in both areas.

The final consideration is the weight of the panel that will subsequently satisfy the design requirements. Weights may be calculated with the following formulae:

Adhesive: $0.05\,\frac{\text{lb}/}{/\text{in.}^3} \times 0.01\text{ in. (thick)} \times 144\,\frac{\text{in.}^2/}{/\text{ft}^2} =$

$0.072\text{ lb/ft}^2\text{ of facing} \times 2\text{ facings/panel} =$

$0.144\text{ lb/ft}^2\text{ of panel}$

Core: (¼ × .003 aluminum honeycomb, 6 lb/ft^3 density, ½ inch thick)

$6.0\,\frac{\text{lb}}{\text{ft}^3} \times \tfrac{1}{2}\text{ in.} \div 12\text{ in./ft} = 0.25\text{ lb/ft}^3$

Facings: (0.016 in. thick aluminum)

$$0.10 \frac{\text{lb}/}{/\text{in.}^3} \times 0.016 \text{ in} \times 144 \frac{\text{in.}^2/}{/\text{ft}^2} =$$

0.231 lb/ft^2 of facing $\times$ 2 facings/ panel $=$
0.462 lb/ft^2 of panel

Splicing of core and face sheets: It is often necessary to join sections of core in an assembly. This is almost always necessary in very large assemblies. It may be easier to splice the core for contoured configurations. Many times it may be desirable to utilize several types of core in one assembly, or it may become necessary to remove a section due to damage.

A stronger splice, and that considered the best splice, would be a scarf joint, which would produce high bond strengths. It would also be easier to pressurize a splice of this configuration. If the splice bond is not a repair, and is considered a normal operation, it is cured at the same time as the rest of the assembly. Liquid or epoxy-based adhesives are usually employed for core splicing because they do not require the pressures that are necessary for many other systems.

The butt splice is used often, and very successfully. This is advantageous because it is the most simple type of core splicing. A small amount of clearance is left between the two sections and is potted with an adhesive compound, usually epoxy, and the potting operation includes a few cells adjacent to the splice.

Mechanical interlocking provides one of the stronger splices. The edges of the two core sections are overlapped and then literally crushed together. The crushed area may then be potted in the same manner as any other type of splice for added strength.

There are other types of splices such as zig-zag, Z-cuts, and dovetail, but they are more difficult to fabricate and are not often used. However, these types of splices do give excellent strength properties.

Assemblies often require face splices to establish continuity in the assembly. Splices are made in the face sheet by butting the two sections of face sheet and covering the mating edges with a strip of metal, essentially producing two lap shear joints. This type of splice may be analyzed as a lap joint. Tooling must be considered in the splice, and provision made not to eliminate or reduce pressure in adjacent areas.

Face splices may be made by simple lap joints and are used frequently with thin-gauge face sheets. When utilizing this type of splice, a flange of the desired length is left on one panel section which will overlap the other. Since this again produces a nonflush surface, tooling must be kept in mind. Splices may be made during the bonding operation, if it is feasible, and with the same adhesive used in the assembly. If the material used for the sandwich is not feasible for the splice, then another system may be used, provided the cure cycle and pressure requirements compare with the sandwich adhesive.

It may be necessary to splice the two assemblies (or more) which are already cured in the subassembly state. This is often accomplished with room-temperature-setting adhesives, and using the lap-type strap. The prime consideration would be adequate bonding area on the strap to ensure strength that meets the design requirements. Some consideration must be given to the possible difference in thermal expansion which could result in warpage. When a strap splice is used that is of sufficient thickness, performance may be improved by beveling or tapering the edges.

Edge members and close-out members: There are many types of edge members for honeycomb design (Fig. 4-19). The simple edge is one in which the faces extend slightly beyond the core in a plane around the panel, and it is simply potted with an adhesive compound. The same configuration may be used with some other fill material such as a metal or plastic block, a piece of wood, or a plastic wet layup laminate. The block inserted as an edge member aids in resisting compressive loads when the panel is attached to the structure. In most cases it would be feasible to use a sheet-metal doubler between the face and the block. This would provide sufficient countersinking depth or bearing thickness for fasteners. The block would be bonded of nonstructural nature but the doubler must be of high quality. The edge member not only absorbs loads but acts as a seal at the same time. Many times they are extruded or machined (Fig. 4-20).

Zee-type edges are the most common of all edge members. The edges are sometimes formed by bending one face of the sandwich to meet the opposite face at the edge of the panel. This is very effective with thin-gauge face sheets and allows the zee to be bonded easily dur-

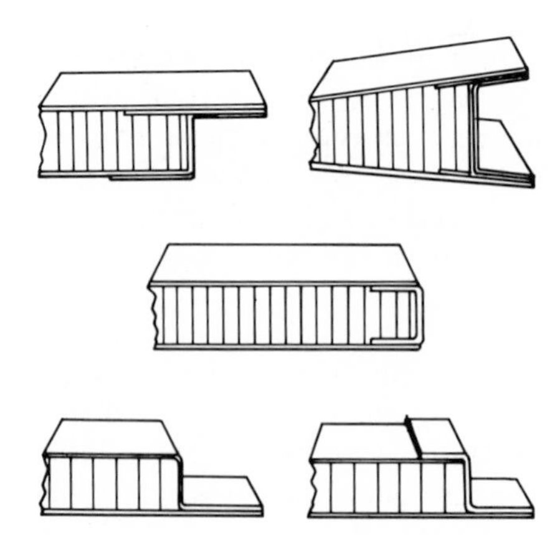

fig. 4-19 *Edge and close-out members.*

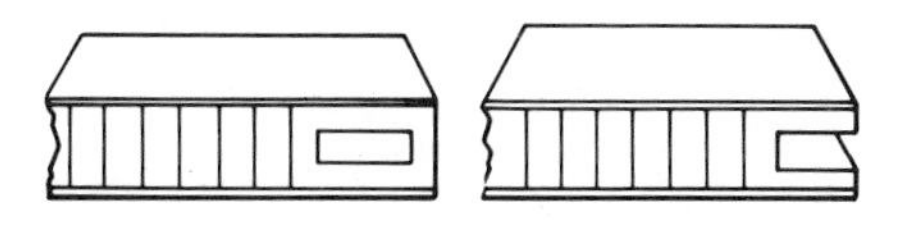

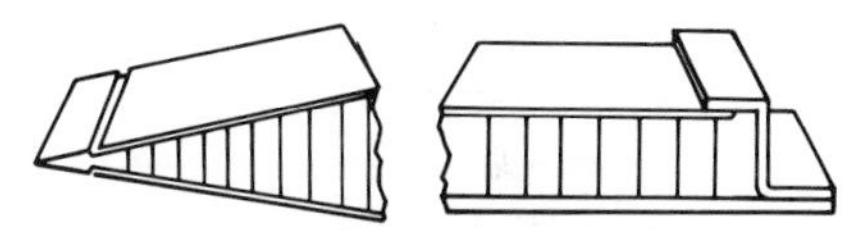

fig. 4-20 *Typical extruded and/or machined edge design.*

ing the virgin curing operation. If this is not feasible, a separate piece of metal is used, a Zee is formed, and then bonded to both face sheets. This can also be accomplished when the panel is bonded. It may be necessary to use a doubler around the edge of the face sheets to increase the edge stiffness, thus reducing the bending moments. The Zee formed by right-angle bends is the more common, and in addition to being formed easily, it permits the core in the sandwich material to be butted up against one section of the Zee, without utilization of extra fill material, or the necessity of core taper at the edge.

A simple technique used by many manufacturers is simply the wet layup of a prepreg around the edges of the sandwich. This is simple and provides an adequate seal, but is not capable of carrying a load of any great magnitude. The core is usually tapered and one face sheet extends slightly beyond the other. The strip of prepreg is then placed on the bottom sheet against the core and lapped around the desired amount on the top face sheet. The prepreg closeout can be cured with the assembly on a "one-shot" basis.

Attachments and inserts: There are so many types of attachments that it would be impossible to do justice to this area with limited space. Many companies design and manufacture their own special attachments or consult experts who make fastening their business. Some typical attachments are shown in Fig. 4-21. Very little can be said concerning the analysis of an attachment or fitting, as this usually requires testing to ensure conformance to the proposed design. In most cases, the loads in the fittings and attachments are transferred to the face sheets of the sandwich assembly and are transferred through the adhesive bond. If the adhesive is sufficiently strong for the expected loads, then the designer must consider the adherends. Many attachments are so complex that it becomes difficult to follow the loads through the joint, from the standpoint of a designer.

Very little information is available for public consumption, because everyone does the testing necessary for his particular design and this information may be worthless to other designers. Many of the manufacturers of this type of hardware can furnish information that is helpful and

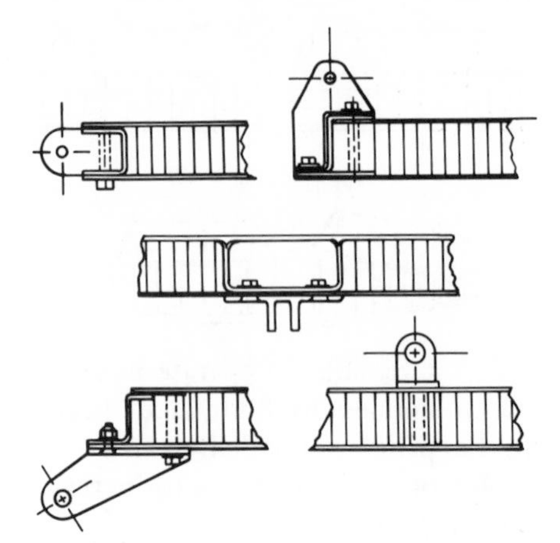

fig. 4-21 *Typical fitting attachments.*

many are prepared to do their own testing. If necessary they will design a special fitting and test it for conformance to the customer's specifications. Inserts are placed within the panel at points of attachments and used to maintain the faces in their original position. It is desirable to bond the inserts during the fabrication of the panel, but it is not necessary because they may be bonded in later, often with room-temperature-setting adhesives. It may be necessary to bond reinforcements in the insert area, depending on the anticipated load. Testing is often necessary to determine the reliability of the insert for the service requirement, but there are specialized manufacturers in this area who can provide valuable information to the designer. Many types are being marketed that are squeezed into a routed hole in the panel after it is fabricated. If the insert is to be subjected to extreme loads, the potted type is most reliable. When the insert is installed, a resin is placed around it creating a bond between the insert and the core. This aids in load distribution, thus preventing concentrations in the face sheets. A densified area may be in order not only for fastener and insert load transfer, but may be necessary in areas of the assembly where heavy stress concentrations are expected (Fig. 4-22).

TABLE 4.1 MMM-A-132 Tensile Shear Requirements (minimum average strength requirements, psi)

	Type I				
Test Conditions	Class 1	Class 2	Type II	Type III	Type IV
10 min @ 500°F	4500	2500	2250	2250	2250
192 hr @ 500°F	2500	1250			
75°F	....	....	2000	2000	2000
10 min @ 180°F	....	....	2000	2000	2000
102 min @ 300°F	....	....	....	1850	2000
192 hr @ 300°F	....	....	....	....	1000

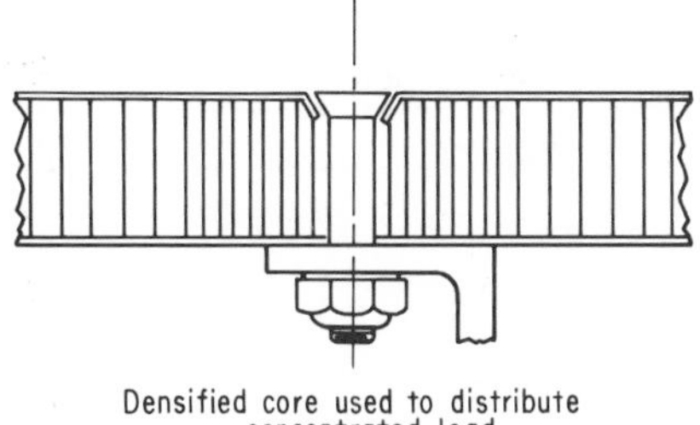

fig. 4-22 *Densified core for load distribution.*

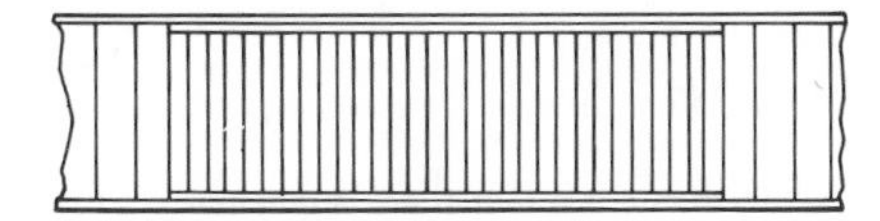

TABLE 4.2 List of Adhesives Qualified to MMM-A-132 "Adhesives, Heat Resistant, Airframe Structural; Metal to Metal" as of March, 1967

Manufacturer's Designation	Qualified	Manufacturer
FM-1000	Type I Class I	American Cyanamid Company
AF-6032 System (AF-32 film) (EC-1660 primer)	Type I Class I	Minnesota Mining & Manufacturing Company
AF-54300 System (AF-300 film) (EC-2254 primer)	Type I Class I	Minnesota Mining & Manufacturing Company
AF-5641 (AF-41 film) (EC-1956 primer)	Type I Class I	Minnesota Mining & Manufacturing Company
Metlbond 408	Type I Class I	Whittaker Corporation Narmco Materials Division
Metlbond 400	Type I Class I	Whittaker Corporation Narmco Materials Division
Epon Adhesive 951	Type I Class I	Shell Chemical Company
FM-238 Adhesive (BR-238 primer)	Type I Class 2	American Cyanamid Company Bloomingdale Dept.

Manufacturer's Designation	Qualified	Manufacturer
Redux Film 775	Type I Class 2	Bonded Structures Limited
Plastilock 620	Type I Class 2	The B. F. Goodrich Company
Plastilock 639	Type 1 Class 2	The B. F. Goodrich Company
AF-6	Type 1 Class 2	Minnesota Mining & Manufacturing Company
AF-5930 (AF-30 film) (EC-1459 primer)	Type 1 Class 2	Minnesota Mining & Manufacturing Company
AF-6030 (AF-30 film) (EC-1660 primer)	Type 1 Class 2	Minnesota Mining & Manufacturing Company
AF-20120 (AF-120 film) (EC-2320 primer)	Type 1 Class 2	Minnesota Mining & Manufacturing Company
Metlbond 406	Type 1 Class 2	Whittaker Corporation Narmco Materials Division
FM-47	Type 1 Class 3	American Cyanamid Company
FM-58	Type 1 Class 3	American Cyanamid Company
FM-61	Type I Class 3	American Cyanamid Company
FM-245	Type 1 Class 3	American Cyanamid Company
FM-250	Type 1 Class 3	American Cyanamid Company
FM-97	Type 1 Class 3	American Cyanamid Company
FM-96	Type 1 Class 3	American Cyanamid Company
BR-92	Type 1 Class 3	American Cyanamid Company
FM-86 with BR-86 primer	Type 1 Class 3	American Cyanamid Company

Manufacturer's Designation	Qualified	Manufacturer
FM-54	Type 1 Class 3	American Cyanamid Company
Geneerco 214	Type 1 Class 3	General Veneer Mfg. Co.
Honey-co Bond 371E	Type 1 Class 3	Honeycomb Co. of America
AF-110 "Scotchweld"	Type I Class 3	Minnesota Mining & Manufacturing Company
EC-1469 "Scotchweld"	Type 1 Class 3	Minnesota Mining & Manufacturing Company
EC-1471 "Scotchweld"	Type 1 Class 3	Minnesota Mining & Manufacturing Company
AS-8233 "Scotchweld"	Type 1 Class 3	Minnesota Mining & Manufacturing Company
AF-5640 "Scotchweld"	Type 1 Class 3	Minnesota Mining & Manufacturing Company
EC-2186 "Scotchweld"	Type 1 Class 3	Minnesota Mining & Manufacturing Company
AF-82110 (AF-110 film) (EC-1682 primer)	Type 1 Class 3	Minnesota Mining & Manufacturing Company
AF-40T	Type 1 Class 3	Minnesota Mining & Manufacturing Company
AF-111	Type 1 Class 3	Minnesota Mining & Manufacturing Company
"Scotchweld" 60204 (AF-204 film) (EC-1660 primer)	Type 1 Class 3	Minnesota Mining & Manufacturing Company
AF-114	Type 1 Class 3	Minnesota Mining & Manufacturing Company
Metlbond 4021	Type 1 Class 3	Whittaker Corporation
No. 328	Type 1 Class 3	Whittaker Corporation
Narmco 324	Type 1 Class 3	Whittaker Corporation
Metlbond 400	Type 1 Class 3	Whittaker Corporation

Manufacturer's Designation	Qualified	Manufacturer
Swallow Adhesive 371W	Type 1 Class 3	Pelham Industries
FM-132	Type 1 Class 3	Minnesota Mining & Manufacturing Company
Bondmaster M224	Type 1 Class 3	Pittsburgh Plate Glass Co.
Bondmaster M602/M611	Type 1 Class 3	Pittsburgh Plate Glass Co.
Bondmaster M-623	Type 1 Class 3	Pittsburgh Plate Glass Co.
Epon 934	Type 1 Class 3	Shell Chemical Company
FM 97-1070	Type 1 Class 3	American Cyanamid Company
Aerobond 430	Type II	Adhesive Engineering Division of Hiller Aircraft
Metlbond-329	Type II	Whittaker Corporation
HT-424	Type II	American Cyanamid Co.
Plastilock 650	Type II	B. F. Goodrich Company
"Scotchweld" EC-1469	Type II	Minnesota Mining & Mfg. Co.
"Scotchweld" AF-5931 (AF-31 film) (EC-1459 primer)	Type II	Minnesota Mining & Mfg. Co.
"Scotchweld" AF-31	Type II	Minnesota Mining & Mfg. Co.
"Scotchweld" EC-1595	Type II	Minnesota Mining & Mfg. Co.
"Scotchweld" AS-9795	Type II	Minnesota Mining & Mfg. Co.
Epon 901-B3	Type II	Shell Chemical Company
Plymaster ACG1031	Type III	Pittsburgh Plate Glass Co.
Epon 422	Type III	Shell Chemical Company
Aerobond 422	Type III	Adhesive Engineering
Aerobond 430	Type III	Adhesive Engineering
HT-424 with HT-424 Primer	Type III	American Cyanamid Company
AF-7431 (AF-31 film) (EC-2174 primer)	Type III Type IV	Minnesota Mining & Mfg. Co.
AF-126	Type 1 Class 2	Minnesota Mining & Mfg. Co.

REFERENCES

Adhesive Bonding, Rohr Aircraft, Chula Vista, Calif.

"Aeroweb Honeycomb Sandwich Construction for Supersonic Aircraft," Ciba Limited, Duxford, Cambridge, England, 1961.

"Aeroweb Honeycomb Sandwich Flooring for Aircraft," The Ciba Company, Duxford, Cambridge, England, 1961.

Epstein, G.: Adhesive Bonds for Sandwich Construction, *Adhesives Age,* August, 1963.

"Handbook of Adhesives," American Cyanamid Company, Havre de Grace, Md., March, 1964.

Herndon, C. F.: "Design of Heavy-Gage Bonded Honeycomb Sandwich," presented at Conference on Structural Bonding, Marshall Space Flight Center, Huntsville, Ala., March 16, 1966.

Honeycomb, Honeycomb Company of America, Bridgeport, Conn.

Honeycomb Bids for High Temperature Use, *Chemical and Engineering News,* July, 1962.

Jacobson, R. E.: Sandwich Construction Consisting of Glass Fabric Reinforced Phenolic Facings and Honeycomb Core for Structural Applications, *Convair Astronautics,* October, 1960.

Kuenzi, E. W., and G. H. Stevens: "Determination of Mechanical Properties of Adhesives for Use in the Design of Bonded Joints," Forest Products Laboratory, September, 1963.

Licari, J. J.: Guide to High Temperature Adhesives, *Product Engineering,* December, 1964.

Ljungstrom, O.: "Design Aspects of Bonded Structures," SAAB Aircraft, Sweden, May, 1959.

McCown, T. E.: "Structural Analysis of Honeycomb Sandwich Construction," AVCO Corporation, Nashville, Tenn., March, 1962.

Merriman, H. R.: Selecting Adhesives for Structural Bonding, *Adhesives Age,* August, 1964.

Norris, C. B., W. S. Ericksen, and W. J. Kommers: "Flexural Rigidity of a Rectangular Strip of Sandwich Construction," Forest Products Laboratory, October, 1958.

Rebeski, H.: "Metal Wings with Pre-Stressed and Adhesive Bonded Skins," The Ciba Company, Duxford, Cambridge, England, May, 1958.

"The Structure of the Comet," The Ciba Company, Duxford, Cambridge, England, 1962.

Why Honeycomb in the First Place? *American Machinist,* March, 1957.

Yurek, D. A.: Adhesive Bonded Joints, *Adhesives Age,* December, 1965.

Zelman, I. M.: "Phoenix Bonding Manual," Hughes Aircraft Company, Report M-304, Culver City, Calif., February, 1964.

CHAPTER 5

Surface Preparation

The newly hatched lobster has one chance in a million of reaching maturity. Those odds could possibly apply to the probability of producing an excellent bond without proper surface preparation. Lobsters raised in a hatchery are protected during their infancy and the odds for survival are increased to one in a hundred. These are the odds one might seek if he were wagering on the success of an adhesive-bonded structure if the surfaces were prepared in any manner other than rigidly controlled conditions.

Surface preparation of adherends for joining adhesively plays a dominant and most important part in the reliability of the finished product. This cannot be overemphasized. Regardless of how well all other operations are performed, if the adhesive does not wet the surface of the substrate, the bond will be substandard, as compared to maximum expectation. This statement should always be remembered, "If it does not wet, it will not bond." Cleaning must be thorough and carefully monitored by all concerned.

It can be safely said that the prime purpose is to develop a bonding surface that will result in an optimum bond and provide the best in service protection possible pertinent to the expected service environment. The target in surface preparation is to ensure that adhesion develops in the joint to the extent that the weakest link in the bonded joint is within the adhesive layer, not at the adhesive/adherend interface. This type of bond fracture is referred to as cohesive failure and a layer of adhesive

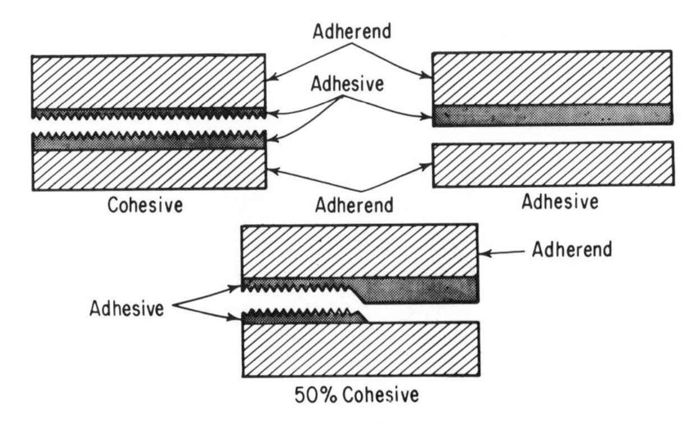

fig. 5-1 *Cohesive and adhesive bond failure.*

remains on both adherends (Fig. 5-1). When fracture occurs at the adhesive/adherend interface, it is referred to as adhesive failure. In analysis of a joint that has been tested to destruction, the mode of failure is expressed as a certain percentage cohesive or adhesive failure. The ideal failure from a surface-preparation standpoint is a 100 percent cohesive failure.

Cleaning procedures depend on the specific substrates employed, the design of the parts, service environment and, in some isolated cases, the adhesive or primer employed. The service environment plays the most important role in the case for proper surface preparation. The aerospace industry is faced with severe environmental problems and the severity grows with each new design. Surface preparation holds the key to the service life of an adhesive bonded structure.

Listed here are several methods of preparing substrates for adhesive bonding.

1. Chemical treatment

Without going into a detailed discussion of the physical and chemical theories of metal surfaces, it can be simply stated that the chemical treatments in some manner alter the surface physically and chemically to increase its free energy level and make it receptive to adhesion. Specific chemical treatments are required for each of the metals in question and are described in detail in the following pages. The surface is usually altered by the reaction of a chemical with some ingredient in the substrate and would possibly result in an oxide formation, dimensional loss, or both.

2. Chemical cleaning

This is sometimes used but should not be confused with chemical etching. In chemical cleaning, there would be relatively no dimensional loss or oxide formation; it would result in the removal of foreign matter from the adherend surface.

3. Abrading

Usually accomplished by sanding or filing, abrading is usually not advisable on metallics unless a chemical cleaner follows. Abrading has a tendency to push foreign particles into the surface of a metallic, which may cause corrosive problems. It may be used satisfactorily on nonmetallics as no corrosion problem exists.

4. Solvent cleaning

This type of cleaning would only be used when there is no other alternative. Very little cleaning can be accomplished in this manner; the foreign matter is only moved around on the faying surface, especially in solvent wiping. Plastic substrates, in most instances, will absorb solvents and, unless they are allowed to remain unbonded until evaporation is complete or the parts are force dried, the probability of porous bond lines is very high.

5. Alkaline cleaning

A strong detergent is used primarily for steels and a pretreatment for aluminum and titanium parts prior to chemical treatments. They are usually heated for more efficiency. In the preparation of metals for bonding, the alkaline cleaning method is usually used as an intermediate cleaning method, i.e. through the process of detergency the alkaline cleaner cleanses the surface by emulsifying the surface contaminants and holding them in suspension, thus producing a chemically clean surface.

The alkaline cleaner is usually a combination of alkaline salt and soap, detergent and/or surfactant. The most commonly used alkaline salts are sodium orthosilicate, sodium metasilicate, trisodium pyrophosphate, or sodium tetraborate. Strong caustic solutions of sodium hydroxide are sometimes used. Alkaline cleaners are available commercially or may be prepared from the appropriate ingredients to existing formulations. Cleaners of balanced alkalinity and those containing inhibitors should be used for steel, magnesium, and titanium. A caustic solution of sodium hydroxide (7 to 8 percent by weight in water) may be used as a cleaner for magnesium. Alloys of high zinc content as ZK60A are attacked by cleaners containing over 2 percent caustic. Inhibited cleaners such as those used for cleaning aluminum should be used with such alloys. Solutions for cleaning should be maintained at a concentration of 3.0 to 4.5 percent by weight in water. Parts prepared for cleaning should be immersed for 8 to 12 min in a well-agitated solution maintained at

170 to 210°F. Parts should be rinsed immediately by immersion and/or spraying with deionized water.

The vapors of alkaline cleaners are not considered toxic, but constant bodily contact may result in skin irritations.

6. Vapor honing

This method is used in special cases but is usually considered too expensive for bonding. The surface finish from this method is considered adequate for adhesive bonding.

7. Wood filing

If wood is to be prepared for adhesive bonding, it should never be sanded with fine grit sandpaper. This tends to fill the pores in the wood and prevents adhesive wetting. Since wood depends largely on mechanical adhesion for strong joints, a vixon or rough file would be adequate for preparing the surface.

8. Conversion coating

This type of surface is used quite extensively on magnesium. It consists of a chemical coating applied to desired thickness, which protects the substrate and provides the adhesive with a good wetting surface. This also prevents corrosion after environmental exposure in service.

9. Caustic cleaner

Caustic cleaners are used as a pretreatment on some metallics before chemical treatment. The milder detergents are usually preferred in lieu of the caustics.

10. Ultrasonic cleaning

This technique is utilized often for cleaning miniature details but is considered impractical and expensive for large assemblies. This method consists of a fluid container into which a high-frequency vibration is induced by vibrating crystals, or some other form of vibrating apparatus.

11. Sandblasting

This is used extensively for preparing the nonmetallics for adhesive joining. Sandblasting is not recommended for metallic substrates unless other cleaning techniques are to follow that will remove the fine particles and foreign matter generated by the blasting operation. If the foreign matter is not removed, oxidation is likely to occur at the adhesive interface. Plastics are stable in this area; therefore sandblasting may be used to a great advantage in preparing them for bonding.

Cleaned parts should be untouched by human hands. They should be wrapped in neutral paper (sometimes primed first), placed in a polyethylene bag or its equivalent, and removed when bonding time approaches. Cleaned parts should always be handled using white, clean gloves; a large sum of money, manufacturing time, and labor can be lost if handling procedures are inadequate. There is absolutely no evidence to support the theory that anything can be bonded satisfactorily if a contaminated surface is present. The new theory, clean as you bond, has not had the time to be tested for endurance. Suppose a missile is stored five years before service? Many of the cleaning techniques being used at present may prove totally inadequate in the future. The age-old question always seems to arise. A detail was bonded without cleaning and the same substrate bonded with etched surfaces and they mechanically tested the same. They may test the same today, but what about a few months from now after the parts have been subjected to heat and moisture factors? The unclean part then becomes unreliable (Fig. 5-2).

12. Distilled water

The cleaning solution mixtures should be prepared and the details rinsed in distilled or deionized water (terms are usually used to indicate

fig. 5-2 *Humidity exposure and its effects on aluminum prepared for bonding by various surface-preparation techniques.*

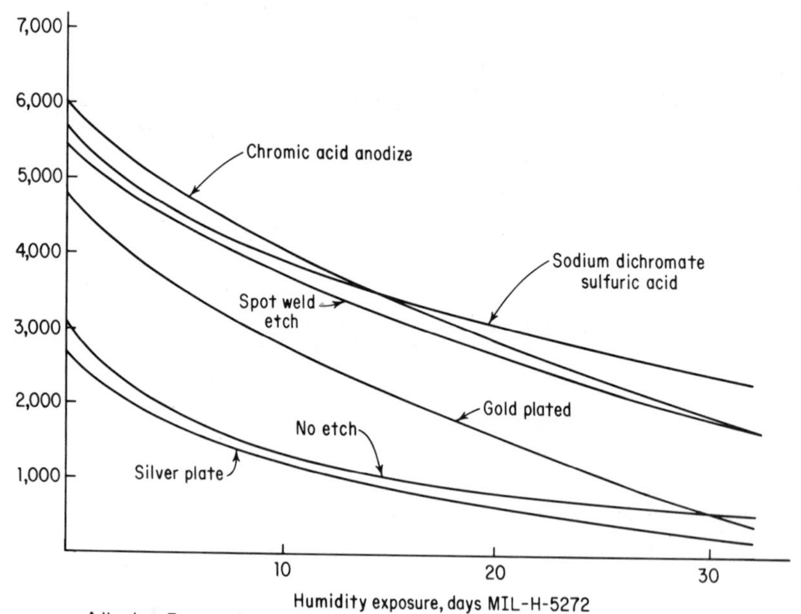

a certain level of purity). The general requirements for acceptable water are:

Resistance	50,000 ohms/cm² at 30°C
Total alkalinity	10 ppm as $CaCO_3$
Chlorides	15 ppm max
pH	6.5 to 7.5

13. Vapor degreasing

One of the terms used frequently in conjunction with cleaning of metals is vapor degreasing, because the first step in preparing a metallic adherend for bonding is degreasing, i.e., the removal of grease, oils, waxes, or other organic matter associated with the forming, handling, and preservation of metals. The degreasing process may be accomplished by solvent washing, but the vapor degreasing technique is preferred, with the exception of titanium.

Vapor degreasing is accomplished by suspending the metal details into a container which contains a chlorinated solvent such as trichloroethylene (boiling point 250°F) that has been vaporized. Because of the difference in temperature between the hot vapors and the cool metal parts, vapors condense on the parts and dissolve the greases, oils, waxes, etc. Most vapor degreasers are also equipped for spraying the parts with clean solvent. It is recommended that the degreaser be cleaned when the boiling point reaches 195°F for trichloroethylene and 256°F for perchloroethylene. These temperatures correspond with the boiling points of a solution of the respective solvent and 25 percent by volume mineral oil. A vapor degreaser to be utilized for precleaning of metals for adhesive bonding should be used for that purpose alone, and not as a general-purpose degreaser, such as breaking down potting compounds, sealants, etc. There is always the possibility it may become contaminated with silicones or some other undesirable material.

The requirements for vapor degreasing are specified in MIL-S-5002, "Surface Treatments (Except Priming and Painting) For Metal and Metal Parts in Aircraft."

Fundamentally, a degreaser is a thermally balanced distillation and should have the following components:

a. Solvent compartment
b. Solvent heating element (steam, gas or electric)
c. Work area—large enough to accommodate the parts to be cleaned
d. Vapor level controller—water-cooled jackets or coils
e. Freeboard area—additional height above the vapor level control to prevent vapors from travelling out of the work area and to provide a space for drying the parts after they are removed from the vapor area

For proper selection and operation of equipment, recommendations from qualified manufacturers of vapor-degreasing equipment should be obtained.

14. Water-break-free test

The term "water-break-free" is used extensively in conjunction with cleaning for adhesive bonding, and actually is a check of the surface tension of the substrate that is to be joined. Distilled water should be used for this test and may be applied with an eye dropper or the equivalent, or even observed during the spray rinse. If the detail is not properly cleaned the water will not spread when placed on the surface but resembles pure mercury when placed on waxed paper. If the detail is clean and surface tension low, the water will spread, wetting a large area (Fig. 5-3).

15. Surface-preparation evaluation

The water-break-free test is by no means the last word in evaluating a surface preparation, although it is considered an adequate method for obtaining information pertinent to the surface condition provided the technique has been validated by other testing media. The only reliable method for determining the bonding quality of a surface preparation is to bond the surface with the candidate adhesive and perform tests under simulated operating conditions. The possibility that a process may produce undesirable effects in the substrate or be incompatible with the adhesive system must be given careful consideration. When subjected to the expected service environment, problems may become apparent that cannot be detected by testing conventional unconditioned test specimens.

The metal-to-metal peel test has been used extensively for determining the bonding qualities of a proposed surface preparation, and the validity of these tests is seldom questioned. However, the problem may be

fig. 5-3 *Water-break-free test.*

encountered with the adhesive system because many adhesives possess poor peel characteristics. If an adhesive with excellent peel properties is used for the evaluation and another system used for the part to be produced, problems may occur because it cannot be assumed that because a surface does or does not meet with one adhesive system the surface will react in a like manner with all adhesives. Peel tests can establish certain limitations, reveal advantages, and suggest guidelines for more extensive future testing.

16. Cleaning area

Before entering into the discussion and methods for preparing surfaces for adhesive bonding, the cleaning area should be mentioned because of the factors associated with the cleaning area that affect the final results of the bonded assembly.

The first objective where chemicals are being used is the safety of personnel responsible for the operation. The cleaning area should be well vented, and exhaust fans should be used to remove any harmful vapors. The humidity and temperature should not reach extreme levels, although close controls are not considered a necessity provided the parts are moved out quickly after cleaning.

The tank locations are important in relation to the sequence of operations, thus preventing droplets of one solution falling into another during the transfer of parts. The cleaning tanks should be well located in relation to the priming area to provide a quick and efficient flow of parts.

The cleaning operation may range from a laboratory basis using beakers and hot plates (or immersion heaters), a large production operation (Fig. 5-4) or out-of-station cleaning which includes field repairs where portable equipment may be utilized. This brings out an important factor, because the prime concern of research and development people is reproducibility of cleaning results. If cleaning is accomplished on a closely controlled laboratory basis, certain variation will occur on a large production basis. The personnel responsible for the production specifications must consider this factor and establish reasonable controls that would tolerate slight deviations on a manufacturing basis, otherwise problems will occur because laboratory results are difficult or impossible to reproduce on a large-scale production basis.

17. Aluminum alloys

Aluminum is presently utilized for adhesive bonding more than any other metal. Consequently, more effort has been expended to develop adequate cleaning techniques for preparing the aluminum surfaces for adhesive bonding. Method A, the sulfuric acid-sodium dichromate

fig. 5-4 *Cleaning tanks. (Douglas Aircraft Company.)*

method was developed by Forest Products Laboratory and is commonly referred to as FPL etch. There are many slight deviations in this method, various companies having modified the basic concept to fit their particular cleaning systems, but the extent of solution modification is minute.

The sulfuric acid-sodium dichromate solution may be utilized on clad or unclad aluminum, with the exception of honeycomb core, and the dimensional loss is negligible when processed properly.

18. Method A, sulfuric acid-sodium dichromate method

a. Remove oil, grease, etc. by vapor degreasing. Where this is impractical, wipe the faying surfaces clean with cellulose tissues soaked in toluene or methyl ethyl ketone.

b. Immerse for 8 to 12 min in a tap water solution of one of the following alkaline cleaners:

Cleaner	*Concentration*	*Temperature*
Turco 4215	6–8 oz/gal	150–160°F
Oakite 61	4–8 oz/gal	160–180°F
Altrex	6–8 oz/gal	160–180°F
Oakite 164	6–8 oz/gal	160–180°F

c. Rinse thoroughly in water.

d. Immerse for 10 to 12 min in a 150 to 160°F solution of the following composition by weight:

Demineralized water	30 parts
Sulfuric acid	10 parts ± 10%
Sodium dichromate	1 part ± 10%

e. Rinse thoroughly in distilled water either by total immersion so that the final pH will be between 8.5 and 5, or by spray rinse to meet the same pH requirement.

f. Air or oven dry parts at temperatures up to 150°F. Should parts remain unbonded or primed beyond the specified limit (usually 4 to 6 hr), they may be returned to the etch tank for a two min "strike," rinsed again, and dried. This will return the surface to a water-break-free condition.

Tank material for this method should be stainless steel lined with polyvinyl chloride, or its equivalent. The tanks should have the equipment necessary to provide a mild agitation to the solution. This is usually accomplished by an air inlet at the bottom of the tank, but filtered or missile-grade air should be used.

In conjunction with a production-type operation, the solutions should be titrated at regular and frequent intervals for proper concentration. A peel specimen is sometimes cleaned, bonded and destructively tested at the beginning of each shift and the results charted. If the chart is closely observed, the indications of an out-of-balance solution will be evident because of the slight decrease in peel values. Water, sodium dichromate, or sulfuric acid may be added when necessary to meet the requirements, but after prolonged use the solution will become dirty or contaminated with foreign matter and must be disposed of and replaced with a fresh mixture.

19. Method B, chromic-sulfuric method

a. Remove grease, oil, etc. by vapor degreasing or solvent cleaning.

b. Immerse 8 to 12 min in alkaline cleaner (same as Method A).

c. Spray rinse.

d. Immerse parts in the following solution for 10 to 13 min, maintained 150 ± 5°F.

Chromic acid .... 5 oz by weight
Sulfuric acid .... 24 oz by weight
Water to make 1 gal

e. Remove and immerse for 3 to 5 min in cold water (75°F or less).

f. Spray rinse with deionized water.

g. Dry in oven, not exceeding 150°F.

Anodized aluminum surfaces (chromic acid, sulfuric acid) are not considered surfaces that will produce bonds of maximum expectation. Sometimes it becomes necessary to bond to surfaces that have been treated by this method. If bonding of anodized surfaces is necessary, a trichloroethylene degrease is adequate. Degreasing should be performed within 4 hr prior to bonding.

If the anodize is sealed it should be removed if at all possible. The following stripper at 195 to 210°F will remove the anodize in 3 to 5 min. (If anodize should be preferred in regions of the parts, that area should be masked with an inert chemical tape.

Chromic acid	160 g
Phosphoric acid (85%)	320 g
Water	to make 1 liter

Before anodizing, the faying surfaces may be primed with a primer system that is not affected by the anodizing process. This eliminates anodize removal prior to bonding because after anodizing the primer may then be removed and the part bonded. A chemically inert maskant tape will serve the same purpose, but usually requires more time in preparation before the anodizing operation.

Aluminum alloy parts which cannot be immersed in acid solutions because of size, incapability, or attached structure, or in field repairs, may be scrubbed with a nonchlorinated cleaner such as Ajax or Bab-O until a water-break-free surface is obtained.

The following mixture may be applied at room temperature for 20–25 min:

Components	*Parts by Weight*
Sulfuric acid	55
Sodium dichromate	10
Water	80
Cab-O-Sil	Sufficient to make a thixotropic paste

During treatment, apply often enough to prevent mixture from turning green or drying out. The etching process may be accelerated by heat and agitation, i.e., the use of a heat lamp and frequent brushing of the paste with a nylon brush. Rinse with distilled water; re-treat as many times as is necessary to get a good water-break-free condition. If accelerated drying is preferred and conventional drying devices are not available, an air gun (hair-dryer type) may be utilized.

20. Steels

In general, most adhesives adhere well to steels. Strong joints can be attained merely by solvent cleaning, especially the stainless series. However, the performance of bonded steel joints is greatly improved under environmental conditions when the surface is treated (Fig. 5-5).

Method A alkaline cleaning of steels

a. Vapor degrease or solvent clean to remove scale, oxides, or other contaminants.

b. Immerse 10 to 12 min in the following solution:

Prebond 700–710 oz to 1 gal. distilled water, and heat 180 to 200°F.

c. Rinse with distilled water.

d. Oven dry at 150°F until dry.

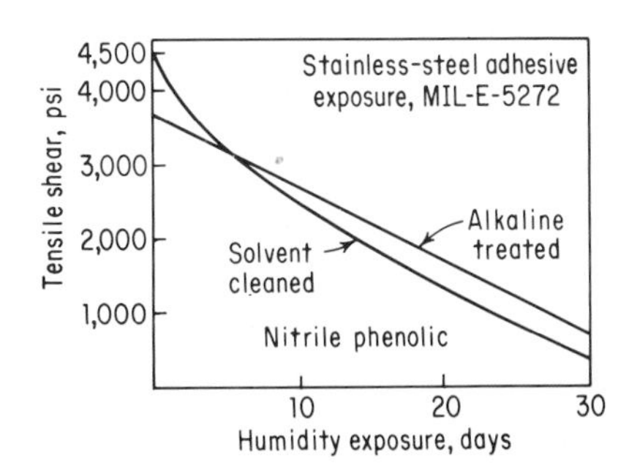

fig. 5-5 *Typical example of the influence of surface preparation on a steel adherend bonded joint when subjected to humidity.*

Check for water-break-free surface. The above treatment may be repeated until a water-break-free surface is obtained.

Note: A quick-drying alcohol rinse following the water rinse will aid in preventing rust during the drying cycle. This may be utilized regardless of the surface treatment used. The steels should be primed as soon as possible after cleaning especially the noncorrosion-resistant steels.

METHOD B CARBON AND LOW-ALLOY STEELS; THE HYDROCHLORIC ACID ETCH

a. Vapor degrease or solvent wipe to remove contaminants.
b. Immerse 10 to 12 min in a solution of prebond 700, 10 to 12 oz to 1 gal. of water, heated to 180°F.
c. Rinse.
d. Immerse 2 to 4 min in room temperature solution of hydrochloric acid (25 to 40 percent by volume) and tap water.
e. Immersion rinse at 160 to 180°F and then spray rinse.
f. Alcohol rinse and dry.

METHOD C THE HYDROCHLORIC-PHOSPHORIC-HYDROFLUORIC ETCH

This process is recommended for the 17–7 PH steels, especially where extreme environmental conditions are anticipated.

a. Vapor degrease or solvent clean.
b. Immerse at 170 to 185°F in the following solution:

Hydrochloric acid (35%)	83.3% by volume
Phosphoric acid (85%)	12.5% by volume
Hydrofluoric acid (60%)	4.2% by volume

c. Spray rinse with demineralized water.
d. Oven dry at 140 to 150°F for 20 to 40 min.

The tank material should be polyvinyl-chloride-lined mild steel.

METHOD D THE OXALIC-SULFURIC PROCESS

This method is sometimes used especially for corrosion-resistant condition-B stainless steels.

a. Degrease to remove scale, oxides, grease, etc.

b. Immerse in the following solution for 8 to 10 min maintained at 180 to 190°F:

	Parts by Weight
Oxalic acid	9
Sulfuric acid	1
Deionized water	80

c. Remove the black smut formation by scrubbing with a stiff brush under running water.

d. Rinse with tap water, followed with a rinse with deionized water.

e. Air dry in a circulating oven at a temperature below 150°F.

Tank material for this etch solution should be polyvinyl-chloride-lined mild steel.

After the etching operation, check the parts for a water-break-free condition. If the water-break-free test is not satisfactory, the entire process may be repeated.

MANUAL METHOD

For field repairs or where the processes listed here are impractical, the faying surface may be scrubbed with a nonchlorinated commercial cleaner, washed thoroughly and dried. A very good water-break-free condition can be produced by this technique, but many times requires a considerable amount of scrubbing.

21. Titanium alloys

METHOD A NITRIC-HYDROFLUORIC ACID

a. Solvent clean.

b. Pickle in the following water solution at room temperature for 30 sec:

Nitric acid (70%)	15% by volume
Hydrofluoric acid (50%)	3% by volume

c. Rinse in distilled water for two min.

Note: This solution should be replenished when the metallic concentration exceeds one part of metal to 1.7 parts of hydrofluoric acid, or when the fluoride content, expressed as hydrofluoric, exceeds 5 percent.

d. Immerse in the following tap water solution at room temperature for 2 to 3 min:

Components	*Percent by Weight*
Trisodium phosphate	5.0
Sodium fluoride	0.9
Hydrofluoric acid	1.6

e. Rinse in distilled water.

f. Force dry in an air circulating oven, maintained at 150 to 165°F for 10 to 20 min, or until dry.

Tank materials recommended for these titanium cleaning solutions are polyvinyl-chloride-lined mild steel.

METHOD B COMMERCIAL SOLUTION FOR TITANIUM

A treatment or process which is commercially available that produces excellent bonds with the majority of titanium alloys is as follows:

a. Solvent clean or vapor blast.

b. Immerse 10 to 15 min in the following solution at room temperature:

Pasa-Jell 107C	5 gal
Water	10 gal

Note: Solution should be stirred or agitated as required to maintain uniformity.

c. Water rinse thoroughly.

d. Air dry at 65 to 90°F.

e. Prime or bond as soon as possible. If the substrate cannot be primed or bonded within 2 hr, store in polyethylene bags. If over 4 hr elapse, recleaning would be in order.

Note: Pasa-Jell should be stored in acid-resistant plastic containers, or plastic-lined containers. The solution should be discarded after treating 800 sq in. of titanium per gal of solution.

METHOD C NONIMMERSION METHOD

In all probability, there will be occasions when the use of an immersion bath is impractical. The jellied version, Pasa-Jell 107M, may be utilized as received from the manufacturer. The titanium surface should be free of foreign matter, which may be accomplished by the use of a nonchlorinated household cleaner. Apply the Pasa-Jell 107M with a stainless steel or nylon bristle brush. Allow to remain on the faying surface for 10 to 15 min. Rinse with distilled water and check for a water-break-free surface. Repeat the above procedure, if necessary. Allow to dry and prime or bond within 2 hr.

Note: Pasa-Jell is a product of Semco, 18881 South Hoover, Los Angeles, Calif.

METHOD D MECHANICAL METHOD

If titanium parts cannot be processed by any of the methods previously mentioned due to size of parts or other reasons, the parts may be lightly abraded, scoured with a nonchlorinated cleaner (Ajax, Bab-O, etc.), rinsed with demineralized water, and dried. This method is recommended for small in-field repairs. Parts cleaned in this manner should be primed or bonded as soon as possible after drying.

METHOD E COLD PHOSPHATE ETCH PROCESS

a. Acetone or methylethylketone wipe.

b. Immerse in the following solution for 2 min at room temperature:

Hydrochloric acid	84 parts/volume
Hydrofluoric acid	43 parts/volume
Orthophosphoric acid	89 parts/volume

c. Rinse with distilled water and force dry for 10 to 12 min at 180 to 190°F.

d. Prime or bond as soon as possible.

22. Magnesium

METHOD A DOW 17 PROCESS

The preferred surface preparation for magnesium for adhesive bonding is the Dow 17 treatment. Tests conducted by several aerospace firms indicate that under extreme environmental conditions, the Dow 17 appears more dependable (see Chap. 8). The Dow 17 treatment must be rigidly controlled and very few companies are equipped to apply this conversion coating. When this treatment is desired, it is usually vendored. When received, it is light green in color if a light coating is present, darker for the heavier; but the light coat is preferred for bonding. After procurement, and 6 hr prior to bonding, it may be vapor degreased and then primed or bonded. The coating is obtained in the following manner:

a. Alkaline clean or vapor degrease.

b. Using an alternating current, anodize the parts for 3 to 5 min at a current density of 5 to 50 amps sq ft in the following solution maintained at 160 to 180°F:

Ammonium acid fluoride	32 oz
Sodium dichromate	13.3 oz
Phosphoric acid (85%)	11.5 oz
Plus water for 1 gal	

c. Follow with a cold-water rinse then a hot-water rinse, and dry below 120° F.

METHOD B CHROME PICKLE PROCESS

The following surface treatment is considered practical for most magnesium alloys and produces a surface capable of producing strong adhesive joints.

a. Vapor degrease or solvent clean (methyl ethyl ketone, acetone).

b. Immerse in the following solution maintained at 160 to 180°F for 6 to 10 min:

Sodium metasilicote	55 parts/weight
Tetrasodium pyrophosphate	35 parts/weight

"Nacconal NR"	5 parts/weight
Sodium hydroxide	15 parts/weight
Distilled water	2400 parts/weight

c. Rinse; preferably spray rinse with cold water.

d. Immerse in the following solution at ambient temperature for 10 to 15 min:

Chromium troxide	13 parts/weight
Calcium nitrate	1 part/weight
Distilled water	75 parts/weight

e. Rinse with cold, distilled water.

f. Force dry 15 to 20 min at 150 to 160°F.

g. Prime or bond as soon as possible, preferably within 2 hr.

Note: "Nacconal NR" is a product of Allied Chemical Company.

METHOD C CHROME PICKLE PROCESS #2

The following treatment, termed modified chrome pickle, is recommended for finely machined parts where close tolerances are essential.

a. Alkaline clean with Oakite 90 maintained at 170 to 190°F or immerse in the following solution for 5 to 10 minutes, maintained at 190 to 210°F:

Sodium carbonate	2.5 lb
Trisodium phosphate	2.5 lb
Soluble soap or wetting agent	10 oz
Distilled water to make 10 gal	

b. Rinse with distilled water.

c. Immerse 30 to 60 sec in the following solution maintained at 195 to 210°F:

Chromic acid	10.5 lb
Water to make 10 gal	

d. Rinse thoroughly in distilled water.

e. Immerse in the following solution 30 to 60 sec at room temperature:

Sodium dichromate	8 lb
Sulfuric acid	17.5 lb
Water to make 10 gal	

f. Allow to drain 10 sec and follow up by a distilled hot-water rinse (140 to 160°F).

METHOD D EBONOL "C" SPECIAL TREATMENT

The Ebonol treatment (same as for copper) has been used successfully for the treatment of magnesium prior to adhesive joining and is considered an adequate process for many alloys.

a. Solvent degrease with acetone, methyl ethyl ketone, etc.

b. Immerse for 12 to 15 min in the following solution maintained at 200 to 210°F:

Ebonol "C" Special	4 lb
Demineralized water to make 3 gal	

c. Rinse in room temperature demineralized water for 1 to 2 min.
d. Rinse in warm water (130 to 140°F) for 90 sec.
e. Oven dry at 130 to 140°F for 45 to 60 min.
f. Prime or bond within 4 hr.

The color will vary with different alloys when this process method is used. Strength values will exhibit more scatter when compared to other methods, which suggests testing for a specific application.

23. Copper and copper alloys (brass, bronze)

METHOD A BLACK OXIDE COATING

Copper and copper alloys are very difficult to bond satisfactorily, especially if high shear and peel strengths are desired. The prime reason for the problems associated with copper bonding is its rapid oxidation. One of the better methods is the chemical blackened technique. This utilizes a chemical product that is commercially available from the Enthane Company, New Haven, Conn., and is marketed under the trade name of Ebonol "C" Special. This method forms a black oxide coating on the copper surface and meets the requirements of MIL-F-495C, "Finish, Chemical, Black, For Copper Alloys." This process is critical and must be carefully controlled, pertinent to time and temperature. Parts should be primed or joined within 4 hr after chemical treating. If this is impossible, parts may be vapor degreased just prior to bonding.

a. Immerse in the following solution for 2 to 3 min; maintain the solution at 185 to 195°F:

Oakite 90	14 oz
Water to make 3 gal	

b. Immerse in the following solution for 20 to 45 sec, solution maintained at room temperature:

Phosphoric acid	9 parts/volume
Nitric acid	1 part/volume

c. Immerse for 10 min in the following solution maintained at 180°F:

Ebonol "C" Special	120 g
Water to make 1 liter	

d. Rinse with distilled water.
e. Force dry at approximately 120°F.
f. Prime or bond as soon as possible.

METHOD B AMMONIUM PERSULFATE PROCESS

a. Vapor degrease.
b. Immerse 1 to 3 min at room temperature in the following solution:

Ammonium persulfate	25 parts/weight
Water	75 parts/weight

c. Water rinse, dry at room temperature, prime or bond as soon as possible.

METHOD C FERRIC CHLORIDE PROCESS

a. Solvent clean (MEK, acetone).
b. Immerse 1 to 3 min in the following solution at room temperature:

Ferric chloride	15 parts/weight
Concentrated nitric acid	30 parts/weight
Water	200 parts/weight

c. Rinse thoroughly with distilled water, dry at room temperature (aid with dry filtered air, if possible). Bond or prime as soon as possible.

METHOD D HYDROCHLORIC ACID–FERRIC CHLORIDE PROCESS

a. Vapor degrease or solvent clean.
b. Immerse 1 to 2 min at room temperature in the following solution:

Concentrated hydrochloric acid	50 parts/weight
Ferric chloride	20 parts/weight
Water	30 parts/weight

c. Water rinse, dry at room temperature, apply adhesive or primer as soon as possible.

METHOD E SODIUM DICHROMATE-SULFURIC ACID PROCESS

a. Immerse 10 min in the following solution, maintained at 145 to 155° F:

Ferric sulfate	12 parts/weight
Sulfuric acid	10 parts/weight
Demineralized water to make 1 gal	

b. Rinse with demineralized water as required.
c. Immerse for 1 to 10 min (as required) in the following solution at room temperature:

Sodium dichromate	6 parts/weight
Sulfuric acid	11 parts/weight
Demineralized water to make 1 gal	

d. Rinse with demineralized water, dry at room temperature (with dry filtered air, if possible). Apply the adhesive or primer as soon as possible.

24. Nickel and nickel base alloys

METHOD A NITRIC ACID PROCESS

a. Vapor degrease or solvent clean.
b. Immerse 4 to 6 sec at room temperature in concentrated nitric acid.
c. Rinse with cold deionized water.
d. Dry at room temperature, aided by clean, filtered air.

METHOD B CHROMIUM TRIOXIDE-HYDROCHLORIC ACID PROCESS

a. Vapor degrease or solvent clean.
b. Immerse in the following solution at room temperature for 60 to 80 sec:

Chromium trioxide .... 15 parts/weight
Hydrochloric acid .... 20 parts/weight

c. Spray rinse with deionized water.
d. Oven dry at 130°F until dry.

Note: If immersion is impossible, the latter solution may be applied with a cheesecloth (after solvent cleaning). Allow to remain on the part for approximately 1 min. Apply to approximately 1 sq ft at a time. Rinse or wipe with demineralized water and air dry.

METHOD C MANUAL CLEANING

Thin nickel-plated parts should not be etched or sanded. A recommended practice is light scouring with a nonchlorinated commercial cleaner, rinse with distilled water, dry under 120°F. Then prime or bond as soon as possible.

One of the major problems associated with bonding plated parts is variation of surface conditions caused by the variations in plating equipment, process methods, and solution concentrations. These variables result in plated surfaces with widespread surface conditions pertinent to surface finish, and inconsistent metallurgical and adhesion properties. Personnel who are not knowledgeable in the art of cleaning plated parts tend to insist on sanding or roughening the faying surface. Sandpaper and emery paper, or equivalents, should never be applied to any type of plated details. Roughening surfaces to obtain good bonds is a completely false impression, as excellent bonds may be obtained with the smoothest finishes, provided they are clean.

Etchants are usually not preferred for plated surfaces, but there are a few exceptions (chrome, chromium). If etchants are impractical, the likely cleaning candidates would be solvent cleaning, vapor degreasing, or soap cleaning.

Primers aid in bonding plated surfaces because they usually improve adhesion, protect the surface against environmental exposures, do not alter component dimensions, but in all probability the most important single factor is the protection provided from cleaning to bonding.

25. Beryllium

METHOD A SODIUM HYDROXIDE CLEANING

a. Vapor degrease.
b. Immerse for 3 to 4 min in the following solution, maintained at 170 to 180°F:

Sodium hydroxide 20%/weight
Distilled water to make 1 gal

c. Rinse with cold water (spray rinse preferable).
d. Oven dry at 275 to 325°F for 10 to 15 min.

METHOD B MECHANICAL CLEANING

Beryllium may be scrubbed with a commercial nonchlorinated cleaner, rinsed with distilled water, and dried at 250 to 300°F. Check for water-break-free and repeat scrubbing operation, if necessary. Beryllium is one of the prime candidates for adhesive-bonded assemblies expected to operate in high-temperature environments. Research in the area of cleaning has been limited, but as beryllium comes into wider use, the necessity for better cleaning techniques will surely bring about more research and better surface-preparation methods.

26. Chromium and chrome plated parts

a. Degrease or solvent clean.
b. Immerse in the following solution for 2 to 5 min, maintained at 190° F:

Concentrated hydrochloric acid	50 parts/weight
Distilled water	50 parts/weight

c. Rinse with cold distilled water.
d. Dry with clean, filtered air.

Note: Cleaning for platings are listed herein because sometimes it becomes necessary to bond to them. When faying surfaces can be stripped, better bond performance would be expected.

27. Zinc and galvanized metals

a. Solvent clean (MEK, acetone).
b. Immerse 2 to 4 min at room temperature in the following solution:

Concentrated hydrochloric acid	15 parts/volume
Distilled water	85 parts/volume

c. Rinse with warm water.
d. Rinse with cold distilled water.
e. Oven dry 15 to 25 min at 140 to 160°F.
f. Apply adhesive or primer as soon as possible.

28. Tungsten and tungsten alloys

a. Solvent clean (MEK, acetone).
b. Immerse 2 to 5 min at room temperature in the following solution:

Hydrofluoric acid	5 parts/weight
Concentrated nitric acid	30 parts/weight
Sulfuric acid	50 parts/weight
Water	15 parts/weight

c. Rinse with distilled water and oven dry 15 to 25 min at 150 to 175°F.

29. Cadmium, silver, gold plating

a. Degrease or solvent clean.
b. Scour with a commercial nonchlorinated cleaner (Bab-O, Ajax, etc.).

c. Rinse with distilled water and dry with clean, filtered air at room temperature.

Adhesive-bonded joints utilizing these surfaces show rapid degradation when subjected to environmental conditions, especially humidity (Fig. 5-1). The results in this figure are not indicative of all adhesive systems. The adhesive choice is very important in bonding any plated substrate, and weathering properties—as stated earlier in this text—may be improved by the use of primers. If a plated part is considered a necessity and is to operate under adverse conditions, the bonded joint may be sealed with a sealant that possesses good weathering resistance.

30. Platinum

Platinum is usually vapor degreased, and primed or bonded. A light scrubbing with a mild commercial cleaner may be used, the surface rinsed, dried quickly, and bonded or primed immediately.

31. Tin

a. Solvent clean.
b. Abrade lightly with 180 grit emery paper.
c. Scour with soap cleaner.
d. Rinse with distilled water and dry with clean, filtered air at room temperature.

Tin is one of the few metals upon which the use of sandpaper is accepted without being followed by an acid etch of some type. The oxide coating must be removed, but after the sanding operation, the faying surface should be scrubbed thoroughly.

32. Lead

a. Abrade lightly with 120 grit sandpaper, using the wet sanding system.
b. Scrub with a mild soap, rinse with distilled water and dry at 120°F.
c. Trichloroethylene degrease just prior to bonding, prime or bond as soon as possible.

33. Iron (cast)

a. Vapor degrease.
b. Scrub with a wire brush, using a commercial cleaner.
c. Rinse with distilled water, dry at 100 to 120°F.
d. Prime, seal, or bond as soon as possible.

34. Galvanized metals

a. Vapor degrease.
b. Immerse in the following solution for 3 to 4 min, maintained at room temperature:

Hydrochloric acid	20 parts/weight
Distilled water	80 parts/weight

c. Rinse with cold distilled water.

d. Dry for 20 min, not exceeding 140°F.

e. Apply primer or adhesive as soon as possible.

The galvanized metals may also be scoured with a nonchlorinated commercial cleaner, rinsed with distilled water, and dried under 140°F. A water-break-free surface should be obtained; if not, repeat the procedure.

35. Aluminum honeycomb core

METHOD A

Vapor degrease (trichloroethylene), allow to dry thoroughly at room temperature. The core should not be submersed in the trichloroethylene.

METHOD B

Spray a cross pattern, passing aliphatic naphtha through each cell as necessary. Dry 30 min at 200 to 240°F.

METHOD C

Sandwich-core materials which cannot be processed by Method A or B may be sanded lightly with 180 grit emery paper, then vacuumed to remove dust particles created by sanding. Particles may also be removed by blasting with clean, filtered air. Vibro sander power tools may be employed provided they do not cause cell distortion.

36. Steel honeycomb core

METHOD A

Vapor degrease with trichloroethylene as necessary, dry at room temperature for a minimum of 2 hr.

METHOD B

Spray with aliphatic naphtha, passing the solvent through each cell and dry at 180 to 200°F.

37. Plastic and paper core

This type of core is normally bondable as received. However, if contaminated, it may be vapor degreased for 20 to 40 sec. Light sanding may be performed provided the dust particles are removed. This type of core should be stored in a dry area, or remain in a sealed container until ready for use. If evidence of moisture is observed, the core should be placed in an oven at approximately 120°F and dried for 20 to 30 min.

38. Molded or cast plastics

Molded or cast plastics may be sanded or lightly sandblasted. Loose particles and dust should be blown off or brushed away. Parts should be free of oil, dust, and moisture before bonding. If molded or cast plastic parts are to enter into a bonding operation, silicone mold releases should never be used during the forming process. If silicone mold releases are used, there is no known practical method to determine if they have been completely removed, even after sandblasting. Even though the silicone release problem is a well-known and established fact, it is still a common occurrence. *Be sure before bonding that a complete history of the part is known.* If the part is silicone contaminated, seek the nearest trash container, and then rush to the nearest cost-savings office and apply for a company cost-savings award. If the company has a zero-defects program, a major contribution has been made to it.

39. Wood

Wood should be filed with a vixon or rough file. It should never be sanded with fine grit sandpaper, as this tends to fill the pores in the wood, resulting in poor wetting. Poor wetting results in substandard joints due to the dependence on mechanical adhesion for satisfactory wood bonding.

All the moisture should be removed that is possible. If evidence of moisture is apparent, dry in an oven maintained at 120 to 130°F for 20 to 30 min.

A vast amount of wood is still used in small aircraft, and many private craft are still in service that utilize wooden wing spares, etc. Bonding operations of this type are naturally critical, and governed by Federal specifications and military-approved bonding materials. A large amount of wood bonding can be associated with the building-construction and furniture-manufacturing industry.

Into the wood bonding area looms the do-it-yourself home economist. Bonding such as repairing furniture and household items are often unsatisfactory because of two prime reasons:

1. Cheap thermoplastic resins that are not adequate for the job; the head of the household rushes out and purchases a 30-cent tube of cement, but good resins cost much more.
2. Inadequate preparation or cleaning; if a repair is being made the old resin must be removed to permit good wetting. The faying surface should then be roughened with vixon file or its equivalent. Naturally the lady of the house would argue about reliability, but the more drastic incident that could occur would not be a fatality but a favor-

ite guest slightly burned in a very undesirable area of the human anatomy, plus a lap of stew, as the dining table comes crashing down.

40. Glass

Method A

a. Solvent clean (MEK, acetone, etc.) or vapor degrease.
b. Light sandblast with fine grit.
c. Vacuum or blow clean with clean, filtered air.

Method B

a. Degrease.
b. Immerse in the following solution 15 to 20 min at room temperature:

Chromium trioxide	1 part/weight
Distilled water	4 parts/weight

c. Wash with distilled water and dry at 180 to 200°F for 20 to 30 min.
Note: Glass to be used for optical purposes should not be etched.

Method C

Clean in an ultrasonic detergent bath, rinse with distilled water and dry at room temperature.

Method D

Use the wet sandpaper or emery cloth method, combined with a mild detergent. Wash thoroughly with distilled water and dry in an oven, not exceeding a temperature of 120°F.

41. Ceramics, pyrocerams, and porcelain

a. Vapor degrease.
b. Sand lightly with 200 grit sandpaper.
c. Vacuum or blast with clean, filtered air.
If a glaze finish is on the faying surface, it should be completely removed for maximum bond strength.

42. Quartz

Method A

a. Vapor degrease or solvent clean.
b. Sandblast with fine grit, or scrub with an abrasive cleaner and steel wool.
c. Wash with distilled water.
d. Dry at 120 to 140°F.

METHOD B

a. Vapor degrease or solvent clean.
b. Immerse for 5 to 10 min in the following solution at room temperature:

Chromium trioxide	2 parts/weight
Distilled water	5 parts/weight

c. Rinse thoroughly with distilled water.
d. Oven dry at 120 to 140°F for 20 to 30 min.

43. Concrete

METHOD A HYDROCHLORIC ACID CLEANING

a. Remove oil, grease and contaminants by vigorous scrubbing with a strong detergent, and utilizing a heavy, stiff-bristle scrub brush.
b. Rinse with cold water.
c. Sand or sandblast the bonding surface completely, removing 0.040 to 0.070 in. of the faying surface.
d. Remove dust particles by vacuum technique or blasting surface with clean oil-free air.
e. Apply the following solution, approximately 1 qt per sq yard of surface area and allow to remain at ambient temperature for approximately 20 min:

Hydrochloric acid	15 parts/weight
Water	85 parts/weight

f. Wash with clean, cold water with as much pressure as is possible, utilizing a hose or equivalent spray technique.
g. Check for acid condition, using litmus paper.
h. As insurance against acid residue, the faying surface may be flushed with the following rinse solution:

Ammonia	2%/weight
Water	98%/weight

i. Flush again with cold water and allow to dry thoroughly before applying primer or adhesive.

METHOD B HYDROCHLORIC ACID PROCESS #2

This method utilizes a slightly stronger solution than Method A and is recommended for old concrete that may be contaminated with grease, oil, etc.

a. Scrub with a strong detergent, using a stiff, bristle-type brush.
b. Rinse with cold water using the highest pressure possible.
c. Apply the following solution at ambient temperature, allow to remain on the faying surface for approximately 15 min; during this time it may be agitated with a stiff nylon brush:

Hydrochloric acid	1 part/weight
Water	2 parts/weight

When this solution is applied, the acid will react with the surface causing small bubbles to form, but after approximately 15 min, the bubbling will subside.

d. Rinse with cold water.

e. Neutralize with the following solution:

Ammonia	3 parts/weight
Water	97 parts/weight

f. Rinse again using the highest possible spray pressure, allow to dry thoroughly before adhesive application.

44. Graphite

a. Solvent clean with acetone, methyl ethyl ketone or equivalent.

b. Abrade with 220 grit sandpaper, remove particles by vacuum or air blast.

c. Clean with mild detergent, rinse with distilled water, and dry in oven, temperature maintained below 120°F. Apply primer or adhesive as soon as possible.

45. Stone work

a. Scour surface with wire brush utilizing a commercial cleaner.

b. Rinse and dry thoroughly. Apply adhesive as soon as possible.

46. Brick

a. Degrease with a cloth saturated with acetone, methyl ethyl ketone, etc.

b. Brush with a stiff wire brush, remove dust with air blast.

c. Be sure the faying surface is dry and then apply the primer or adhesive.

47. Jewels and precious stones

These items need only to be degreased by the use of solvents and then bonded. They should not be handled with bare hands after cleaning because of the probability of contamination with skin oils.

48. Leather

a. Degrease with a clean cloth saturated with acetone, methyl ethyl ketone, etc.

b. Abrade lightly with 200 grit sandpaper.

c. Apply a blast of clean air or vacuum the faying surface.

d. Apply primer or adhesive.

49. Plaster

a. Abrade with 200 grit sandpaper or wire brush.

b. Remove dust and apply adhesives.

Solvents are not recommended for cleaning a plaster surface because of its rapid absorption properties, especially if the finish has been sanded away. If solvents are used, be sure that enough time has elapsed for the solvents to evaporate before the application of the adhesive system.

50. Painted surfaces

If high strengths are required, painted surfaces are always stripped. A bond with high structural properties would be impossible because the joint could not exceed the strength of the paint, which in most cases would be inferior to the bonded joint. If the adhesive system utilized solvent systems, the paint would be subject to degradation in most cases.

If a nonstructural bond is desired and stripping the paint is undesirable, the painted surface may be washed with a detergent solution and then sanded lightly with wet sandpaper (20 to 240 grit). Wash again with clear cold water and dry thoroughly.

51. Paper and paper laminates

Normal paper that is not waxed does not require any special preparation prior to bonding. Paper laminates may be abraded lightly with dry sandpaper (200 grit) and the dust particles removed.

52. Asbestos and fiber board

Abrade with sandpaper (200 grit), remove dust particles, and bond.

53. Laminated plastics

METHOD A TEAR PLY METHOD

If practical, remove a ply of the laminate. Slide a thin knife blade under a ply, peel desired distance and trim. Remove loose particles by brushing or blowing surface with clean filtered air. This does not provide positive insurance that all mold releases and contaminants have been removed, but the probability is increased.

METHOD B SANDBLASTING OR SANDING

The surface of laminated plastics should be sanded lightly or sandblasted to remove contamination and release agents. Solvent wipe with methyl ethyl ketone and allow ample time for all solvents to evaporate. If the surface is porous, Method C would be preferred because of solvent absorption.

METHOD C MANUAL SCOURING

The faying surface is scrubbed with an abrasive household-type cleaner to remove contamination and release agents. The surface is then rinsed with distilled water and dried under 150°F. The parts should exhibit a water-break-free surface, if not the above procedures are repeated.

METHOD D SOLVENT SOAK AND ABRADING

If water-break-free surfaces cannot be attained by Methods A, B, or C the plastic laminate material should be treated as follows:

a. Soak the laminate for 48 hr in reagent-grade acetone.

b. Dry for 3 to 4 hr at 190 to 220°F.

c. Sand lightly with 200 grit sandpaper, remove particles by air blast or vacuum and check for water-break-free condition.

d. If water-break is exhibited, parts may be bonded. If not, resoak the laminate for an additional 24 hr minimum and repeat steps *b* and *c*.

e. Parts that do not meet the requirements of the water-break-free test should be rejected.

54. Rubber (natural)

METHOD A SANDING

When bonding with rubber-type cements, sanding lightly and a toluene or alcohol clean are adequate.

METHOD B SULFURIC ACID PROCESS

a. Immerse in the concentrated sulfuric acid for 5 to 10 min (until a brittle surface is formed), rinse thoroughly with distilled water, and dry at room temperature. The rubber can then be flexed, resulting in a surface laden with fine cracks. The surface cracks aid in forming an optimum bond because mechanical adhesion plays an important role in bonding rubber.

METHOD C NONIMMERSION SULFURIC PROCESS

Prepare a thixotropic paste using sulfuric acid and Cab-O-Sil. Brush on the faying surface and allow to remain for 10 to 20 min. Rinse thoroughly, dry and flex to form cracks.

Note: When using Methods B or C, residual acid may be neutralized by either a 5-min soak or wipe-on application of a dilute solution (5 to 10 percent) of ammonium hydroxide, followed by another distilled-water rinse and room-temperature dry.

55. Rubber (synthetic)

METHOD A ABRADING OR SANDING

a. Solvent clean with methyl alcohol.
b. Abrade with 200 grit emery cloth.
c. Solvent clean again with methyl alcohol, allow to dry thoroughly to ensure complete solvent evaporation.

METHOD B NITRIC ACID PROCESS

a. Wipe with methyl alcohol.
b. Immerse for 4 to 8 min in concentrated nitric acid.
c. Rinse with distilled water.
d. Immerse for 5 to 10 min in the following solution at room temperature:

Ammonium hydroxide	20%/weight
Water	80%/weight

e. Rinse with distilled water, dry at room temperature and flex to form fine cracks in the faying surface.

Note: Method C under natural rubber may be used if immersion is impractical. It may be necessary to allow the parts to remain on the surface a longer time on the synthetics. The amount of acid exposure depends on the grade of rubber.

56. Polycarbonate

METHOD A SOLVENT CEMENTING

The polycarbonates may be bonded to themselves by solvent cementing. This is accomplished by the use of methylene chloride. The minimum amount of solvent should be used, contradictory to most cementing procedures. One bonding area should be softened while the other faying surface remains dry and ready to mate. Soaking both bonding surfaces or using an excessive amount of methylene chloride may cause a substandard joint because of air bubble formations or crystallization due to moisture condensation from the air. Before solvent fusion, the parts should be free of contaminants. This may be accomplished by a mild soap cleaner, wiped clean with a tissue, then a thorough drying at room temperature.

Note: The drying rate of the methylene chloride may be reduced by adding 3 to 5 percent polycarbonate resin to the solvent. This method is not recommended if the parts do not mate perfectly.

METHOD B ABRADE AND SCRUB METHOD

a. Clean with isopropyl alcohol, heptane or naphtha.
b. Abrade lightly with 200 grit emery cloth or sandblast lightly.

c. Scrub with nonchlorinated commercial cleaner (Ajax, Bab-O).
d. Rinse with distilled water and dry at room temperature.

57. Fluorocarbons

Polytetrafluoroethylene (TFE), Polyfluoroethylene propylene (FEP), Polychlorotrifluoroethylene (CFE), Polyvinylfluoride (PVF), Polymonochlorotrifluoroethylene (Kel-F).

METHOD A SODIUM-NAPHTHALENE PROCESS

a. Solvent clean with acetone or MEK.
b. Abrade lightly with 200 grit emery paper.
c. Immerse for 10 sec to 1½ min at room temperature in the following solution:

Sodium metal (chips, wire or ribbon)	46 g
Naphthalene	128 g
Tetrahydrofuran	1 liter

During immersion, the etching action will be faster if the part or the solution is agitated.

d. Neutralize with isopropyl alcohol. (This will dispose of small sodium particles that may be on the part before rinsing.)
e. Rinse with distilled water and dry at 110 to 120°F. When properly etched, the part will be a light to dark brown color depending on the immersion time and strength of the solution. Unsatisfactory etched surfaces may be returned to the solution for additional etching if necessary. This solution slowly depletes itself due to the reaction of the sodium with the fluorine or chlorine atom of the substrate. As the solution grows weaker, more time is required for adequate etching. Therefore, the immersion time is largely dependent on the knowledge and experience of the operator. The surface may be inspected by the use of an eye dropper partially filled with distilled water. Allow one drop to fall approximately one-half inch onto the surface to be checked. Use the dropper to pick up the water drop in one slow suction. Inspect the surface for the presence of water. If the surface is properly etched, it will show a continuous wet spot similar in size and shape to the original drop of water. This test, when conducted on an unetched surface, will result in little or no water visible on the surface. A partially etched surface will show a discontinuous film of moisture after removal with the eye dropper.

Preparation of this solution must be accomplished with extreme caution. The sodium metal produces a strong caustic compound when it comes in contact with the air. It is usually supplied in an inert atmosphere or immersed in a hydrocarbon liquid. Do not allow the sodium metal to come in contact with water, chlorinated hydrocarbons, or carbon dioxide. *Use caution* in the preparation and use of this surface treatment. One company alone sent five people to the hospital in a 6-week period with severe burns, one individual permanently scarred over 70 percent of his body, simply because of negligence or lack of understanding of the dangers of this solution. Can you imagine a technician carving a 2-lb ingot of sodium on an open table, underneath a fire sprinkler system, one door in the room and the sodium between him and the door? This actually happened! The young man was permanently disabled physically. The solution should be prepared in a dry box, if possible. Pour the tetrahydrofuran into a wide mouth glass container having a solvent-resistant screw top. The screw top should have a fluorocarbon film liner. Cut the sodium metal into chips or pieces finer than $\frac{1}{16}$-in. thick and immerse them quickly into the tetrahydrofuran and add the naphthalene slowly. Allow the solution to stand approximately 6 hr, shake occasionally, or until the sodium and naphthalene have completely dissolved. The solution will be a dark brown to black when ready for use. Never pour the solution in an open container for use. Replace the cap on the container at once after use. Do not process parts rapidly over a long period of time as the exotherm from the chemical reaction will result in elevated temperatures in the solution and approach a danger point.

After the solution has served its purpose it must be disposed of in a safe manner. The chances are high that small particles of undissolved sodium are still present in this solution. Many companies have disposable teams for this purpose. This solution may be disposed of by pouring isopropyl alcohol into the solution, thus neutralizing the sodium. It may then be disposed of by pouring down a water drain or a disposal sump.

METHOD B PROPRIETARY PROCESSES

Proprietary compositions may also be used. Four of the better known are listed, not necessarily in order of preference.

Trade name	
Tetraetch	W. L. Gore Associates, 487 Paper Mill Road, Newark, Del.
Fluoroetch	Action Associates, 1180 Raymond Boulevard, Newark, N.J.
Fluorobond	Joclin Manufacturing Company, 15 Lufbery Avenue, Wallingford, Conn.
Bondaid	W. S. Shamban & Company, 11543 West Olympic Boulevard, Los Angeles, Calif.

These solutions are not considered dangerous, but should be handled with caution according to the manufacturer's recommendations. The fluorocarbons may also be purchased in a bondable form which is termed "bondized." They are available from most manufacturers who market this type of product.

METHOD C CASING PROCESS

A new process for treating certain plastics in the fluorocarbon area for bonding has been developed by Bell Telephone Laboratories. The surface treatments given in Methods A and B change certain characteristics of Teflon and other fluorocarbons, including strength, color, and—more importantly—the dielectric properties. This new process called "casing" supposedly does not adversely effect the above mentioned properties. "Casing" also applies to the polyethylene.

This process is accomplished by exposing the plastic material to an electrically activated inert gas, such as helium or neon, and utilizing a glow discharge tube. This forms a tough, bondable skin on the surface of the substrate. This skin is compared to the layer that is formed on paint when exposed to the atmosphere, and not only produces a good service for bonding, but for printing and painting as well.

The "casing" process was used on the two samples of polyethylene (unfoamed and foamed) shown at the top of Fig. 5-6. All samples in this photograph were placed on a thick aluminum plate and heated above the melting point. The molten plastic in both "casing-treated" samples were contained within their molecular shells, while the untreated samples flowed freely.

fig. 5-6 *"Casing" treated plastics versus untreated (Bell Laboratories, Murray Hill, N. J.)*

58. Polyvinyl chloride

a. Degrease with trichloroethylene.
b. Abrade with 200 grit sandpaper.
c. Clean again with trichloroethylene, wipe with soft clean cheesecloth and bond.

59. Polyethylene tetraphthalate and linear polyesters (Mylar)

METHOD A ALKALINE CLEAN

a. Degrease with acetone or MEK.
b. Immerse in the following solution maintained at 180 to 190°F for 5 to 10 min.

Prebond 700	1 lb
Distilled water	1 gal

c. Rinse with distilled water.
d. Dry at 120 to 140°F for 20 to 30 min.

METHOD B

a. Clean with acetone or MEK.
b. Scrub thoroughly with nonchlorinated commercial cleaner (Bab-O, Ajax, etc.).
c. Rinse with distilled water.
d. Dry at 120 to 130°F for 20 to 30 min. Check for water-break-free and if the faying surfaces do not meet the requirements of this test, repeat the above procedure.

METHOD C SODIUM NAPHTHALENE PROCESS

Mylar may be treated as specified in Method A under Fluorocarbons, but the Mylar will not show a visible change in color after treatment but should meet the surface-tension requirements defined in the same section.

60. Phenolic, polyester laminates and molded phenolics (Durez)

a. Clean with acetone, MEK.
b. Abrade with 200 grit emery cloth.
c. Wipe with cheesecloth saturated with acetone, MEK.
d. Scrub with Ajax, Bab-O, etc.
e. Rinse with distilled water.
f. Dry at 130 to 150°F for 20 to 30 min.

61. Polyurethanes

a. Clean with acetone or MEK.
b. Abrade with 200 grit emery cloth.

c. Wipe faying surface with acetone or MEK. Dry thoroughly to evaporate solvents and bond as soon as possible.

62. Polyethylene and Polypropylene

METHOD A SODIUM DICHROMATE-SULFURIC ACID PROCESS

a. Degrease with acetone, MEK or xylene.
b. Immerse in the following solution maintained at 150 to 160°F for approximately 2 to 3 min:

Sodium dichromate	15 parts/weight
Concentrated sulfuric acid	250 parts/weight
Distilled water	25 parts/weight

c. Rinse with distilled water and dry at 110 to 120°F for 15 to 30 min.

This process is more effective for polypropylene if the temperature of the sodium dichromate-sulfuric acid solution is raised to 175 to 185°F, and it may be necessary to extend the immersion time for a few seconds.

METHOD B OXIDIZING FLAME METHOD

Polyethylene and polypropylene may be prepared for bonding by the oxidizing flame method which utilizes an oxyacetylene burner passed over the faying surface until it appears glossy. To insure that too much oxide is not on the surface, a light scouring with soap and water is in order. Wash with distilled water and dry at room temperature. These materials may also be purchased from the suppliers in bondable form.

63. Polyether

a. Wipe with clean cheesecloth saturated with MEK or acetone.
b. Immerse in the following solution maintained at 155 to 165°F for 8 to 12 min:

Sodium dichromate	15 parts/weight
Concentrated sulfuric acid	300 parts/weight
Distilled water	30 parts/weight

c. Rinse with distilled water.
d. Dry at 90 to 120°F until dry.

64. Polyformaldehyde

a. Clean with acetone or MEK.
b. Immerse for 15 to 30 sec at room temperature in the following solution:

Sodium dichromate	12 parts/weight
Concentrated sulfuric acid	300 parts/weight
Distilled water	25 parts/weight

c. Rinse with distilled water.
d. Oven dry at 100 to 300°F for 20 to 30 min.

65. Acrylics

a. Wipe with clean cheesecloth saturated with methyl alcohol and dry.
b. Abrade with 200 grit sandpaper and remove dust particles with blast of clean air or vacuum.
c. Check for water-break-free and repeat procedure, if necessary.

66. Polyamide (nylon)

a. Solvent clean with acetone, methyl ethyl ketone, etc.
b. Scrub parts with nonchlorinated abrasive cleanser.
c. Rinse with distilled water and dry at a temperature below 150°F.
d. Check for water-break-free surface and repeat procedure if necessary.

67. Polyolefins

METHOD A

a. Solvent clean with naphtha by immersion, if possible, wipe with clean cheesecloth, and dry.
b. Abrade lightly with 180 to 200 grit sandpaper, remove particles with a blast of clean air or vacuum. Repeat if water-break surface is not obtained.

METHOD B OXIDIZING FLAME TECHNIQUE

a. Solvent clean with naphtha and dry.
b. Subject the faying surface to a hot flame for a split second, or as necessary to obtain a glossy-appearing finish.

68. Polystyrene

METHOD A ABRADING OR SANDING TECHNIQUE

a. Degrease with a clean cheesecloth saturated with methyl or isopropyl alcohol, and allow to dry at room temperature.
b. Abrade with 100 grit sandpaper and remove dust particles. A water-break-free surface should be obtained or the entire procedure repeated.

METHOD B SODIUM DICHROMATE-SULFURIC ACID PROCESS

a. Degrease with methyl or isopropyl alcohol.
b. Immerse for 3 to 4 min in the following solution maintained at 220 to 230°F:

Sulfuric acid	90 parts/weight
Sodium dichromate	10 parts/weight

c. Rinse thoroughly with distilled water, dry below 120°F and check for water-break-free condition.

If parts fail to exhibit water-break-free surface, then repeat item *b*, rinse, and dry again.

METHOD C NONIMMERSION PROCESS

If polystyrene parts are to be used in high-frequency electrical applications it may be desirable that only the faying surfaces be treated, thus the following procedure may be used.

a. Degrease with methyl or isopropyl alcohol.

b. Apply to the faying surface the following thixotropic paste:

Sulfuric acid	3 parts/weight
Powdered potassium	1 part/weight

Add Cab-O-Sil as required to obtain a thixotropic paste as desired.

c. Heat parts to 200°F and hold for 2 to 3 min with the paste on the parts.

d. Rinse thoroughly with distilled water, dry, and check for water-break-free surface. Parts may be re-etched if necessary.

Polystyrene parts should not be dried above 150°F.

69. Cellulosics

a. Degrease with a clean cheesecloth saturated with methanol or isopropanol.

b. Abrade with 200 grit sandpaper or emery cloth.

c. Remove dust particles by clean air blast or vacuum.

d. Scour with abrasive commercial cleaner, rinse with distilled water, and dry at 200°F for 30 min. If possible, apply adhesive or primer immediately, preferably while the parts are still warm.

70. Epoxy and epoxide resin parts

a. Degrease with clean cheesecloth saturated with acetone, methyl ethyl ketone, and dry.

b. Abrade the surface with 200 grit sandpaper and remove dust particles with blast of clean air.

c. Place in an oven or apply heat not to exceed 150°F to insure that solvents are evaporated, prime or bond as soon as possible.

TABLE 5.1

The following reference specifications are applicable to materials and processes listed in this chapter.

Federal Specifications

O-A-51	Acetone, technical
O-C-303	Chromium trioxide, technical (chromic acid)
O-H-795	Hydrofluoric acid, technical
O-M-232	Methanol (Methyl alcohol)
O-N-350	Nitric acid, technical

O-O-670	Orthophosphoric (Phosphoric) acid, technical
O-S-595	Sodium dichromate, technical grade (sodium dichromate), Sodium dichromate dihydrate
O-S-642	Sodium phosphate, tribasic, technical, anhydrous, dodecahydrate and monohydrate
O-S-809	Sulfuric acid, technical
O-T-634	Trichloroethylene, technical
P-C-436	Cleaning compound, alkali boiling vat (soak) on hydrosteam
TT-M-261	Methylethylketone (for use in organic coatings)
TT-N-95	Naphtha, aliphatic
TT-T-548	Toluene, technical
PPP-B-0026	Bag, plastic, polyethylene (general purpose)

Military Specifications

MIL-F-495	Finish, chemical, black, for copper alloys
MIL-M-3171	Magnesium alloy, processes for pretreatment and prevention of corrosion on
MIL-S-5002	Surface treatments and metallic coatings for metal surfaces of weapon systems
MIL-C-5541	Chemical films and chemical film materials for aluminum and aluminum alloys
MIL-A-8625	Anodic coatings, for aluminum and aluminum alloys
MIL-A-13528	Acid, hydrochloric, inhibited, rust-removing
MIL-F-14580	Ferric chloride, anhydrous
MIL-P-17667	Paper, wrapping, chemically neutral (non-corrosive)
MIL-P-20693	Plastic molding material, polyamide (nylon), rigid
MIL-M-45202	Magnesium alloys, anodic treatment of

Military Standards

MIL-STD-171	Finishing of metal and wood surfaces

REFERENCES

Adhesive Engineering, Technical Bulletin AE-100, San Carlos, Calif., June, 1964.

"Cleaning and Conditioning Metal Surfaces," Semco Sales, Los Angeles, Calif.

DeLollis, N.J., and O. Montonya: "Surface Treatments for Difficult to Bond Plastic Materials," The Scandia Corporation, Albuquerque, N.M., May, 1962.

Eccobond Brochure, Emerson & Cumings, Canton, Mass., September, 1965.

Eickner, H. W.: "Adhesive Bonding Properties of Various Metals as Affected by Chemical and Anodizing Treatments of the Surfaces," Forest Products Laboratory, Madison, Wis., August, 1960.

"Handbook of Adhesives," American Cyanamid Company, Havre de Grace, Md.

Jacoby, H.: Bonding Polycarbonate Resin Parts, *Adhesives Age,* October, 1963.

Kieth, R. E., R. E. Monroe, and D. C. Martin: "Adhesive Bonding of Titanium and Its Alloys," Marshall Space Flight Center Technical Memorandum X53313, Huntsville, Ala., August, 1965.

Levine, M., G. Ilkka, and P. Weiss: Wettability of Surface-treated Metals and the Effect of Lap Shear Adhesion, *Adhesives Age,* June, 1964.

McNutt, J. E.: Wetting and Adhesion, *Adhesives Age,* October, 1964.

"Magnesium Finishing," Dow Chemical Company, Midland, Mich.

"Magnesium Finishing, Military Applications," Dow Chemical Company, Midland, Mich.

Monoleyer Bonds Metals to Thermo-Plastics, *Chemical and Engineering News,* December, 1964.

"New Process Prepares Plastics for Stronger Bonding," Bell Telephone Laboratories, New York, March, 1966.

"Preparation of Substrates for Adhesive Bonding," Hughson Chemical Company, Erie, Pa.

"Processing–Surface Preparation–Dow 17 Anodize Process–Bonding HK-31 Magnesium–Thorium Alloy–Process Variables, Investigation Of," Report FGT-2000, General Dynamics, Fort Worth, Tex., October, 1962.

Ruling, R. C.: "Ablative Shield Development Testing–Adhesive Evaluation and Elevated Temperature Properties," Report A472, McDonnell Aircraft, St. Louis, Mo., March 10, 1964.

Schonhorm, H.: "Approach to Adhesion via Interfacial Deposition of Amphipathic Molecules," Bell Telephone Laboratories, Murray Hill, N.J., April, 1964.

Snogren, R. C.: "Surface Treatment of Joints for Structural Adhesive Bonding," Hughes Aircraft Company, Culver City, Calif., presented at American Society of Mechanical Engineers, May, 1966.

Thelen, Hallinger, Haigh, Drew, Varker, Frank: Treatment of Metal Surfaces for Adhesive Bonding, *Bulletin of the Franklin Institute,* February, 1958.

CHAPTER 6

Adhesive Tooling and Fabrication Techniques

Adhesive-bonding tools may be described as special fixtures, jigs, or materials designed specifically for each configuration of part (or parts) to be held together during assembly and cure. Tools hold the details in place, apply the necessary pressure, and sometimes transfer or apply the necessary heat for curing. When designing tools or specifying fabrication procedures, the following items must be considered: (1) the cost involved, (2) how many parts are to be fabricated, (3) amount of pressure needed, (4) the cure conditions, (5) tolerances allowable, and (6) shop performance. Shop performance includes two primary areas: the ease of handling, i.e., if the tool must be transported, weight may be a prime factor, consequently the production tool should be designed as light as possible to facilitate handling. The second item is the simplicity of the tool, i.e., the ease of "setup." One of the advantages of adhesive bonding is the ability to fasten several details at the same time, yet the tooling must be designed to do the job with as few tooling components as possible. The more components incorporated into a tool, the more probability of poor setup or pressure deviations during the bonding operation.

There are two evident extremes associated with tooling techniques relative to adhesive bonding. These are overdesign and underdesign. The tool or curing techniques should be selected on the basis of the amount of pressure and heat required, and the simplicity by which the parts can be satisfactorily produced. In the planning stages, tooling that is currently available that may be utilized must be considered.

Tolerances, life expectancy, and cure temperature govern the choices of materials for a bonding fixture, i.e., plastic, plaster, clay, aluminum, steel, etc. A wise choice in this area is important from the economy angle. It would be unwise to select a metallic tool that would require costly machining if a plastic tool would be adequate, and by the same token it would be foolish to select a plastic if high pressures and temperatures were required for curing the adhesive and long production runs were anticipated.

A prime example of overdesign would be the fabrication of a large, bulky matched die tool when the simple vacuum-bag technique would be adequate.

The tooling engineer may select one of many approaches, but should always work in conjunction with the materials engineer to be sure all fabrication aspects and properties of the adhesive system are taken into consideration. The production personnel responsible for fabricating the assembly should be consulted to eliminate scheduling problems. The quality-control engineer should be prepared at this time to specify in-process controls and inspection requirements. Special fixtures may be necessary for control coupons or specimens. It may be desirable to incorporate the control specimen fixture on the tool, especially if a cavity-type tool is being designed.

When a new tool or fixture reaches the manufacturing area, the shop man must be briefed on its use, and a dry run is usually used to check the bonding operation. As an example, if a new bonding tool is designed for the fabrication of a honeycomb panel, the details may be "laid-up" utilizing a polyvinyl chloride (PVC) film between the core, edge members, inserts, and outer face sheets. The thickness of the PVC film should be equivalent to the adhesive film thickness, plus the primer (if primer is used). The assembly is then run through the normal cure cycle, pulled apart, and the PVC is checked for visual impressions. This will indicate areas where the pressure may be inadequate. During the dry run, the fixture should also be equipped with thermocouples to check for universal heating, heat-up rate and cooling factor.

A substandard bond due to insufficient pressure is difficult to detect because it usually affects the cohesive properties of the adhesive system. This would not result in a complete void, but a joint below maximum expectation. A discrepancy in this area can be detected by the use of ultrasonics (see Chap. 10).

After all the precautions have been taken to produce a satisfactory part from a new tool or technique, the first assembly is usually destructively tested. Samples are taken from the assembly and mechanically tested. The type of samples and strength-level acceptance requirements usually depend largely on the function of the part, design configuration, operating environment, and load-bearing requirements. If the equipment is available, the assembly should be ultrasonically tested or x-rayed before destructive tests. The destructive-test data can then be equated to the nondestructive-test values, and used for development of nondestructive inspection tests.

DEAD WEIGHT

This is a simple approach that is sometimes used for bonding small details, especially if a room-temperature-setting adhesive is utilized and high pressures are not required. This method may be used as a quick check for press accuracy or the relationship of pressure to bond-line thickness for an adhesive system. A simple lap-shear tensile specimen may be fabricated with dead weight, bond-line thickness measurements obtained, and mechanically tested. The same procedure may be followed utilizing a press, thus allowing a bond-line measurement and mechanical test comparison. The dead-weight method is also used for fabrication of large honeycomb, balsa-sandwich, and plywood assemblies. This technique has been used to a great advantage in bonding heat sinks to electrical terminal boards that were bonded with slow-curing systems and required a high production rate. This eliminates the necessity for several presses or other equipment. In a dead-weight operation of this type, a silicone slip sheet is used to insure a more even transfer of pressure, and also serves as a release agent. The pressure applied in pounds per square inch is calculated simply by dividing the contact area of the part or assembly by the weight of the object used for pressure.

MAGNETS

Magnets may be used in a special or unusual situation. This technique can be used for bonding thin-gauge materials where other methods are impractical. A steel backup plate is used underneath the lower surface detail and a magnet of the desired configuration on top of the assembly. Approximately 10 psi is the maximum expected pressure that may be obtained with a common magnet. The primary disadvantage is the life of magnets being limited, due to the loss of magnetism when subjected to elevated temperatures. This technique has been utilized for rather large assemblies by aerospace firms. An example is an aircraft missile launcher that was bonded with a polyamide system. First attempts were made with the pressure-bag technique which proved unsatisfactory. The design configuration of the assembly eliminated the vacuum bag technique. The press or dead weight was impractical. The tool was developed utilizing a steel cavity; a heavy base and the magnets were inserted through an end opening, and the assembly cured in an oven. Anyone familiar with adhesive bonding would concur that there is a better approach. This type of tooling may be eliminated if design, materials, and tooling engineering are thinking on the same wavelength.

The magnet method may be used for small field repairs or detail attachments where clamps or elaborate tooling would be impractical.

CLAMP AND SPRING DEVICES

Spring devices are undesirable for elevated curing bonded assemblies due to the loss of spring tension when subjected to heat. Clamps tend to loosen as the adhesive flows and results in loss of pressure. The pressure calculations in this parameter of tooling are only an educated guess and would never be used where accurate pressures are essential. Clamps are still used by many firms for production bonding, but in most cases, other techniques could be utilized that would produce better parts. The prime reason the clamps stay around is production cost. Clamps may be used to an advantage for repair jobs and one-shot parts and assist in holding other tooling or fixtures, but should be eliminated whenever possible. It could be termed depressing to see three or four technicians shuffling in the direction of an oven, carrying a base plate on which an assembly rests, stepping on thermocouple wires, three hundred "C" clamps that completely cover the part, and a shop foreman standing hopelessly to one side wringing his hands. To quote an old cliché, "no amount of planning will ever replace dumb luck."

THE PRESS

This is the simplest and most widely used of all tools for adhesive bonding and also the oldest type of equipment used in structural adhesive bonding. The press is versatile and may be utilized in many ways. Sizes range from the small 10×10 in. platen laboratory press to the giant 40- or 50-ft presses designed for fabrication of large assemblies (Fig. 6-1). They are usually hydraulic but some models also utilize an air ram on one platen for low-pressure bonding, with a "pop-off" or balance valve to maintain a constant pressure as pressure builds up due to thermal expansion of the parts or is reduced by the flow of the adhesive. As a modification for bonding contoured or odd shaped parts, a formed or machined die may be attached to one or both platens which then takes the status of a flat platen. The die then becomes the heat source (if any) and

fig. 6-1 *Large press utilized for adhesive bonding.*

the pressure medium (Fig. 6-2). The heat may be applied to the platens by means of steam, water, or electric heating, but steam and water are preferable for the large presses. This is because steam or water heating is more universal and not likely to result in hot spots. Press platens can usu-

fig. 6.2 *Flat honeycomb-panel press tool.*

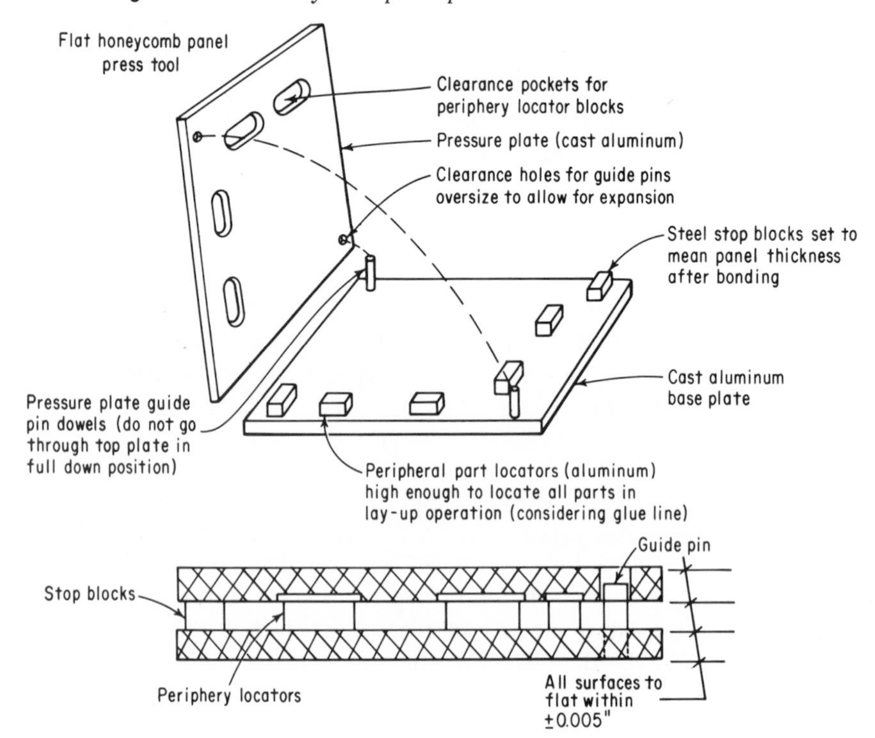

ally be held within ± 5°F. Operating temperatures can be reached very quickly and a rapid cooldown may be obtained by circulating cold water through the platens. Water-cooled platens may also be obtained in the electric-heated models.

Pressure gauges on hydraulic presses in most cases read in hydraulic pounds. When parts are being bonded, this must be converted to pounds per square inch. If the ram diameter is calculated into the instrumentation, this is accomplished by dividing the area of the part in contact with the platen into the gauge reading. If not, this formula will suffice:

$$\frac{\text{psi desired} \times \text{contact area of part}}{\text{Area ram cross section}} = \text{hydraulic pounds}$$

$$\text{Ram cross section} = \text{ram squared} \times 0.785$$

If a press does not have a balance valve, it should be monitored during the early stage of the operation. Pressures usually increase due to the thermal expansion of the substrates, and then decrease as the adhesive begins to flow.

The press is not only the most effective, but also the most economical approach for bonding flat surfaces where high pressure is required. The laboratory-type press is the most common of all adhesive tooling used in research and development, qualification testing, certification of adhesives, etc. These tests must be as accurate as possible; thus the press must be maintained in excellent operating condition pertinent to platen mating surfaces, control instrumentation, and gauges. The area of major concern is the mating surfaces of the platens, because they will not only fade easily into an unsatisfactory mated condition, but it is usually very difficult to detect this uneven distribution of pressure. An easy method of performing a quick check is the scattering of small metallic spheres (preferably lead beads). Over the platen, the pressure is applied, released, and micrometer readings taken on the flattened spheres.

Many finger-type lap shear specimens are bonded in small presses on a laboratory basis without the use of fixtures; thus pressure becomes of utmost importance because a ½-mil bond-line thickness variation may be important. The stock used to shim the upper coupons should be given considerable attention, because the shim stock must be the same thickness as the coupons on the platen plus the desired bond-line thickness. Mill sheet stock as received from the manufacturer may vary from 1 to 3 mils or more in thickness as received. The coupons should be measured for thickness and matched as nearly as possible because it is easy to visualize what can happen if two panels on the high-tolerance side are mated and two on the low side mated, and bonded at the same "shot." The result could be several mils difference in bond-line thickness if two panels bonded simultaneously, and caul sheets will not completely eliminate this problem.

SHRINK AND LAGGING TAPES

Shrink and lagging tapes employ the heat-and-heat-shrink factor, and may capitalize on the thermal-expansion properties of the adherends. Shrink tapes may be purchased from various sources and a number of types are available, but a cellophane or Mylar tape is usually used in a lagging operation. The tape is wet, wound securely around the mated parts, and shrunk, either from a heat source or allowing to dry at room temperature. The commercial shrink tapes are usually prestretched plastic tapes that contact on heat application. Pressure obtained in this manner is effective for bonding small parts and especially odd configurations on a one-shot basis, where development of tooling would be impractical. The shrink-tape concept may also be used with small tooling to obtain pressure.

The amount of pressure that may be obtained is limited; the maximum is approximately 15 psi. An excellent example of this method would be a thin substrate bonded to a tubular part, with a low-pressure adhesive system. The part can be lagged, placed in an oven, and cured. In the event of a room-temperature cure requirement, cellophane tape may be used, applied wet, and slightly stretched; thus drying will shrink the tape and

fig. 6-3 *Positive pressure tool utilizing fire-hose technique. (Hughes Aircraft Company, Culver City, Calif.)*

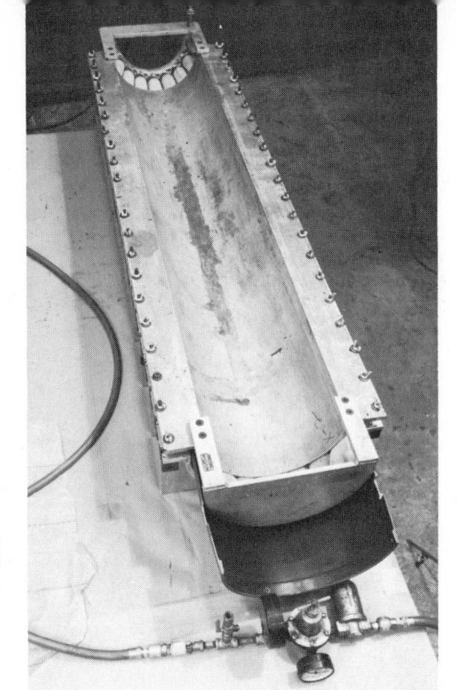

fig. 6-4 *Positive pressure tool assembled with part located and ready for cure. (Hughes Aircraft Company, Culver City, Calif.)*

produce pressure. Shrinkage of the tapes may be accelerated by the use of a hair-dryer type of air gun. They are available in a variety of forms and a wide range pertinent to weight, width, and thickness. The shrink-tape method is inexpensive, fast, and is a clever approach when applicable.

AIR-PRESSURE TOOLING

The air-pressure type of tooling is sometimes used employing solid cast or machined tools that are pressurized by the expansion of air-inflated fire hoses, metallic diaphragms, or plastic bags by means of a positive air-pressure source. The primary problem with this type of tooling is the uneven distribution of pressure. Figure 6-3 is an example of positive-pressure-type tooling utilizing the fire-hose technique. Note the aluminum pressure medium on the female tool which is used in an effort to produce an even transfer of pressure to the faying surface. Figure 6-4 shows the same tool assembled with a half-section of a plastic missile into which a thin-gauge steel torsion doubler is being bonded, and Fig. 6-5 shows the same approach for bonding the hinges to a plastic missile structure. The assemblies utilize a nitrile-phenolic adhesive system and are cured in an oven at 250 to 350°F with 40 to 50 psi bonding pressure. A direct air-pressure reading is obtained from the air-pressure gauges shown in the photographs. This type of tooling is usually heavy and difficult to handle.

fig. 6-5 *A positive pressure tool, utilizing the fire-hose technique for bonding the hinges to a plastic missile structure. (Hughes Aircraft Company.)*

AUTOCLAVE

The autoclave is the most practical of all tooling for bonding adhesives on a production basis. It would be difficult to justify the cost of an autoclave on short production runs, but the quality is usually better, especially in the fabrication of large-contoured and odd-shaped parts. Practically all companies with a substantial amount of bonding utilize this method.

Heat and pressure can be controlled and applied simultaneously with an autoclave. This may be accomplished by the use of special jigs or fixtures which are heated in an oven (or contain their own electric heat sources), but this may be expensive and uneconomical from a tooling standpoint. By using an autoclave, pressure and heat may be applied without the use of expensive, matched dies (Fig. 6-6). An autoclave consists of a cylinder; doors open at one or both ends, allowing the loading and unloading of work. There are several means of applying heat and pressure. Air and steam may be used for pressure; electricity and steam for heat. The parts are usually placed on a fixture, a carriage, or both, and the assembly enclosed in a vacuum bag which is evacuated before and during cure. This allows the autoclave to receive positive pressure and pressurize the assembly without removing it from the vacuum. Several parts, of various heights and contours, may be bonded at one shot without pressure problems. Autoclaves are available from small clam-shell type to large units 18 ft in diameter and 40 ft long (Fig. 6-7).

Autoclave pressure can be maintained up to 200 psi and many have maximum operating temperatures up to 1000°F. The platens are usually aluminum, but many of the later models are equipped with steel platens for curing the polyaromatic systems which require high-pressure and temperature cure cycles.

One of the difficult bonding operations utilizing the autoclave is a simple, flat honeycomb panel, especially if it is a composite structure. The difference in thermal expansion and production of a universal temperature rise in the assembly always poses a problem. Many times a rubber blanket is needed underneath the assembly to allow slippage during heat-up. This approach is used when aluminum skins are in contact with steel platens.

The vacuum is pulled on the bagged assembly to hold the details in place but is usually reduced or released completely as the internal positive pressure is increased. This decreases the possibility of pulling air through the assembly or causing the adhesive to foam or become porous if the seal is imperfect.

Many autoclaved assemblies are defective because of poor bleeding. Fiberglass cloth is usually used for a bleeder cloth. The glass cloth is draped over the entire assembly and bagged. By increasing or decreasing the layers of glass cloth, the heat-up rate on the top surface of the assembly may be controlled to some extent. If several assemblies are being

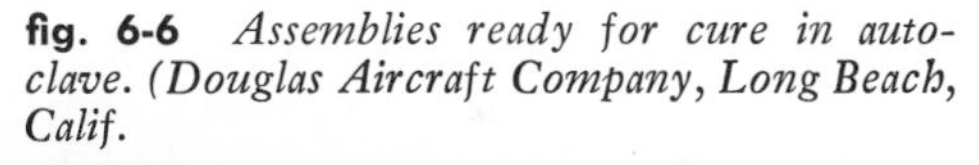

fig. 6-6 *Assemblies ready for cure in autoclave. (Douglas Aircraft Company, Long Beach, Calif.*

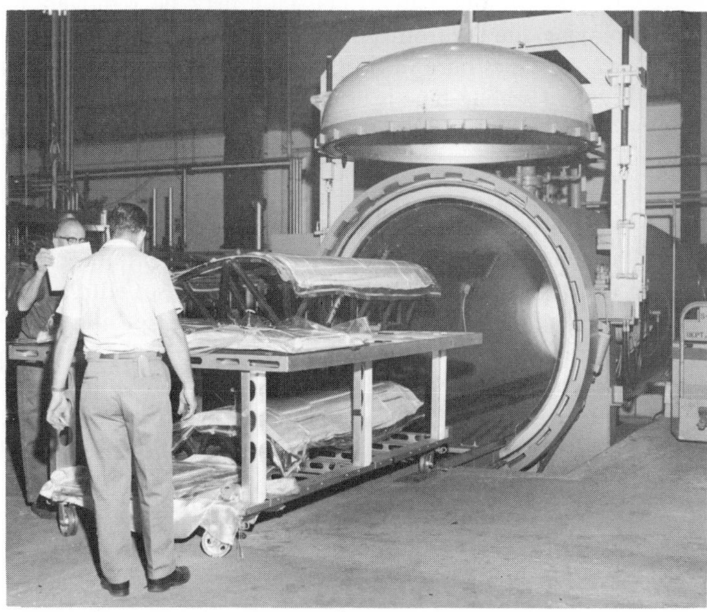

fig. 6-7 *37 x 17 ft autoclave utilized at Marshall Space Flight Center, Huntsville, Ala.*

fabricated at the same time, care must be exercised in positioning the assemblies properly to be positive each assembly will have an avenue for gas escape to a vacuum port. Cotton or nylon rope placed properly around the units will guarantee this condition and also prevent panel sealing. When utilizing an adhesive that flows readily, the glass cloth sometimes becomes saturated at the edge of the assembly and seals off the panel or assembly, thus preventing further gas evacuation.

Many autoclave platens are equipped with an outer ring assembly which allows the blanket or bag to be sealed with a rubber rope. The bag is tucked underneath and a flange tightened down on the rope to compress it and effect a seal. If dichromate paste is utilized for sealing, a maskant tape (vinyl or polyester) may be placed around the outer parameter of the platen. The tape may be removed after curing, thus removing the dichromate paste and expediting platen cleanup. Some of the sealing tapes cure during the bond cycle and are easily removed, but this type of sealing tape is usually limited to 400°F.

VACUUM-BAG BONDING

This method of applying pressure for adhesive bonding may be utilized at room temperatures, or in an oven or an autoclave as discussed earlier (Fig. 6-8). This technique consists of a sealed plastic bag placed over the part, and a vacuum pulled to meet the specified or desired pressure requirement (Fig. 6-9). The primary problem is the pressure limitation, which restricts this type of operation to the low-pressure curing systems because 12 to 15 psi is usually the pressure limit. This technique is especially economical on a one-shot part, because very little tooling is required.

The vacuum-bag method is used often for room-temperature bonding.

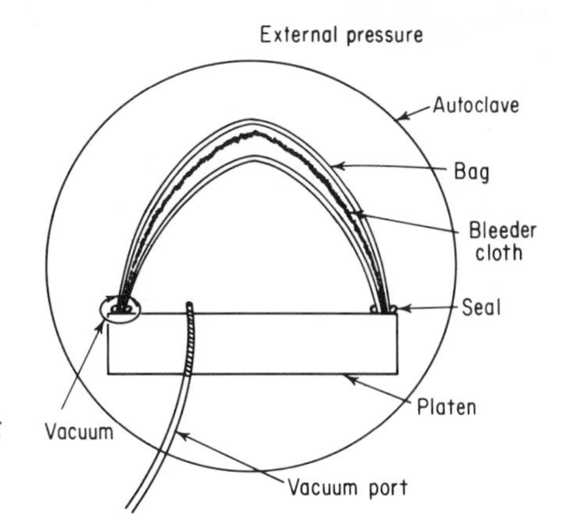

fig. 6-8 *The vacuum-bag theory.*

The assembly or panels may be placed on a table or fixture in the layup room, bagged, placed under vacuum, and allowed to cure. For this type of bagging operation polyvinyl chloride or polyvinyl alcohol films are utilized. They are economical and may be sealed with heat or dichromate paste.

Figure 6-10 shows the vacuum technique being utilized to encapsulate an expanded copper screen to the exterior of a missile structure. The assembly is warmed at this point until the adhesive is fluid and the air bubbles, which are visible in the photo, will be worked to the edges with a rubber "squeegie" and the assembly is then oven cured. Very large as-

fig. 6-9 *Assembly bagged and ready for cure. (Douglas Aircraft Company.)*

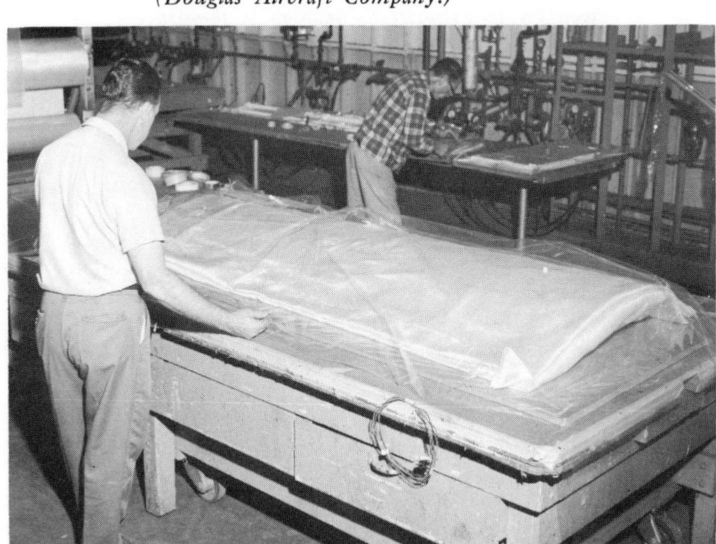

fig. 6-10 *Vacuum-bag technique used for encapsulation of components on a missile structure. (Hughes Aircraft Company, Culver City, Calif.)*

semblies are cured in this manner utilizing large walk-in ovens equipped with vacuum lines.

A heating pad or blanket may be used in conjunction with the vacuum bag, which exhibits a very high performance level, especially on large honeycomb or sandwich assemblies. Figure 6-11 shows a contoured honeycomb sandwich panel that has been bagged in a manner that will hold the panel to the tool and produce the desired contour. Figure 6-12

fig. 6-11 *Bagging of honeycomb sandwich panel. (North American Aviation, Columbus, Ohio.)*

fig. 6-12 *A heating blanket is placed over a sandwich assembly for curing. (North American Aviation.)*

shows the same assembly as the heating blanket is positioned. Figure 6-13 shows the assembly during the cure cycle. Thirty-two thermocouples are attached to the assembly, and the recorder in the left of the photo monitors the temperature throughout the cure cycle.

fig. 6-13 *Honeycomb assembly during cure, utilizing vacuum-bag and heating blanket. (North American Aviation, Columbus, Ohio.)*

In any vacuum-bagging operation one point cannot be overemphasized, that is, the proper application of the bleeder system. This may be accomplished in most operations by draping the assembly with fiberglass cloth, or creating an avenue for gases to escape by placing cotton rope or stockinette at proper points throughout the assembly and directing the bleed avenue to the evacuating outlet.

BAGGING MATERIALS

A six-month survey of one company involved in adhesive bonding disclosed seven vacuum-bag ruptures that scraped or damaged 56 assemblies. Bag materials are available for any operation, but often poor selections are made. Thick rubber bags may be used for simple layups, but this type of material would not conform closely enough to part configurations where complex configurations and sharp contours are bonded. It is also difficult to predict the service life of a bag. When should it be discarded? Several questions should be answered before specifying or using a bagging material. (1) What effect does heat aging have upon its physical properties? (2) What is its elongation and shrinkage factor at room temperature and at elevated temperatures? (3) What is its tear resistance? (4) Is it flammable—what is the flash point? (5) Is it compatible with solvents? (6) What is the cost—would cheaper material suffice? (7) Is it a one-cycle or multiple-cycle material? (8) Last but not least, what is the service temperature and how long will it be subjected to that temperature? Listed below are some tips.

1. Polyvinyl chloride film is an excellent one-shot material when not subjected to temperatures above 180°F, and may be heat sealed.
2. Polyvinyl alcohol is ideal for one-shot material when the bonding temperature does not exceed 250°F. This film also has the advantage of being water soluble, which means it can be in contact with the adhesive and if adhesion occurs it may be removed with water. This property can be a valuable asset, especially when utilized in an encapsulation-type bonding operation. A smooth surface can be obtained by placing the bag over the adhesive (glass bleeders usually leave impressions). The work or adhesive is visible through the film and the air bubbles may be worked out to the edge of the part and picked up by a bleeder that is located at the edges of the faying surface. Polyvinyl alcohol also has very high tear resistance and may be heat sealed effectively. PVA is available in a variety of weights and widths, and the cost is quite low.
3. Polyvinyl fluoride is a good bagging material that may be used up to 450°F. It has excellent tear resistance and is easily sealed.

fig. 6-14 *Bonding wings to a missile fuselage with rigid-type tooling.*

These three materials listed are very popular as one-shot materials because they perform well and are not expensive.

For multiple-shot operations, especially autoclave production operations, a heavier bag material should be selected. For information concerning other bagging materials, see Tables 6.1 to 6.7 at the end of this chapter.

RIGID PRESSURE TOOLS

Rigid tooling falls into a category that is considered restricted to a limited number of adhesives and belongs in the "clamping" category. Fixtures for fabricating test specimens often use this approach, but it is considered a poor tooling choice. Pressure may be obtained by threaded-screw devices, heavy clamping devices, or a rigid fixture that is locked in place and capitalizes on the thermal expansion of the substrates.

Figure 6-14 shows wings that are being bonded to a missile structure utilizing a "walk-in" oven for curing. Note the "C" clamps and steel bars used for alignment of the wings. No one could argue with the tool designers that tooling of this nature will not do the job because it does, at the present time; but enough money has been wasted by many firms on expendable "Rube Goldberg" tooling of this type to purchase fine autoclaves and associated equipment that would perform better in this area of adhesive bonding. The autoclave could be used for the next or subsequent contracts—but they cannot afford an autoclave!

fig. 6-15 *The use of aluminum beads as a pressure medium. (Fairchild Hiller Corporation, Hagerstown, Md.)*

CAVITY-TYPE TOOLING

The cavity-type tool is designed to hold mating parts during a bonding operation. This tooling concept is common in the production of aircraft and missile assemblies. The cavity tool may be constructed of metal or plastic; the metal is usually machined to the desired configuration and the plastics are cast or molded.

Assemblies are cured in the cavity tool utilizing autoclaves, ovens, built-in heating devices (electric or steam), heating blankets, or presses. Figure 6-15 illustrates a method that has merit but is not widely used. This concept consists of a cavity tool that utilizes aluminum beads as a pressure medium. The photo shows the details in place and the beads are being placed in the cavity. Figure 6-16 shows the cavity filled with beads

fig. 6-16 *The pressure plate is placed on the beads as a pressure medium. (Fairchild-Hiller Corporation.)*

and the pressure plate being located. This concept may be used with small details that are cured via the vacuum bag method where the contours are sharp or several details are bonded to one-shot and the bag tends to bridge and cause nonuniform pressure transmission. This simple technique and the aluminum beads provide even pressure transmission and distribution plus excellent heat transfer. The beads are inexpensive and may be used indefinitely. They do not seal an assembly; consequently, improper bleeding is highly improbable when cured under vacuum pressure if a bleeder is placed between the assembly and the aluminum beads.

PLASTIC TOOLING

Plastic tooling is considered a wise choice for curing adhesive-bonded assemblies and certainly offers special features that are advantageous. The plastics came into widespread use in World War II as tooling materials in the aircraft industry because of the high production requirements and the scarcity of skilled tool-and-diemakers. The early plastic tooling was based on cast phenolics and polyesters. The epoxies have replaced the earlier materials because they are easier to use, the shrinkage factor is lower, they have better adhesive and cohesive properties, they are less corrosive to metals and in general, are stronger and tougher.

There are several factors that must be considered in the selection of plastic materials for adhesive bonding tools.

1. Life expectancy of the tool

This determines the type of plastic construction, i.e., foamed systems, cast-resin systems, or reinforced-plastic systems. Plastic tooling is subject to creep or cold flow, consequently this is considered a major problem to contend with if many thousands of parts are to be fabricated utilizing the same tool, especially if an elevated-temperature cure is necessary. Plastic tooling is also subject to fatigue over an extended period; consequently, only a minor fraction of the maximum expected load applicable to a "single shot" should be exerted on the tool in multiple-cycle operations. If the life expectancy of the plastic tool is shorter than the metallic due to adverse operating conditions, it may still be cheaper than its metallic counterpart.

2. The operating temperature

Considerations must be given to the continuous and intermittent operating temperature. The plastics will show signs of degradation under prolonged usage at elevated temperatures and a rapid heatup and cooldown rate tends to cause deterioration due to thermal shock. The other prime factor associated with cure cycles is the low thermal conductivity of plastics as compared to metals. If the requirements of the tool demand a thick-walled mass construction, the heatup rate would be slow. This problem area may be minimized by incorporating heat-conductive fillers into the base resin such as metallic fibers, metal screening, metallic oxides, etc.

3. Machinability

Metal-filled systems offer the best machining characteristics. If finished machining is required, materials should be avoided that contain silica or other abrasive materials.

4. Compound contours

The fabrication of contoured curvatures is considered one of the advantages of adhesive bonding, thus tooling must be fabricated to produce these parts. They are more difficult to tool for; consequently a cast plastic is a simple approach provided that other properties are acceptable.

5. Tooling weight

The weight and mobility of the tool must be considered. The plastic tools offer outstanding features in this area.

6. Modification

The ease with which modifications, alterations, and repairs may be accomplished with plastic tools is one of the largest single factors favoring this type of tooling. Design changes may be incorporated quickly into a working tool without serious production interruptions.

In summary, economic gains can be realized by the utilization of plastic tools, provided the properties are acceptable but quality cannot be sacrificed for quantity. The plastic tool must be consistent in producing acceptable parts for the expected duration of the tool. When considering plastic tooling, it is advisable to consult experts in the field. The companies responsible for marketing these materials can offer helpful suggestions in selecting the proper resins and fillers or can formulate a system to meet a special design requirement.

HONEYCOMB CONFIGURATION TOOLING AND TECHNIQUES

There are many methods for shaping honeycomb core and sandwich materials to the desired configuration. The method utilized depends on equipment capabilities, desired tolerances, quality or parts, core density and thickness, contour complexity, detail size, type of core material (metallic, plastic, paper, etc.), and the quantity of parts.

Core material is usually purchased from the manufacturer in flat panels to a customer specification. It may be procured in the unexpanded condition, shaped to a desired contour, and then expanded. (HOBE indicates *Ho*neycomb *B*efore *E*xpansion and is a registered trademark of Hexcel Products, Berkeley, Calif. CUE indicates *C*ore *U*nexpanded and is a registered trademark of American Cyanamid Company).

HAND CUTTING

The manual method is used if the density, thickness, and quantity of parts are small, but would be undesirable for complex contours and if close tolerances are essential. This type of carving is usually accomplished with a serrated knife or a sharp instrument such as a putty knife with a razor-sharp edge. The core may be attached to the table and held in position with double-backed tape, held manually or placed in a fixture and cut to the planned dimension. A template or guide fixture may be practical especially if test specimens are being fabricated. This type of cutting is common in laboratory test specimen fabrication.

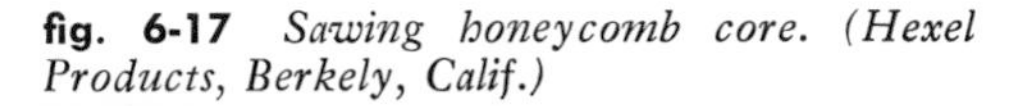

fig. 6-17 *Sawing honeycomb core. (Hexel Products, Berkely, Calif.)*

fig. 6-18 *Contouring honeycomb core utilizing the vertical saw method. (Hexel Products.)*

SAWING

This method is used throughout the aerospace industry. It is usually accomplished utilizing a circular bandsaw but may incorporate several types of saw guides, tilt tables, and fixtures. This technique is used for very large panels and high production rates may be attained by this method. Figure 6-17 shows core being sliced utilizing the common band-

fig. 6-19 *Hand router. (Douglas Aircraft Company.)*

saw and Fig. 6-18 illustrates a more complex sawing operation utilizing a vertical saw combined with a movable table for contouring a larger section of core material.

Special blades are available designed to saw aluminum honeycomb core and they are designed to slice rather than remove chips, thus producing a smoother and more uniform finish. A blade speed of 5 to 7 thousand FPM is considered a prime speed, but much slower or faster speeds may be used, depending on the core material. It is recommended that the core be sawed slightly oversize and if final trim is necessary, it may be accomplished during layup. A core detail that is slightly oversize (0.020 to 0.050 inches) will conform under pressure and produces a satisfactory bondment. Many times a slight crush is desirable, especially the thin cell wall core.

The most important factors in sawing honeycomb core are the knowledge and experience of the saw operator.

DIE-CUTTING METHOD

This technique is used very little in the aerospace industry, although high production rates may be attained with this method. It may be used on a laboratory or low-production basis, utilizing a special die and a hydraulic press. Cores with small cells and minimum density of 4 lb/ft produce the best results.

ROUTING

Routing is considered the most satisfactory method for contouring and shaping honeycombs, and closer tolerances may be realized in the finished product.

The equipment may be an elaborate router with multiple spindles or a simple hand router (Fig. 6-19). Special machines are fabricated for shaping honeycomb but standard shop equipment, such as planers and shapers, is commonly used (Figs. 6-20, 6-21). These machines are usually equipped with valve-stem-type cutters.

SANDING AND GRINDING

A small amount of sanding and grinding may be possible without damage to the core, but care must be exercised to prevent damage to the core from heat or oscillation. Very close tolerances may be held utilizing a horizontal belt sander or its equivalent. A modest amount of sanding may be necessary during the prefit operation. After a forming or

fig. 6-20 *Shaping honeycomb core utilizing valve stem-type cutter. (Hexel Products.)*

shaping operation, a honeycomb section may be placed in a fixture as shown in Fig. 6-22 for dimension check, preform, or a prefit check. Cores of high density may be ground unsupported but low-density cores should be stabilized.

fig. 6-21 *Routing honeycomb. (Hexel Products.)*

fig. 6-22 *Honeycomb in fixture for dimension check. (Douglas Aircraft Company, Long Beach, Calif.)*

CRUSHING

Aluminum honeycomb may be crush formed during the bonding operation, taking advantage of the heat and pressure and a cavity die or fixture. This operation is usually accomplished in a press and utilizes crush stops to limit platen travel in the desired areas. The adhesive is applied prior to assembly of the part, and as the press is closed the core is crushed in the desired areas. The crushed or high-densified areas are stronger and may be used for fitting attachments. This would eliminate fillers for attachment points. A large crush-formed assembly would naturally be slightly heavier than one contoured by other methods.

The core may be crushed to the desired contour and the bonding operation performed at a later time. A press and set of dies would be utilized in a similar manner as the bond-crush form technique.

MACHINING UNEXPANDED HONEYCOMB

This method eliminates the stabilization problem, as the desired contour or shape is attained before the core is expanded into its final form. Figure 6-23 shows a piece of unexpanded honeycomb during the machining operation, and Fig. 6-24 shows the core after expansion. Unexpanded core may be sawed, milled, or ground, and this type of panel production usually produces a part with fewer core defects. If tolerances are to be close, several factors must be considered: (1) the

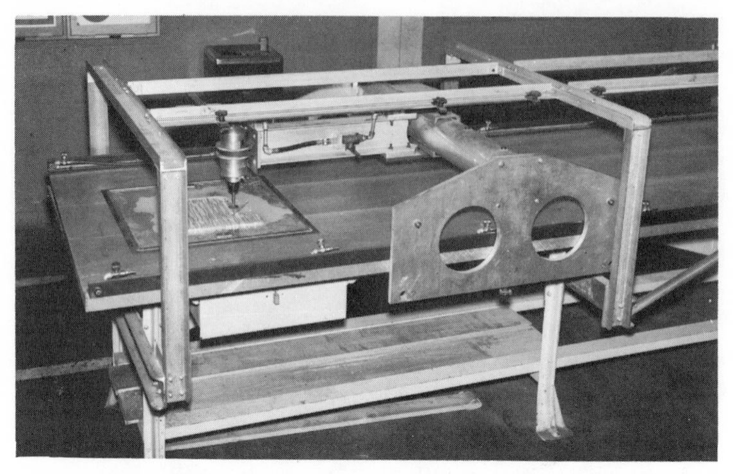

fig. 6-23 *Machinery; unexpanded honeycomb core. (Douglas Aircraft Company.)*

expandable direction, (2) ribbon direction contraction, and (3) accuracy of expansion and the severity of the contour.

BRAKE FORMING

Simple contours of fairly light radius may be formed in honeycomb core with a simple press brake (Fig. 6-25). The gap in the "U"-shaped die block should be approximately 1.75 times the core thickness. Both sides of the honeycomb core should be protected with an unbonded face sheet of aluminum.

fig. 6-24 *Aluminum honeycomb after machining and expansion.*

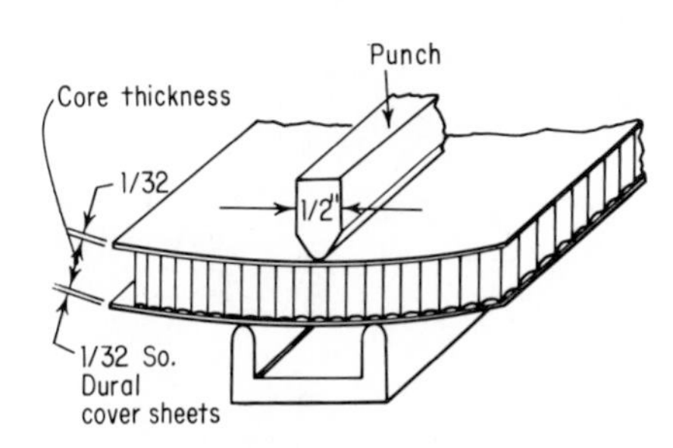

fig. 6-25 *Forming honeycomb with a press brake.*

ROLL FORMING

Cylindrical shapes are best formed by this technique. Both simple and compound curves in metal honeycomb can be mass produced using this method. Cylindrical and spherical radii down to approximately five times the honeycomb thickness are produced without serious problems or inherent loss in strength. The slightly flat surface on the face of the core is usually advantageous to the bonding operation because of a slight increase in surface area and the "locking" that occurs if proper node filling is realized.

The operation is performed on a double-roller piece of equipment (such as Farnham Roller, Slip Roller, etc.). The opening between the rolls must permit the entering and passage of the full thickness of the core without damage, using a step approach with slightly tighter radii until the desired curvature is achieved. A thin slip sheet of light-gauge aluminum should be used to prevent damage to the core. In tight-roll forming, a piece of sandpaper, cork, or other gripping material is used to aid in the resistance to node splitting or damage. A slight amount of node delamination can be tolerated because the adhesive will fill the "split" when the assembly is bonded. When roll forming very heavy honeycomb it may be advisable to bond a face sheet to one side of the core and, if necessary, remove it after the operation. The skin should be thin to facilitate peeling during removal. If it is not feasible to form a large heavy core assembly it may be cut, rolled in small sections, and then bonded together. If the assembly is too thick to roll successfully it may be sliced, roll formed, and then laminated with a structural-type adhesive system. This type of structure is stronger than the original, but is also slightly heavier.

FORMING DURING THE BONDING OPERATION

Where only mild contours are desired and low-density core is utilized, the forming may be accomplished during the bonding operation. The

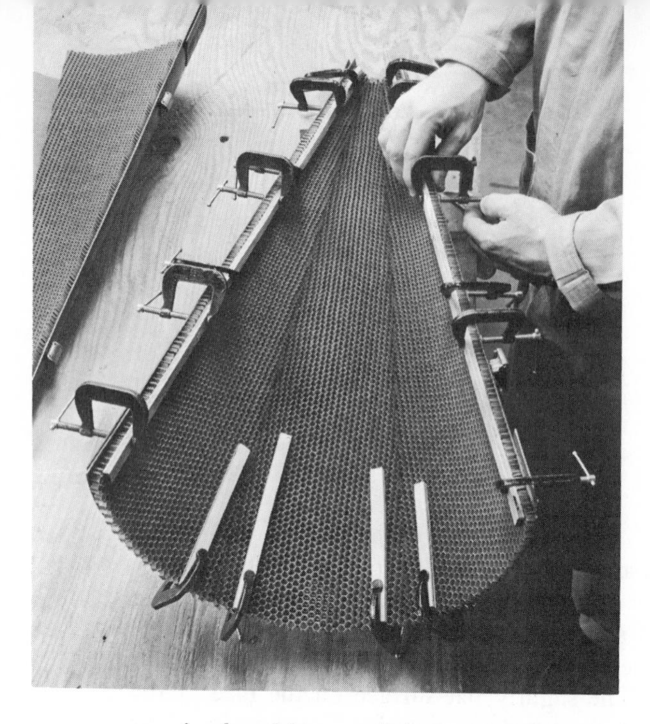

fig. 6-26 *Forming HRP core. (Hexel Products, Berkeley, Calif.)*

core is placed in a tool that meets the contour requirement and is forced into place with the application of bonding pressure and heat. This is a common practice and the general consensus among adhesive personnel is that "locked-in" stresses are minute. If thin force sheeting is used, it will also conform and requires no preforming.

PLASTIC HONEYCOMB CORE

HRP (Honeycomb Reinforced Plastic) can also be formed to mild contours by forming during the bonding operation. The most efficient method of forming HRP is by the utilization of heat. Plastic honeycombs are usually brittle at room temperature; thus the heating tends to soften the resins and allow the core material to conform to a forming dye or fixture. The amount of heat and exposure time naturally varies with the type, density, and thickness of the core. The core is heated in a matched die, or heated and moved quickly into the forming tool and allowed to cool with enough pressure to hold the core to the desired contour.

It may be desirable to slice the core for easier forming. When using this approach, the core is cut into sections, placed into the fixture, and butt joined as shown in Fig. 6-26.

The anticlastic or "saddling" characteristic may be used as an advantage. A spherical or other compound curvature can be made by simply forming the core beyond the desired radius in one direction. The radius is then opened up so that the anticlastic forms a compound curved surface. A rough approximation of the contour is all that is usually required, since the core will conform to the bonding tool during the joining operation.

A tight radius can be achieved with HRP core of high density by "slashing" the core, which consists of small cuts, one-third to one-half the wall thickness in depth, made on the inner radius of the material. The mechanical properties are not decreased to any large degree as the cuts will be filled during the bonding operation. The inflicted cuts will allow the core to be more easily formed, and this operation is sometimes performed and the core placed directly into the bonding tool.

Because the sandwich materials exhibit high strength-to-weight ratios, a vast amount of research is being performed in this area. Better plastic cores are being developed, and studies are being made to fabricate lighter and stronger metallic core. Figure 6-27 is a photograph of a tool designed by Battelle Memorial Institute for the manufacture of magnesium-lithium honeycomb core on an experimental basis, under a prime contract from NASA. This type of core shows great promise as a structural material in future space vehicles.

CHUCKING AND STABILIZING HONEYCOMB CORE

Honeycomb core should be carved or machined in the unstabilized condition whenever possible. Many times this is impossible because the core must be held or gripped firmly for the cutting operation as the cell walls are very thin and would otherwise be distorted or torn. If the foil wall core is snugged in a vice-type apparatus, the probability of damage is obvious. There are a number of methods used to stabilize or rigidize the core for chucking.

fig. 6-27 *Tool designed by Battelle Memorial Institute for manufacturing magnesium lithium honeycomb core.*

A. Polyethylene glycol

"Polyglycol," as it is normally referred to, is obtained in the form of a flaky solid that will melt at 120 to 150°F. It may be obtained in a variety of molecular weights, but the molecular weight is of little importance as long as the above melting range is obtained. Dow Chemical Corporation supplies several types to fit the need of any machining operation.

A common method is a chuck that is provided with a circulatory system for hot and cold water. This provides for quick melting and freezing. The chuck is heated and the polyglycol applied. The cold water is then circulated and the core becomes stable when the polyglycol becomes solid or rigid. The core actually becomes bonded to the chucking plate. It is then removed by heating the chuck. If very thin foil is being machined it may be advantageous to completely fill the cells to prevent damage during the cutting operation, or to provide complete stabilization in desired areas. The polyglycol may be removed after the operation by a hot-water rinse.

If specially designed chucks are not available, the core may be stabilized or chucked to a solid metallic plate and the plate gripped or mechanically fastened to a desired fixed position. The polyglycol may be melted by the use of an electric heat gun or equivalent, allowed to rigidize, and then removed by the same method.

B. Hot melt adhesives

On special occasions when heavy-cell-wall materials are being used, this method may be utilized for bonding the core to a solid plate for machining. The "hot melts" may be obtained with melting points as low as 170°F. They can be melted with the heat gun in the same manner as polyglycol, but pose a clean-up problem. After machining the core may be solvent cleaned or vapor degreased. One using this type of chucking should be aware of the "hot melt" properties and the solvent that would be a soluble agent for the adhesive. The only remaining question would be whether or not the solvent would damage the core node bonds. This method is seldom used but a very strong and rigid chuck may be attained.

C. Tape chucking

Limited machining may be accomplished using this technique. A double-back tape is utilized, both sides covered with an adhesive. The core is taped to a plate or feed table. This method would not be used for an extensive or heavy machine operation, but may be utilized to a great advantage in conjunction with light carving.

D. Ice chucking

Water is sometimes used to freeze the core to a chuck. The core is placed on a platen that is equipped with refrigeration capabilities. The platen is chilled, the core located, and water poured into the cells. The ice formed will hold the core in place during the machining operation provided the temperature is not raised above the freezing point of water during the operation. After machining, the ice is allowed to melt and the core removed. This method is not advocated by many technical people due to rust or corrosion problems. After ice chucking, the core should be moved quickly to the surface-preparation area and prepared for bonding.

E. Vacuum chucking

This is a very satisfactory and efficient method for machining honeycomb core. The machines are equipped with vacuum parts in the chuck, feed table, or platen. A temporary skin may be attached (polyglycol method, tape, etc.) to one side of the core and then removed after the operation. A temporary skin may be attached to the machined side and the other side machined. Permanent skins have been bonded to the machined surface in some operations. This is usually difficult due to the contamination problem. Recleaning after one skin has been bonded is impractical.

CONCLUSION: SIMPLICITY

Simple tooling should be the guideline for adhesive-bonding tools especially for high production capabilities. Tools should be designed to permit rapid setups and to be as idiot-proof as possible. This is not an implication that the average shopman is not capable of handling complex tooling, but rather a suggestion for expediency and reliable assemblies. Figure 6-27 is a good example of tooling for simple contoured honeycomb assemblies, geared for high production rates. Figure 6-28 is a simple tooling fixture for bonding airfoil assemblies for aircraft production. This type of tooling has proven its reliability pertinent to long-run and high-speed production.

If suddenly a new employee should be assigned to a bonding shop, regardless of his previous experience, the individual would view some weird looking contraptions described as bonding tools. The employee, not being familiar with the tool, would probably panic. It is possible that the archaic looking device may hold 25 parts in place for a one-shot assembly fabrication. The tool may have had its beginning as a rela-

fig. 6-28 *A simple production tool, bagged and ready for cure. (Douglas Aircraft Company, Long Beach, Calif.)*

tively simple tool but as the learning curve advanced, a clip was added, then several clips, angles, T-bars and other assemblies or components were included in the same bonding sequence. If an individual has grown in experience with the tool, he may be the only one who can make a satisfactory setup because he has knowledge of the problem areas. If one observes an experienced technician in the assembly and setup of a complex tool, certain movements would be noted, i.e., a small shim incorporated, a little piece of rubber or Teflon inserted at some point or possibly a "C" clamp attached at a particular point. The reason is obvious; if the tooling were perfect, which never occurs, the probability is extremely high that the detail parts would leave something to be desired, i.e., a joggle may not mate properly; an insert may not be the proper length; clips may have slight imperfections; or the tolerance buildup may have taken its toll, etc. Consequently, these areas that may receive substandard pressurization will need some encouragement from the operator, and being familiar with the operation he knows exactly what to do. As was stated earlier, regardless of the efforts made to this point to produce a reliable assembly, the assembly technician will have the final word in determining the quality of the bonded assembly. Pressure deviations result in substandard joints, and joints that are below maximum expectation relative to mechanical strength are difficult, and, in many cases impossible, to detect. Substandard joints due to tooling deficiencies result in a bond that will be subject to more rapid degradation when subjected to environmental conditions, and the service life expectation has been curtailed beyond any reasonable doubt. The very next time you board an airline, look to your lucky stars and pray that the people who were responsible for the bonded assemblies were knowledgeable and conscientious. The life that an unheralded assembly man in some remote vapor infested corner of a large company saves–may be yours!

As adhesive-bonding techniques advance, the demand for more complex and efficient tooling will tax the tool designer more and more. The shopman must be trained and ready to accept new challenges. He must be prepared to utilize new tools and techniques to produce reliable assemblies with longer service life expectations. It is the duty of the proponents of adhesive bonding to strive for excellency. This means to educate the masses, advocate more research, propose better working conditions for manufacturing, indoctrinate the nonbelievers, and last, but far from least, push for better tooling concepts and knowledgeable personnel to use them.

fig. 6-29 *Fixture for bonding air foil assemblies, designed for aircraft production. (Douglas Aircraft Company, Long Beach, Calif.)*

TABLE 6.1 Description of Materials Evaluated for Vacuum Bagging

Manufacturer	Material designation	Material thickness, in.	Description
F. B. Wright ..	Wrightex 920	0.063	Butyl rubber sheeting with high temperature additives
F. B. Wright ..	Wrightex 920 "Improved"	0.125	Butyl rubber sheeting with alleged higher temperature additives
Permacel	DL 5435	0.0322	Glass cloth coated both sides with fully vulcanized self-bonding silcone rubber
Permacel	ES 5009	0.0211	Glass cloth coated one side with unvulcanized silicone rubber and opposite side with fully vulcanized silicone rubber
Permacel	ES 3969	0.014	Glass cloth coated one side with fully vulcanized silicone rubber
Noland Paper	Vac-Pak H.S. 8171	0.0025	Green tinted, transparent nylon film with additives for heat stabilization
DuPont	Tedlar PVG200SG40TR	0.0025	Transparent polyvinyl fluoride film
DuPont	Tedlar PVF100SG30TR	0.0011	Transparent polyvinyl fluoride film
Monosol	PVA	0.008	Transparent polyvinyl alcohol film
DuPont	Cellophane 150PD*	0.0013	Uncoated transparent cellophane film
DuPont	Mylar 200A †	0.0022	Transparent polyester film
Mosites Rubber	Mosites 1440	0.125	Silicone rubber sheeting

* Manufactured from regenerated cellulose, a softening agent, and water.
† Manufactured from polyethylene terphthalate.

TABLE 6.2 Average Results of Aging at 375°F upon Tensile Values of Multiple-Cycle Vacuum-Bagging Materials

Material	Initial tensile strength, psi (1)		Tensile strength after aging, psi					
			Aged 24 hours		Aged 48 hours		Aged 72 hours	
	at R.T.	at 375° F	at R.T.	at 375° F	at R.T.	at 375° F	at R.T.	at 375° F
Wrightex 920, 1/16 in. thick	845	176	880	176	564	176	564 (2)	176
"Improved" Wrightex 920 1/8 in. thick (3)	1500	....	1470	....	1240	....	865	....
Permacel DL 5435	2000	2350	2350	2550	2210	2210	2210	2210
Permacel ES 5009	2370	3320	2210	2630	2840	2840	2690	2840 (4)
Permacel ES 3969	2210	1900	2370	2210	2370	1740	1740	1740
Mosites 1440 1/8 in thick (5), (6)	1420	....	....	....	....	....	1420	....

NOTES: 1. Tensile strength determined per Fed. Test Std. No. 601, Method 4111, Die 111.
2. Wrightex 920 (1/6 in. thick) was somewhat brittle and hard after being aged 72 hr.
3. "Improved" Wrightex 920 could not be tested at elevated temperatures in the laboratory oven due to its high elongation.
4. The unvulcanized rubber coated side of the glass cloth gradually cured due to elevated temperature aging.
5. Mosites 1440 could not be tested at elevated temperatures in the laboratory oven due to its high elongation.
6. Mosites 1440 was not tested after 24 and 48 hr aging due to lack of material.

TABLE 6.3 Average Results of Aging at 375°F upon Tensile Values of One-Cycle Vacuum-Bagging Materials

Materials	Initial tensile strength, psi (1)		After aging 5 hours, tensile strength, psi	
	at R.T.	at 375° F	at R.T.	at 375° F
Vac-Pak H.S. 8171	5,570	4,950	4,950	4,950
Tedlar PVF 200SG40TR	4,430	2,980	5,310	1,770
Tedlar PVF 100SG30TR	10,500	6,030 (2)	(3)	(3)
PVA (8 mil) (4)	4,430	1,110	7,200 (5)	1,110 (5)
Cellophane 150PD	8,500	(6)	4,430	(6)
Mylar 200A	15,000	12,600	13,000	7,300

NOTES: 1. Tensile strength determined per Fed. Test Std. No. 601, Method 4111, Die 111.
2. Shrinkage began when placed in 375° F oven.
3. Tensile strength could not be measured due to shrinkage.
4. Tensile strength of hygroscopic PVA varied with humidity.
5. After aging, PVA was brittle at room temperature; the film was flexible at 375°F.
6. Tensile strength was too low to be measured.

TABLE 6.4 Effect of Aging at 375°F upon Ultimate Elongation (1) and Shrinkage of One-Cycle Vacuum-Bagging Materials

Materials	Initial test		Aged 24 hours		Aged 48 hours		Aged 72 hours	
	R.T.	375° F	R.T.	375° F	R.T.	375° F	R.T.	375° F
Wrightex 920, 1/16 in.	210%	100%	200%	103%	160%	113%	140%	113% (3)
"Improved" Wrightex 920, 1/8 in (6)	310%	—	400%	—	305%	—	285%	— (3)
Permacel DL 5435 (4)	10%	10%	10%	10%	10%	10%	10%	10%
Permacel ES 5009 (4)	10%	10%	10%	10%	10%	10%	10%	10% (5)
Permacel ES 3969 (4)	10%	10%	10%	10%	10%	10%	10%	10%
Mosites 1440 (7)	650%	—	—	—	—	—	650%	—

NOTES: 1. Ultimate elongation determined per Fed. Test Std. No. 601, Method 4121.

2. No multiple cycle bagging materials had shrinkage after aging at 375°F for 72 hr.

3. Wrightex 920 (1/16 in. thick) was somewhat brittle and hard after being aged 72 hr; Wrightex 920 (1/8 in. thick) was quite hard after being aged 72 hr.

4. True elongation did not take place in these materials; the glass fibers were distorted and the silicon rubber elongated to result in a 10 percent ultimate elongation before breakage.

5. The unvulcanized rubber coated side of the glass cloth gradually cured due to elevated temperature aging.

6. "Improved" Wrightex 920 could not be tested at elevated temperatures in the laboratory due to its high elongation.

7. Mosites 1440 could not be tested at elevated temperatures in the laboratory oven due to its high elongation.

TABLE 6.5 Effect of Aging at 375°F upon Ultimate Elongation (1) and Shrinkage of One-Cycle Vacuum-Bagging Materials

Materials	Initial elongation (1)		Elongation (1) after being aged 5 hours		After being aged 5 hours	
	at R.T.	at 375° F	at R.T.	at 375° F	Shrinkage (2)	Appearance
Vac-Pak H.S. 8171	130%	180%	20%	130%		No visible change
Tedlar PVF 200SG40TR	160%	227%	80%	130%	4%	No visible change
Tedlar PVF 100SG30TR	68%	80% (3)	(4)	(4)	60%	Brittle
PVA (8 mil) (5)	200%	230%	10%	200%	7%	Brittle and hard, cloudy brown
Cellophane 150PD	15%	(6)	10%	(6)	10%	Brown, wrinkled
Mylar 200A	20%	23%	20%	37%	3%	No visible change

NOTES: 1. Ultimate elongation determined per Fed. Test Std. No. 601, Method 4121. Distance between center of the knife edges was 1.000 ± 0.003 in.

2. Shrinkage was determined by difference in bench marks before and after aging per Fed. Test Std. No. 601, Method 4121. Samples were aged at 375°F for 5 hr.

3. Shrinkage began when samples were placed in 375°F oven.

4. Elongation should not be measured due to excessive shrinkage; sample was quite brittle at room temperature.

5. Elongation of hygroscopic PVA varied with humidity.

6. Elongation could not be measured due to zero strength at 375° F.

TABLE 6.6 General Vacuum (1) and Pressure Test of Unaged Materials

Materials	Failure at room temperature under pressure (2)		Failure at 375° F under pressure (3)	
	Air pressure, psi	Type of failure	Air pressure, psi	Type of failure
Wrightex 920, 1/16 in.	45, 45, 46	After some elongation, tears in all directions from center of hole	36, 38. 40	Lengthwise tear
Wrightex 920, 1/8 in.	95	No failure	95	No failure
Permacel DL 5435	95	No failure	95 (1)	None
Permacel ES 5009	95	No failure	95 (1)	None
Permacel ES 3969	50, 76, 80	Tear at edge of outer plate	55, 61, 73	Tear at edge of outer plate
Mosites 1440	95	No failure	95	No failure
Vac-Pak H.S. 8171	21.5, 27, 28	Failure in center after high elongation; "blown effect"	30, 48, 52	Failure in center after some elongation; "blown effect"
Tedlar PVF 200SG40TR	41, 42, 44	Small hole in center	21, 23, 25	Lengthwise tear
Tedlar PVF 100SF30TR	19, 21, 26	Lengthwise failure in center	17, 20, 20	Lengthwise tear
PVA (8 mil)	47, 47, 50	Failure in center after high elongation; "blown effect"	28, 30, 35	Failure in center after some elongation; "blown effect"
Cellophane 150PD	30, 31, 35	Expanded, then shattered	.8, 1, 1.2	Shattered immediately
Mylar 200A	46, 47, 48	Failure in center; some "blown effect"	38, 40, 43	Failure in center; "blown effect"

NOTES: 1. All samples withstand the vacuum of 28 in. of mercury applied at room temperature. The vacuum was applied at a rate of 5 in. of mercury per min.
2. Pressure was applied at the rate of 5 psi per min.
3. The film and text fixture were heated to 375° F for 10 min. prior to application of pressure.

TABLE 6.7 Initial Properties of Selected Vacuum-Bagging Materials

Material	Tear resistance lb/in. thickness (1)	Flammability	Solubility in boiling water (4) soluble matter lost, percent	Adhesives which adhere to films Before Bonding	After bonding (7)
Wrightex 920, 1/16 in.	32–159	Burns very slowly	0%	F-141, F-161	F-161, HT-424, HT-430, FM-1000
Wrightex 920, 1/8 in.	145–152	Burns very slowly	0%	F-141, F-161	F-161, HT 424, HT-430, FM-1000
Permacel DL 5435	342	Self-extinguishing	0%	None	None
Permacel ES 5009	429	Self-extinguishing	0%	None	None
Permacel ES 3969	142–190	Self-extinguishing	0%	None	None
Mosites 1440	312–320	Self-extinguishing	0%	None	None
Vac-Pak H.S. 8171	600–800	Slow burning to self-extinguishing	2.81%	None	FM-61, FM-86, FM-96, F-141, HT-430, FM-1000
Tedlar 200SG40TR	280–400	Burned slowly until film charred; then burned rapidly	2.335%	FM-86, F-141, F-161	FM-86, HT-430
Tedlar 100SG30TR	636–726	Burned slowly until film charred; then burned rapidly	.336%	FM-86, F-141, F-161	FM-86, HT-430
Polyvinyl Alcohol	938–1250 (2)	Burns slowly	100% (5)	FM-61, FM-86 F-141, F-161	FM-1000
Cellophane 150PD	780	Burns rapidly	23.34% (6)	FM-61, FM-86, F-141, F-161, HT-424	HT-430, FM-1000
Mylar 200A	363–455	Burns slowly	6.06%	FM-86, F-141, F-161, HT-424,	FM-96, F-161, HT-430

NOTES: 1. Tear resistance specimens prepared and tested per Fed. Test Std. No. 601, Method 4211, Die C.

2. Tear resistance increased as the moisture absorption increased.
3. Flammability was determined by placing specimen in flame until kindling point was reached. Specimen was then removed from flame. In relative classification of burning rate, paper would "burn slowly."
4. Five in. sq specimens were weighed to the nearest 0.1 mg before and after being boiled in demineralized water for 5 hr.
5. Polyvinyl alcohol dissolved immediately upon contact with boiling water.
6. Cellophane was brittle and wrinkled upon contact with boiling water.
7. All adhesives except FM-86 were cured at 340 ± 10° F for 1 hr under 40 psi; FM-86 was cured at 180° F for 2 hr under 25 psi.

REFERENCES

"Carving and Forming Honeycomb Materials," Hexel Products, TSB-117, Berkeley, Calif.

Douthitt, C. A., and D. F. Weyher: "Curing Adhesives in Production Runs," ASME Publication 66-MD-34, June 13, 1965.

"Handbook of Adhesives," American Cyanamid Company, Havre de Grace, Md.

Hull, H. R.: Production of Contoured Honeycomb, *Western Machinery and Steel World*, October, 1958.

"Magno-Ceram, A New Concept in Plastic Tooling," Magnolia Plastics, Atlanta, Ga.

Nelson, P. K., and E. S. Howell: "Limited Evaluation of Various Materials for Bagging in Bonding Operations." AVCO Corp., Nashville, Tenn.

Personal correspondence with Douglas Aircraft Company, Long Beach, Calif.; Hughes Aircraft Company, Culver City, Calif.; Hexel Products, Berkeley, Calif.; North American Aviation, Columbus, Ohio; Lockheed-Georgia Company, Marietta, Ga.; Northrup-Norair Corporation, Hawthorne, Calif.; The Rohr Corporation, Riverside, Calif.; The Boeing Company, Renton, Wash.; The AVCO Company, Nashville, Tenn.; Bell Helicopter Company, Fort Worth, Tex.; The Lockheed-California Company, Sunnyvale, Calif.; Honeycomb Corporation of America, Bridgeport, Conn.

"Polyethylene Glycols," The Dow Chemical Company, Midland, Mich.

"Pre-fitted Parts for Aircraft Construction," *Adhesives Age*, January, 1966.

"The SPI Handbook," The Society of Plastic Industries, published by the John P. Watkins Company.

"Tooling with Plastics," Magnolia Plastics, Inc., Atlanta, Ga.

CHAPTER 7

Testing Adhesives

A test may be defined as the method or process by which the quality, presence, or genuineness of anything is determined; a means of trial. Someone once said, "One test is worth a thousand expert opinions." The latter statement has never had a more realistic application than in the area of adhesive bonding. If it could be correctly assumed that the adhesive chosen to perform a particular service would always retain consistent and universal quality throughout the same batch, roll, etc., testing would be unnecessary. If physical properties were not sensitive to varying conditions, mechanical testing would be a waste of time. Adhesives deteriorate under environmental conditions, necessitating environmental testing.

An adhesive test may measure a single property or several properties at the same time and it may be a simple test or an extensive complex study. In any case, tests for adhesives have been devised to measure a particular property as accurately as possible. A majority of tests now in use are products of years of experience and research. In general, present standard test methods are considered adequate; nevertheless, further improvements are sought constantly. When standard tests will not produce the necessary information, special tests must be devised to satisfy the situation and promote a sound decision by the cognizant individual or responsible function.

The American Society for Testing Materials (ASTM) regards testing as a dynamic science, always receptive to further improvement. The military services require testing and qualification of adhesives for military programs, but progressive companies in the field of adhesives test far beyond these basic requirements. The adhesive manufacturers or formulators are faced with an acute problem of producing a product of universal and consistent physical and mechanical properties. During the formulating and development of an adhesive system an enormous amount of testing is necessary, not only to validate the general properties, but also, in many cases, to produce data that comply with a customer procurement, military specification, or both.

The customer or consuming organization finds the data reported by the manufacturers very helpful but cannot, or will not rely entirely upon the reports of the sales agency. Neither will one company rely on another company's research reports, even though they may prove valuable and accurate, because everyone prefers to do his own testing. There are probably more duplications in adhesive testing than all other fastening modes combined. The reason is obvious: everyone will utilize the same system and possibly the same adherends but with entirely different requirements and manufacturing conditions. One hardware manufacturer may place more emphasis on one testing mode than another manufacturer, due to a normal reaction that stems from a previous failure.

A wise approach for the materials engineer, when screening for an adhesive to do a particular job, is to search the vendor data sheets for information in the area of major concern, look for unfavorable or poor data in certain areas, or search for missing information. This would permit one to perform cursory tests in areas of doubt, and thus to accept or eliminate the candidate. One should always test any area where data are missing. Does it seem reasonable that a vendor would trouble himself to report three pages of data from an assortment of tests and leave out, for instance, peel data or moisture resistance?

After tests have been conducted and the adhesive qualified for production, other areas of testing must be considered. This includes such items as incoming or receiving inspection, material control, quality control, and certification.

GENERAL PROPERTY TESTING

There are a number of factors that should be known concerning the general properties of adhesives. These are factors that govern the behavior of an adhesive from the time it is received by the user until it is incorporated into a final cured joint. They are commonly referred to as the working properties which include viscosity, storage life, weight loss, pot life, cure rate, tack, flow rate, and penetration characteristics.

The adhesive manufacturers are unable to control the general properties to a desirable standard; therefore, testing by both manufacturer and purchaser is necessary for quality assurance.

The general properties are seldom given the consideration that is given mechanical tests during the selection of an adhesive for a production operation. The trend seems to be reversed, that is, to determine the mechanical properties in the laboratory under ideal conditions, and then write a manufacturing specification that is surely geared to control the general properties.

VISCOSITY

The viscosity is an important property of the adhesive resin, especially if it is to be applied by a metering device or semiautomatic dispenser. Viscosity tests also reveal information pertinent to age, quality of the resin, and, in many cases, the wetting characteristics which may be directly equated to the age factor. The relationship between viscosity and other product dimensions often makes these flow-property measurements the most convenient method of detecting changes in density, stability, solids content, and molecular weight.

A number of tests exist for evaluating the flow or consistency properties of liquid adhesives. The more common of these is the utilization of

fig. 7-1 *The Brookfield viscometer. (Brookfield Engineering Laboratories, Inc., Stoughton, Mass.)*

an instrument known as the viscometer (ASTM D1084–60 Method B and ASTM D1286–57). A viscometer usually utilizes a stand, a mounted electric motor, and a rotating shaft to which a spindle is attached (Fig. 7-1). There are a variety of spindles available for any testing need. An instrument is incorporated into the device to register the results in centipoises (cps), the fundamental unit of viscosity measurement. This measurement may be compared to a water reading or a viscosity standard solution which is supplied by the viscometer manufacturer. If a viscosity standard fluid is utilized for adhesive tests, be absolutely positive that it does not contain silicone oils.

Another test for determining the consistency of liquid adhesives is defined in ASTM D1084–63 which utilizes "consistency cups." The cups are designed to deliver 50 ml of the sample in 30 to 100 sec under controlled temperature (75°F) and controlled relative humidity (50 ± 2 percent). The consistency cup is mounted on a test stand with a receiving container underneath it. The outlet in the bottom of the consistency cup is closed by placing a finger over it and the adhesive is poured into the cup. The number of seconds from the time the finger is removed from the orifice is measured by the use of a stop watch and reported as the consistency of the adhesive material.

When testing the more viscous materials, especially the RTV's, the extrusion rate is recommended. A Semco 440 nozzle or its equivalent is attached to a standard Semco cartridge filled with the material. The cartridge is then placed in an air-operated sealant gun and the gun attached to a constant air supply of 90 ± 6 psi. Two to three in. of the material should be extruded to clear trapped air and at least 1 in. extruded prior to each observation. The material is then extruded into a cup that has been previously weighed and labeled. The material is extruded for 10 sec and then the cup reweighed. Three cup contents are averaged and the extrusion reported in grams per minute.

Flow-rate tests also apply to film adhesives. The procedure utilizes a stainless steel plate that is covered with cellophane or thin Teflon sheeting. Three specimens, approximately 1½ in. in diameter are cut or stamped out of the roll of film to be tested. The specimens are placed on the first prepared plate and covered with another plastic sheet, followed by a second steel plate. The specimens are then cured utilizing the recommended cure cycle and pressure (press, autoclave, etc.). After the specimens are cured, they are removed from between the metal plates and taped to a white or other light-colored table top. The cured specimens are measured by the use of a compensating polar planameter or venier calipers and the percent flow calculated as follows:

$$\text{Percent flow} = \frac{\text{final area} - \text{original area}}{\text{original area}} \times 100$$

Results are reported as flow rate, which is the average of a minimum of three specimens.

STORAGE LIFE

When an adhesive is stored for a considerable length of time, changes may occur. The storage life of an adhesive is the length of time it can be stored and remain suitable for use. Storage life may be determined by the viscosity test previously described, especially in the area of liquid resins, but for the structural film-type tape, mechanical tests are utilized—in most cases, the simple lap shear tensile test. The lap shear results may be compared to a known "fresh value," utilizing a minimum pounds per square inch value for qualification. This is the prime reason preproduction tests and qualification tests are conducted on stored adhesives at regular intervals to determine their strength properties. The timing of these tests is established from information obtained from the vendor, military specifications, or practical experience.

POT LIFE

The pot life or working life of an adhesive is the time lapse between the time the adhesive is ready for use and the time when the adhesive is no longer usable, and naturally applies to the liquid or paste systems. The working life is usually established by experience or is accepted according to the manufacturer's recommendations. The adhesive age, shipping environment, storage application environment, and exactness of catalyzation influence the working-life factor. ASTM D1338 established two procedures that are applicable to checking pot life but are seldom used in the aerospace industry. In manufacturing assembly operations pot life sometimes poses a problem because if the adhesives are used in the gel state, proper wetting may not occur, thus resulting in a poor joint. The most satisfactory control in this area is the quality-control-technician check sheet. The mix time is recorded and signed off on the check sheet and repeated when the application has been completed. With the pot life specified with a 20 percent safety factor, the use of "over-the-hill" adhesives is eliminated.

SOLIDS CONTENT (WEIGHT LOSS)

The solids content is usually determined by weighing a small amount in a clean container, heating or curing until a constant weight is obtained, and weighing again. The percentage of solids may be calculated as follows:

$$\text{Percent solids} = \frac{\text{sample weight after curing}}{\text{sample weight before curing}} \times 100$$

The data reported should include (1) drying temperature, (2) number of specimens, (3) solids content of each specimen, and (4) average of all samples. The percent-solids tests are important because they indicate the amount of volatiles released during cure.

CURE RATE

Structural adhesives usually require curing by the application of heat or the addition of a hardener, or both of the above, with or without pressure. Since it is often desirable to be aware of the variation in bond strength with rate of cure, the strength developed in the adhesive joint when cured at various time cycles may be determined by the lap shear tensile specimen. Specimens are fabricated from the same batch (and mix, if applicable), cured utilizing a standard cure, a one-half cure, two-thirds cure, and post cure. They are then tested destructively and compared, the results reported in pounds per square inch. A recommended method is outlined in ASTM D1144.

PENETRATION (WETTING)

This may be determined by placing a measured amount of adhesive (usually 0.020 to 0.030 g) on the top layer of a stack of 4 to 6 sheets of Whatman's No. 4 laboratory filter paper. The entire stack is then placed in a press and pressurized to 100 psi for 10 to 15 sec. The sample is then removed from the press and the filter paper examined for penetration. Only a comparison can be made of adhesives with various wetting characteristics.

TACK

Tack is the property of an adhesive to adhere to another surface on contact. This may be referred to as the "stickiness" of an adhesive and may be determined by the utilization of a tensile testing machine and test blocks. The results would be the pounds per square inch required to part the test blocks. Although tack is an important factor, especially in honeycomb layup, tack tests are not given prime consideration in the aerospace industry.

DENSITY (FOAMED ADHESIVES)

The density of standard adhesives (liquid, paste, and films) is available from the manufacturer. Many foamed adhesives are being used in the aerospace industry, usually as a filler, which, in many cases, adds stiffness and reduces the "K" factor. The density test is accomplished by foaming a section and cutting 2.0-sq-in. cubes from the sample. The dimensions of the specimens are measured to the nearest 0.1 in. and the volume calculated to the nearest 0.01 in. The specimen is then weighed to the nearest 0.1 g and the density calculated using the following formula:

$$\text{Density (lb/ft}^3\text{)} = \frac{\text{weight in grams}}{\text{volume in inches}^3} \times 3.8$$

RESISTANCE TESTS (CONDUCTIVE ADHESIVES)

The resistivity of a conductive adhesive tends to increase with age, therefore tests must be conducted frequently to determine if the adhesive is suitable for usage. A strip of adhesive $\frac{1}{16}$ in. wide is deposited onto a sandblasted surface of a glass epoxy slide specimen $1 \times 12 \times 0.125$ in. The test specimen may be masked off to ensure a smooth and consistent bead of adhesive. The conductive material is then cured in the recommended manner and the tape removed. A 50-mA current is then passed through the adhesive strip, utilizing an ohmmeter or equivalent, and the total resistance determined. The resistivity is then determined by the following formula:

$$P = \frac{RA}{L}$$

where P = resistivity, microhm-centimeters
R = resistance, ohms $\times 10^6$
A = cross-sectional area, cm^2 (average of 3 locations)
L = length, cm

The resistivity is reported in ohms per square centimeter.

AMOUNT-OF-SOLIDS-APPLIED EVALUATION

In production operations where the amount of adhesive applied is important, it becomes necessary to determine the quantity of adhesive applied in a spreading, spraying, or coating operation. A procedure is defined in Federal Test Standard 175 and is performed in the following manner:

> The test adherend should be the same material as the production part that is to be coated. The size of the test specimen is unimportant but should be simple in geometric shape to facilitate accurate measurements and large enough to give good weight indications of the cured adhesive system. The adherend is weighed and the adhesive applied in the same manner as is anticipated for the production part. The adhesive is cured with the identi-

cal cure cycle that is, or will be, specified for the actual part. After cure, the specimen is weighed again and solid adhesive applied is calculated as follows:

When W_2 and W_1 are expressed in grams:

$$D = \frac{317.5\ (W_2 - W_1)}{NA}$$

When W_1 and W_2 are expressed in grams:

$$D = \frac{9000\ (W_2 - W_1)}{NA}$$

Where D = weight of adhesive solids applied, expressed in pounds per thousand square feet of surface area

W_2 = weight of the test specimens after application of the adhesive and cure

W_1 = weight of the specimens prior to coating

A = area of test specimen (sq in.)

N = number of specimens coated

Several specimens should be tested to obtain an accurate determination.

MECHANICAL TESTING

The method used to determine the actual strength of adhesives is the mechanical test. The more important and more widely used methods will be discussed in the following paragraphs.

TENSILE TESTS

An adhesive is in tensile loading when the acting forces are applied perpendicularly to the plane of the adhesive. The tensile strength of an adhesive joint is the maximum tensile load per unit area required to break the joint. There are various types of tensile tests that are currently being used in the adhesives industry.

LAP SHEAR

The lap shear test is the most common of all adhesive tests because it is easily fabricated and simple to test. Standard lap shear coupons are 1 in. wide and 4 in. long. Two of these are overlapped ½ in. and joined. They are cured and then mechanically tested utilizing a loading rate of 800 to 1400 lb per min (1400 lb per min is equivalent to 0.05-in.-per-min jaw separation). The bond-line thickness, width, and overlap measurements should be taken and recorded before tensile destruction. The glue-line thickness is obtained by subtracting the total adherend thickness from the total specimen thickness across the glue line. Lap shear tests are defined in MM-A-132 and ASTM D1002.

Often, too much emphasis is placed on the lap shear test, especially in the area of production control, because only useful information obtained from the lap shear specimen that accompanies a part during cure pertains to the surface condition. The pressure (even in an autoclave) would seldom be the same as the assembly. The cure cycle would not be comparable because of the difference in heatup rate.

There are many factors that effect the final result obtained from the lap shear tensile test:

1. Temperature (ideal usually considered 75 ± 5°F for room-temperature testing)
2. Jaw alignment of the testing machine
3. Alignment of the specimen itself in the joined area
4. The length of overlap (see T/L curves, Chap. 4)
5. The rate of loading

The latter is very important and is usually underemphasized. Table 7.1 indicates what may happen if the rate of loading varies.

After testing has been performed, the failure must be analyzed. This may be adhesive (failure of adhesive from substrate or primer), cohesive (failure within the adhesive layer), primer failure (failure of the primer from the substrate), adherend failure (failure of the substrate), or the primer may fail cohesively, which seldom happens if the specimens are processed properly. If a substrate has a conversion coating (magnesium, plating, etc.), the failure could occur within this parameter. The fact is evident that the analysis of the failure is of prime importance. If substandard bonds are tested, the weak link may be established and the necessary corrections made.

The report should include the type of adherends, adhesive and primer used, loading rate, bond-line thickness, area of bond, load, pounds per square inch, conditioning, type of failure, and temperature environment during test procedure.

BLISTER DETECTION TENSILE TESTS

The blister-detection tests are also tensile tests; the prime difference is that the panels are bonded, which will yield, when sawed, approximately 11 specimens (1.0 × 12.0 in.). The lap shear tensile is usually bonded singly or the finger-type panels, but the blister type is bonded in flat panels and then tested in the same manner as the lap shear tensile specimen. The prime advantage of the blister specimen over the normal lap shear is the observation for gas that may be trapped in the panel and a comparison of the flow medium or bond-line thickness. (Test defined in MMM-A-132.) One precaution that must be taken here rests in the sawing of the specimen. If the cut is too deep at the lap joint, the yield strength of the material may be affected, thus producing poor results. Care must also be exercised during the sawing operation to prevent overheating of the test panel. Reporting would be the same as for lap shear tensile tests.

BUTT TENSILE (ROD)

This type of tensile specimen is usually produced by the utilization of ⅜ to ½ in. solid cylindrical rods. They are butted and bonded, usually utilizing a "V"-type block that produces a joint with excellent alignment, the alignment during cure being of utmost importance.

The fabricated specimen should be a minimum of 7 in. in length, allowing 1½ in. for gripping and a 4-in. distance between the jaws when the specimen is gripped. The test is performed utilizing a standard tensile testing machine equipped with "V"-type jaws or equipped with a fixture for holding a cylindrical test specimen. The load rate should be applied 1000 to 1500 lb per min or jaw separation at 0.005 in. per min. The results are reported in pounds per square inch obtained by dividing the breaking load by the area of the faying surface. The report should include the type of failure, load rate, test temperature, and information pertinent to the adhesive and adherends.

It is sometimes desirable to report the standard deviation and coefficient of variation for butt tensile tests (or others) as follows:

$$S = \sqrt{\frac{2X^2 - nX^2}{n - 1}}$$

$$n = 100 \frac{S}{X}$$

where S = estimated standard deviation
$\overline{X}$ = value of a single observation
n = number of observations
$\overline{X}$ = arithmetic mean of a set of observations
n = estimated coefficient of variation

TUBULAR SHEAR TESTS

It may be necessary or desirable, in an effort to approach realism, to test tubular type specimens. In a test of this type one tube is smaller than the other, and the smaller is inserted and bonded into the larger. For example, some spacecraft utilize joints of this type for landing-gear mechanisms and they may be plastic, metallic, or a combination. They are tested in the same manner as other tensile-type specimens, but it may be necessary to plug the gripping area with a solid material to prevent collapse during loading. The testing device must be equipped with the "V"-type jaws and perfect alignment is essential. Test specimens should be a minimum of 6 in. in length allowing 1½ in. for gripping and 3 in. distance between the jaws after the specimen is gripped. The testing speed should be 1000 to 1500 lb per min and applied at a constant rate. When fabricating specimens of this type with liquid or paste adhesives, a sprinkling of glass beads of the correct size for the desired bond line is recommended to insure a universal bond line and satisfactory alignment of the specimen.

The report should include shear strength reported in pounds per square inch, which is obtained by dividing the bonded area into the breaking tensile load. The report should also include information pertinent to the adhesive and adherends, type of failure, loading rate, temperature of surroundings during test, and average of the values for the number of specimens tested.

"DOG-BONE" TENSILE SPECIMENS

This type of specimen is seldom used, but may be the best of all tensile specimens for environmental testing, as changes in color, cracking, etc., may be visually observed during the test. This specimen resembles the metallurgical specimen in appearance and is usually cast in a Teflon die if liquid resin is being used. If structural tapes are being laminated, the die may be open at both sides, thus allowing the specimen to be pressurized by means of a press or other pressure means. The cross-section measurements are taken in the small area and tested in the tensile testing machine. The data is reported in pounds per square inch obtained by division of the cross-section (breaking area) into the tensile load. The major objection to the "dog-bone" tensile specimen is the difficulty in fabrication, and engineering would usually prefer a specimen that would employ the substrates that are to be used in the assembly for which the adhesive is being tested.

DOUBLE LAP SHEAR TENSILE TEST

Tensile tests may be performed using any type of shear specimen shown in Chap. 4, Fig. 4-6. When testing a simple lap shear tensile specimen, the stress load is not true shear due to the loading effect; consequently, an unknown amount of peel enters into the test. In an effort to combat or minimize this situation, the double lap shear specimen is used. This type of specimen may better represent the stress loading of a production-type structural joint (Fig. 7-2).

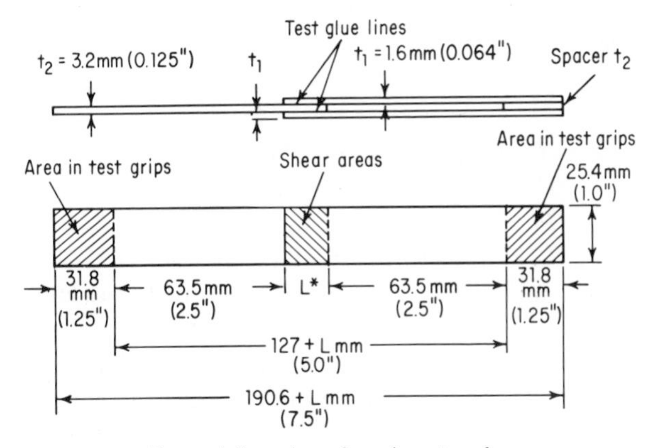

Form and dimensions of specimen type A

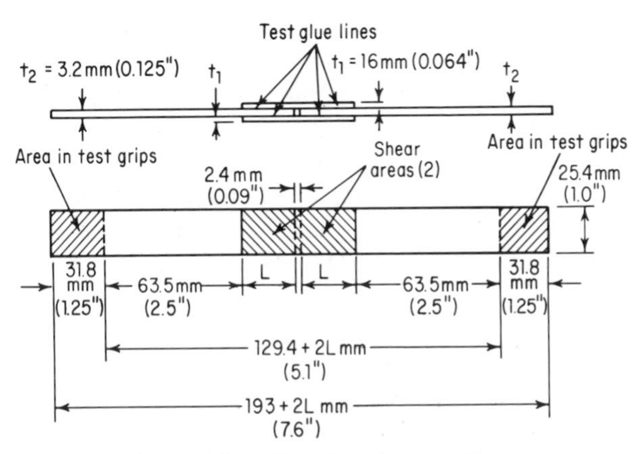

Form and dimensions of specimen type B

fig. 7-2 *Double lap shear specimen.*

Variations in adherend metal thickness and the length of overlap normally influence the test values and make direct comparison of test values questionable. Therefore, it is preferable for comparative tests to standardize on the adherends and specimen configuration. It would also be undesirable to exceed the yield strength of the adherend during the test, which would adversely affect the test data. Obviously the double-lap tensile specimen would produce considerably higher shear strengths than the simple-lap configuration.

The overlap may vary with the adherend type, adherend thickness, and the general level of strength expected with the adhesive being tested. To prevent the yield factor from influencing the test data, the maximum permissible length of overlap may be computed from the following relationship:

$$L = \frac{F_{ty} T_1}{\pi} \quad or \quad L = \frac{F_{ty} T_2}{2\pi}$$

where L = length of overlap, in.
T_1 = thickness of metal, in.
T_2 = thickness of metal, in.
F_{ty} = yield point of metal, psi
π = 150 percent of the estimated average shear strength of the adhesive being tested, psi

Test panels should be fabricated in a large panel and then sawed to the desired size because a more universal bond-line thickness would be realized. A fine-tooth typesetter's circular saw is recommended for cutting the specimens to size, but caution must be exercised to prevent overheating during the sawing operation. The test is performed utilizing a standard tensile testing machine applying a load rate of 1200 to 1400 lb per min (0.005-in.-jaw separation per min).

The report should include information pertinent to the adhesive and substrates employed plus the type of failure, testing temperature, load rate, and shear strength in pounds per square inch, which is obtained by dividing the faying surface (bond area) of both bond lines into the tensile failing load. This test in a more refined state has been proposed as a standard by ASTM D14 West Coast Committee on Adhesives.

CROSS LAP SHEAR SPECIMENS

This test was especially designed for testing glass bonded to itself or to other materials. The adherends are cut to 1 × 1½ × ½ in. The glass should be annealed to prevent breakage. The test coupons are bonded together at right angles in such a manner that 1 sq in. of bonded area extends with a ¼- in. overlap on all four sides of the square bonded area. The specimens are bonded, placed in a special gripping fixture, and tested on a universal testing machine with a load rate of 500 to 700 lb per min. The values are calculated in the same manner as other shear specimens. This is not a widely used test, but the need may arise for testing in this manner. The procedure is defined in detail in ASTM D1382-64. Glass and other materials may be tested in shear by cutting specimens 1 × 0.5 × 0.050 in. and bonding into a simple lap joint (Fig. 7-3). This method is commonly used for testing thin plastic sheeting, cork, rubber, and other materials that do not possess the tensile strength necessary for testing alone utilizing a conventional coupon. The loads induced during

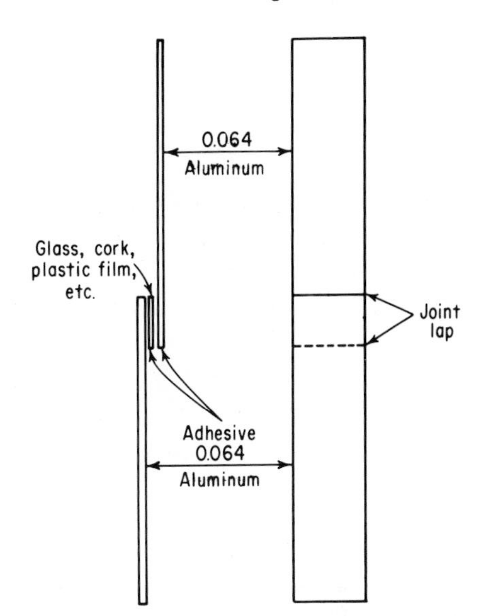

fig. 7-3 *Specimen for testing materials possessing poor tensile characteristics.*

the test are not exactly desirable due to thick joints, but useful data can be obtained pertaining to adhesion, surface cleaning, etc.

FLATWISE TENSILE TESTS

This type of tensile test is usually applicable but is not limited to sandwich or honeycomb testing. It has been used for testing substrates such as cork, laminates, and ablative coatings. This test is widely used in the destruct testing of honeycomb production parts as it is simple to saw the specimen from the actual part. Average values for sandwich constructed parts range from 20 to 800 lb per sq. in., depending on the adhesive system and the type of core. The ideal situation would naturally be a core failure during the test indicating that the bonded joint is stronger than the substrate. The rate of loading is usually 800 to 1500 lb per min.

A standard sandwich specimen is cut to size (2 $\times$ 2 in.) and bonded to special solid metal blocks designed to fit the jaws of a tensile testing machine (Fig. 7-4). The specimens are then tested similarly to the lap shear specimens and the results are reported in pounds per square inch (obtained by dividing the area into the load). The adhesive employed for bonding the specimens to the metallic blocks should be sufficiently rigid so that large amounts of deformation will not occur during the test. A room-temperature setting adhesive is suitable, which can be removed, permitting the blocks to be used again and again. The report should include the mode of failure, information pertinent to the adherends and adhesives, load rate, and test temperature.

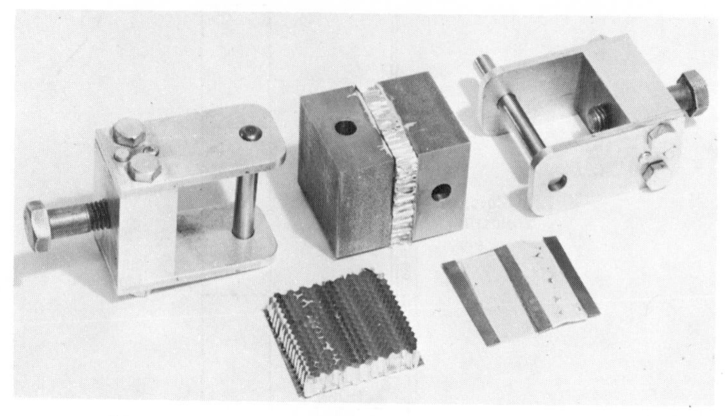

fig. 7-4 *Flatwise tensile specimen and test fixture.*

CORE SHEAR TESTS (DELAMINATION)

Core shear strength tests are used primarily as an acceptance test for honeycomb core. The test may be performed on a universal tensile testing machine maintaining a steady loading rate of 300 to 600 lb per min. The specimen should be approximately ⅝ in. thick, 5 in. in width and 10 in. in length (Fig. 7-5). The core specimen should be monitored so the fixture pins engage all the cells of a continuous longitudinal row with

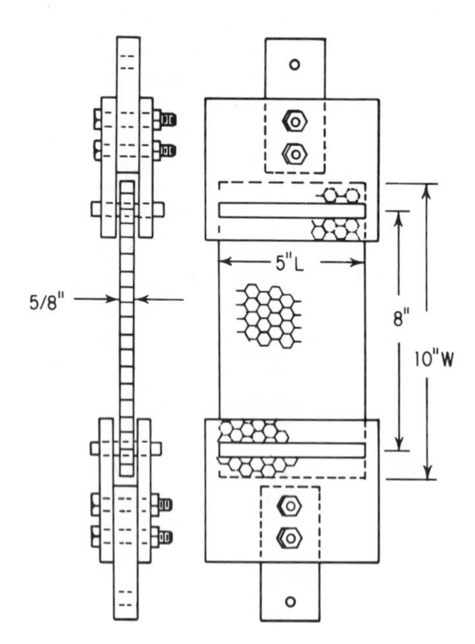

fig. 7-5 *Core delamination test, fixture, and specimen.*

an even extension of cell rows at each end so that opposite pins in opposite plates engage cells in the same transverse row. The specimen should be mounted to allow at least 8 in. between the holding points. The pins should be of round stock and suitable diameter to permit attachment of the core without distortion or damage to the foil or node bond. If a specimen exceeds 12.5 lb, it is considered acceptable. The report should include the failing stress load, testing speed, test temperature, and a sketch pin-pointing the area of failure. There is a minimum amount of emphasis placed on a test of this nature because the core is never loaded in this manner after bonding and during service. This test is detailed in MIL–C–7438C, "Core Material, Aluminum, For Sandwich Construction."

COMPRESSION SHEAR TEST

The compressive test is not widely used in the aerospace industry simply because practically the same information may be obtained from the tensile shear test. The standard lap shear specimen may be used for compression tests provided a fixture is available to hold the specimen upright between the compression chambers. Another specimen that may be used is prepared from metallic blocks (0.075 in. thick and 2.0 × 2.0 in. square). They are bonded with a ¼-in. mismatch in the upright position, allowing the compression to be placed on one block. The compression-load rate should be applied utilizing a head travel 0.015 in. per min and should be applied as uniformly as possible.

The compressive failure may be calculated in pounds per square inch, based on the actual bonded area measured to the nearest 0.01 sq in. Reporting data are obtained by dividing the area of the sample into the compression load. The test temperature, adhesive, primer-cure cycle, cure pressure, and the mode of failure should be included in the test report.

ASTM D2182-63T, "Strength Properties of Metal-to-Metal Adhesives by Compression Loading (Disk Shear)," defines another method which is highly regarded if compression shear tests are considered necessary. The test specimens are fabricated with a disk that has approximately 1 sq in. bonding area which is bonded to a 4 × 1.5 × 0.050 in. adherend. A special shearing tool is necessary for the test, utilized with a universal testing machine equipped with compression cells.

PEEL TESTS

A well-designed assembly would not have any joints loaded in peel, but peel forces do arise, hence peel tests are very important. The utilization of this type of test contributes information as to the behavior of the adhesive to an adherend, and indicates the surface condition of the substrate. The peel test involves stripping or tearing away of a flexible member of an assembly that has been bonded with an adhesive to another member, which may be flexible or rigid, and the specimen is usually loaded 90 or 180°. The rate of loading is not as important as in lap shear tensile testing, but should be known and controlled as closely as possible. The bond-line thickness plays a more important role in the values, especially in the elastomeric adhesives. Average "T"-peel values may vary from 8 to 70 lb, the latter being considered a very high value. Adhesives are generally considered poor in peel strength but values have been obtained in the 300-lb range by increasing the bond-line thickness to ¼ in. or more. The prime reason for these high values is the mode of loading, and is not considered indicative of the values normally expected. The data are always reported in pounds rather than pounds per square inch due to the X dimension in one direction. Peel values will normally vary more (percentagewise) than other mechanical tests due to the one-point loading over a peel distance of 1 to 10 in. Peel tests are widely used in production control for information pertinent to cleaning-tank titration levels.

T-PEEL TESTS

The T-peel test is the most common of all peel tests because of the simple configuration of the test specimen and its ease of fabrication. T-peel tests are defined in MMM-A-132 and ASTM D1876-61T. The objective of these tests is to determine the relative peel resistance of adhesive bond between two flexible adherends. T-peel specimens are usually 1 in. wide, 8 to 10 in. long, and 0.010 to 0.020 in. in thickness (Fig. 7-6). The test is performed on a standard tensile testing machine and readings are taken directly, then reported as pounds per linear inch. The peel resistance may be reported in three figures: high, low, and average. Peel resistance may vary considerably over the entire peeling range of the specimen, and the highest stress point is usually the initial breaking point. A machine equipped with an autographic chart device is advantageous, presenting a chart reading that can be interpreted in terms of inches per minute of jaw separation as one coordinate and applied load as the other (stress versus strain). The ideal method of calculation would be the average taken with a planimeter. If a planimeter is not used, the average may be calculated as the average of load readings taken at fixed increments of crosshead motion. The load may be recorded at 1-in. intervals (or 0.5-in. interval jaw separation) following the initial peak, until at least 10 readings have been obtained.

The loading rate may vary from 0.5 to 20 in. per min depending on

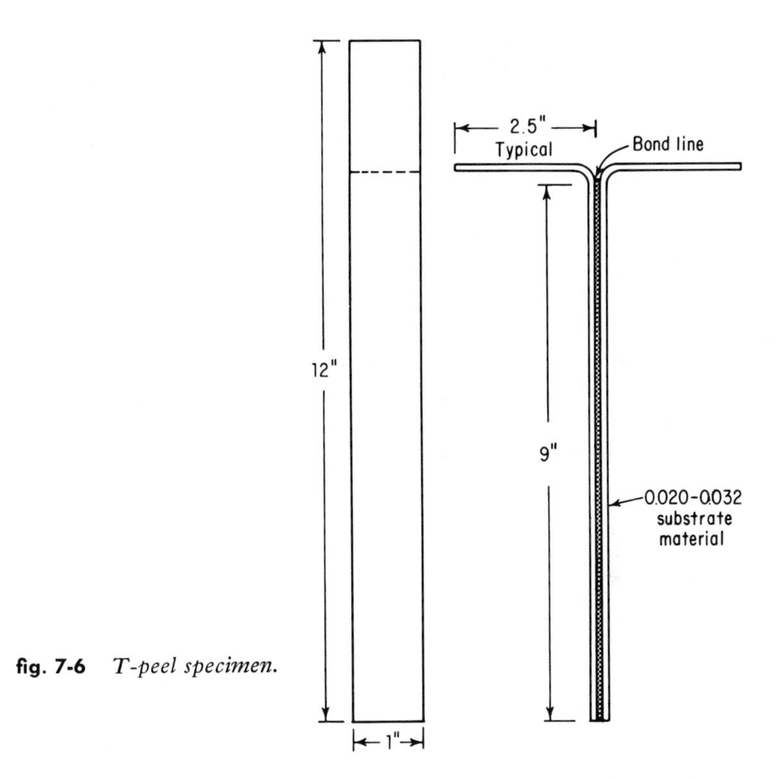

fig. 7-6 *T-peel specimen.*

the adhesive system being tested, but approximately 10 in. per min is usually preferred.

The report should include the loading rate, peel resistance in pounds, number of specimens tested, temperature during tests, and the most important factor of all—the mode of failure.

THE BELL "T" PEEL

Engineers at Bell Helicopter Company of Fort Worth, Tex., have devised a peel test to satisfy the needs of their company. The test employs one flexible member (0.016) and a more rigid member (0.063), and the specimen is ½ in. wide and approximately 10 in. long. Test panels are fabricated in much the same manner as a standard T peel; i.e., a large panel is bonded and the specimens are sawed from it, thus aiding in being more representative of production parts. The Bell peel test differs from the standard T peel because the angle of peel is slightly different. The specimen is peeled through a spool arrangement, the spools being allowed to rotate during the load application (Fig. 7-7). The following components are identified in the test apparatus as follows: (1) Tinis Olsen stress/strain recorder, (2) bottom clamp for peel specimen, (3) peel specimen, (4) Bell's metal-to-metal peel head, (5) environmental test chamber, and (6) temperature-recording monitor for environmental test chamber.

Values obtained are normally slightly higher when the Bell test is utilized as compared to the standard T peel, and more consistent values are obtained because the angle of peel is more consistent. The loading rate is approximately 20 in. per min and calculations are made much the same as for a standard T peel.

The report should include information pertinent to the adhesive system and adherends, peel values obtained, number of specimens tested, test temperature, and the mode of failure.

fig. 7-7 *The Bell T peel. (Bell Helicopter Company, Fort Worth, Tex.)*

1, Tinius Olsen stress/strain recorder; 2, Bottom clamp for peel specimen; 3, Peel specimen; 4, Bell's metal-to-metal peel head; 5, Environmental test chamber; 6, Temperature-recording monitor for environmental test chamber.

CLIMBING DRUM PEEL

This technique is intended primarily for determining the peel resistance of adhesive bonds between metal facings and sandwich cores and between metal or plastic laminates (or combinations) when one member is flexible. The test apparatus consists of a flanged drum, flexible loading straps or cables, suitable clamps for holding the test specimen, and a testing machine capable of peeling the facing at a rate of 3 to 5 in. per min.

The specimen should be a minimum of 10 in. long and 3 in. wide. The core thickness is not important, except that the specimen should not bend during the test. These specimens are commonly taken from production panels for destruct test evaluations and may be sawed or machined from larger panels of any type as long as care is exercised during removal (Fig. 7-8). The test is performed with a climbing drum peel apparatus illustrated in Fig. 7-8, coupled with a standard tensile testing machine equipped with a recording device. The jaw-separation rate of the machine should be 1 to 2 in. per min.

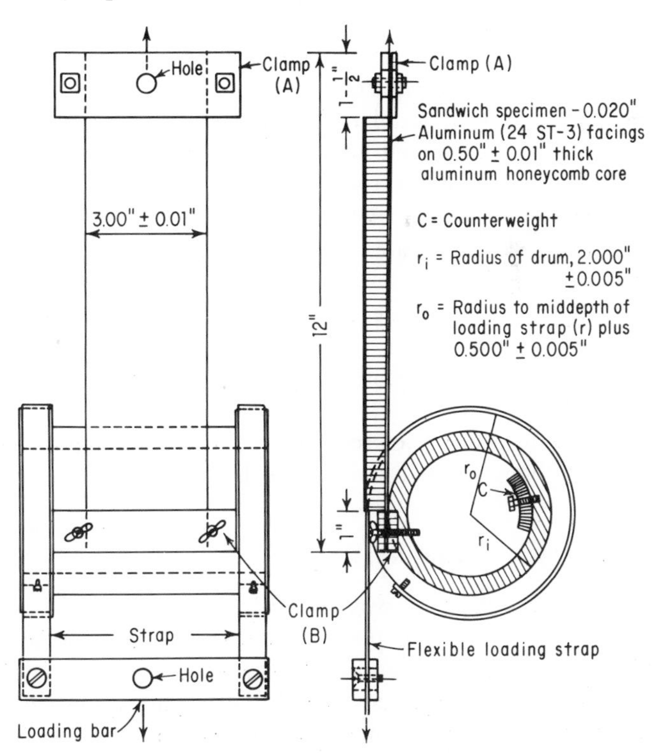

fig. 7-8 *Climbing drum peel specimen and test fixture.*

The results obtained from this test include the torque required to bend the facings as well as to peel the face from the core, plus the weight of the drum apparatus. This must be taken into consideration; therefore, after the apparatus is attached to the test machine, a face sheet identical to the one to be tested is rolled through the device and the results tabulated. The forces imposed on the adherend during the peel test usually result in greater bending than is obtained in bending the unbonded adherend around the drum. More torque is required, therefore, to bend a bonded rather than an unbonded material; thus, this compensation is only approximate.

The peel values are obtained by taking readings from the chart in one inch increments, excluding the first inch, and using an average of these plugged into the following formula:

$$T = \frac{(r_o - r_i) \quad (F_p - F_o)}{w}$$

where T = average peel torque, in.-lb per in. of width

r_o = radius of flange, including one-half the thickness of the loading straps, in. (see Fig. 7-8)

r_i = radius of drum plus one-half the thickness of the adherend being peeled, in.

F_p = average load required to bend and peel adherend plus the load required to overcome the resisting torque, lb

F_o = load required to overcome the resisting torque, lb (using corrected load obtained by false face sheet)

w = width of specimen, in.

The report should include information pertinent to the adhesive and adherends, loading rate, number of specimens tested, values in pounds for each specimen (minimum, maximum and average), temperature during test, and the mode of failure, i.e., cohesive failure within the adhesive, adhesive failure at the facing (interface), adhesive failure to the core, core failure, or combinations of any of these in percentages as near as can be estimated.

Climbing-drum peel tests are defined in Federal Test Standard 175, Military Standard 401 and ASTM D1781-62.

IMPACT PEEL (TOUGHNESS TEST)

This peel method was devised by General Dynamics/Convair, San Diego, Calif., and was proposed to ASTM as a standard method in 1964. The test employs a standard T-peel specimen and a device to suspend it in a normal T-peel position (Fig. 7-9). The top portion of the specimen is attached in a permanent manner, and to the lower portion (90°) is

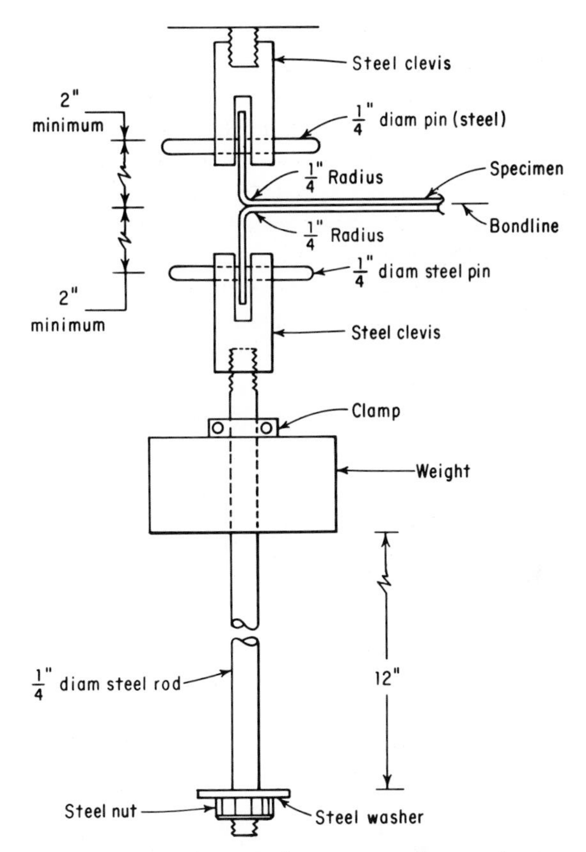

fig. 7-9 *Specimen and apparatus for conducting toughness test.*

attached a rod with a stop at the desired distance, a weight that is dropped that distance (1-lb weight, dropped 1 ft = ft-lb). The results are recorded as the amount of peeling that occurred from the weight being dropped.

It has been discovered that adhesives which exhibit very high normal peel strengths may be very poor when impact loaded, and naturally the opposite also occurs. When testing a specimen, the weight should be dropped at least three times, the specimen removed and measured for total separation (peel) and divided by 3. The following data should be reported:

1. The total bond-line separation in inches for each specimen tested, divided by 3.

2. The impact load, weight used, and distance dropped, expressed in foot-pounds.

3. The toughness of the adhesive for each specimen expressed in inches of bond-line separation/foot-pounds of force. This should incorporate data from at least four specimens for each adhesive tested.
4. Type of adhesive, cure cycle, cure pressure, substrate used, and primer, if any.
5. Width of the specimen and bond-line thickness.
6. Conditioning procedure used prior to testing, if any.
7. Test temperature (critical).
8. Nature of failure, including the average estimated percentages of failure in cohesion or adhesion, contact failure, or substrate failure.

This test indicates very clearly that, testing in a conventional manner, the proper data may not be obtained if adverse loading conditions exist. An improper test results in poor design and very often failure. The impact peel test was devised for a special purpose after failure had occurred in a part that had been designed around conventional tests. If standard tests are not indicative of the environment or function of the assembly, then tests should be devised to simulate that condition as nearly as possible.

STRIPPING STRENGTH OF ADHESIVES

A stripping test is actually a peel test, but differs in load application because it was designed for testing flexible materials, i.e., rubber, fiber materials, screen, plastic sheeting, etc. One adherend must have proper flexural modulus of elasticity and thickness to permit an approximate 180°-angle turn near the point of load application without breakage or excessive loss of mechanical properties (Fig. 7-10). The test is performed with a standard tensile testing machine employing a jaw separation rate of 6 to 12 in. per min. The specimens are 1 in. in width and 8 to 10 in. in length. Inconsistencies or scatter may be anticipated with porous materials (cloth, screen) because the degrees of penetration of the adhesive and the thickness of the bond line may vary, which will cause peel-strength variations.

Tests of this type are defined in Federal Standard 175, Method 1041.1, and ASTM D903-49.

The report should include information pertinent to the adhesive and adherends, loading rate, stripping values (reported in pounds per linear inch), number of specimens, test temperature, and mode of failure.

FATIGUE TESTS

Fatigue testing is defined as the repeated application of a given load and is a measure of the ability of an adhesive-bonded specimen to resist

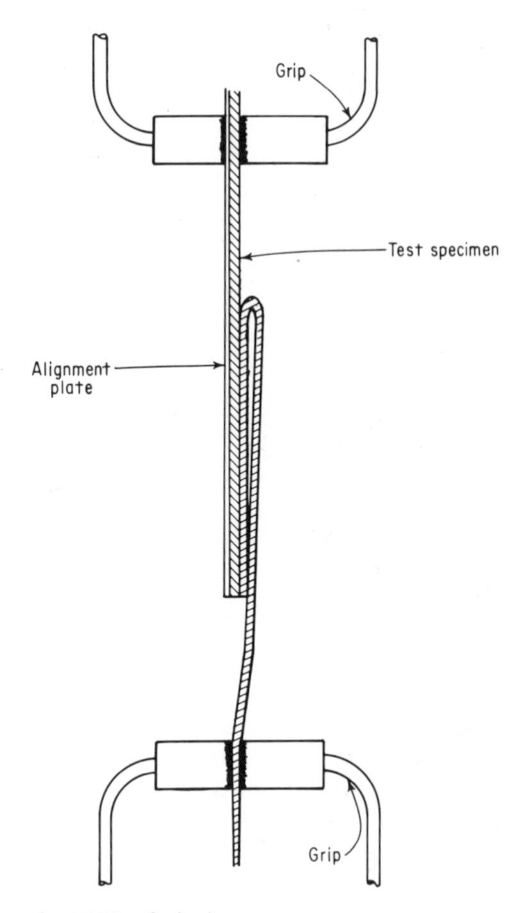

fig. 7-10 *Stripping test.*

failure. The specimens may be of various configurations but the standard lap shear as specified in ASTM D1002 and MMM-A-132 is commonly used. If plastic substrates are utilized it may be desirable to fabricate double-lap shear specimens, thus reducing the probability of adherend failure. This test is accomplished on a fatigue testing machine; the specimen is held in a suitable manner to induce cyclic axial load or an alternating bending stress into the adhesive joint. Since fatigue life is dependent on frequency, amplitude, temperature, mode of stressing, and magnitude of stress, these variables must be specified and controlled. Data are reported in number of cycles to failure with corresponding load calculated in pounds per square inch. They may be plotted on curves termed stress-lag-cycle coordinates, and fatigue strength of the adhesive is reported in cycles for a known load.

Adhesives exhibit excellent properties in fatigue, but testing in this area is still essential due to the heavy fatigue loads imposed on assemblies in aircraft, missiles, and spacecraft. One of the problems posed in fatigue testing is simulating the loads of the actual assembly.

A large amount of work has been accomplished in the area of predicting the service life of an adhesive subjected to heavy fatigue loads. The time is near at hand when a technician may take an instrument, test a panel, and remove it because it is near fatigue failure. This will be a Herculean step forward for the adhesives industry.

CREEP

The deformation or dimensional change occurring in an adhesive-bonded specimen under stress and over a long period of time is called "creep." There are several tests advocated for creep determination, but all that are in use leave something to be desired, especially in the creep-deformation area. Creep rupture tests are more accurate and are obviously easier to report and interpret. Creep tests are of prime importance but are not emphasized as much as other mechanical tests. They should be, for the simple fact that creep is a weak area in most if not all adhesive systems. They all tend to creep; the question is how much? It seems apparent that extensive tests should always be conducted in anticipated problem areas but this is not always the case. Several examples could be cited here—large assemblies connected with critical defense programs where adhesives are being used and proper creep tests have never been performed. The general thought in most cases seems to be, "Let them creep!" The point that should be made here is the fact that it is almost impossible to trace a failure to creep. The failure can always be attributed to another influence.

MMM-A-132, "Adhesives, Heat Resistance, Airframe Structural, Metal-to-Metal," specifies the use of a standard lap shear specimen tested utilizing a dead-weight load. A lap shear specimen is used with a scribe line drawn through the center of the joint, plainly marking each adherend (Fig. 7-11). The specimens are gripped or suspended by drilling a hole in one end of the specimen and inserting an 0.25-in. diameter pin through the hole into a stable or permanent fixture and a dead weight (1600 psi at room temperature, 800 psi at 180°F) to the opposite end. When performing an elevated-temperature test, the specimen may be suspended in an oven or "hot box." To meet the requirements specified in MMM-A-132, the deformation must not exceed 0.015 in. deflection in a 192-hour test in either the room or elevated-temperature tests. If the creep deformation is reported, it is reported in thousandths of an inch slippage under a known load. If the specimens are tested to destruction, the report would

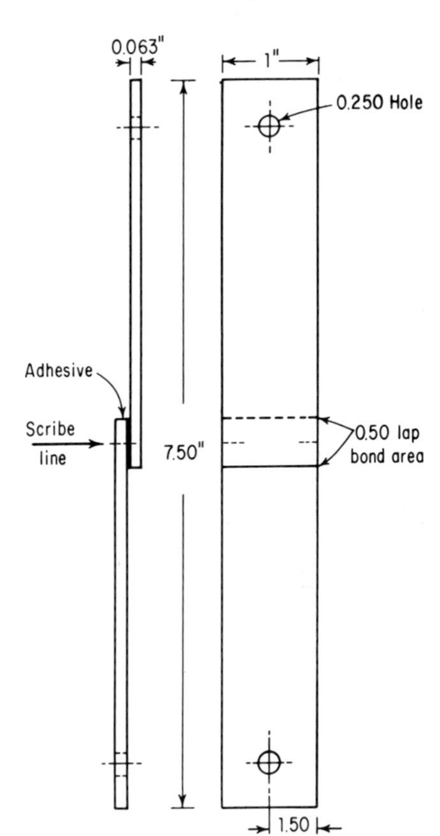

fig. 7-11 *Creep specimen.*

contain the total amount of time (in hours) the specimens remained under a known load before rupture occurred.

ASTM D2293-64T defines a test procedure for determining creep properties of adhesives in shear by compressive loading. This test utilizes the lap shear specimen which is sawed ¼ in. beyond the lap, resulting in a specimen 1 × 1 × 0.063 in. The test apparatus consists of a slotted bolt, nut, steel spring, and washers (Fig. 7-12). The spring may be compressed utilizing a tensile testing machine, until the desired load is attained and the length of the spring measured with a pair of venier calipers or its equivalent. The specimen is inserted into the slot, between two washers (scribe lines on specimens as in weight test described earlier), and the nut tightened, compressing the spring to the proper dimension to achieve the desired load (600 lb recommended). The specimen should be monitored at regular intervals for slippage, and the use of a calibrated microscope will aid in assessing the condition of the specimen with more detailed accuracy. If buckling occurs in the adherends due to compressive

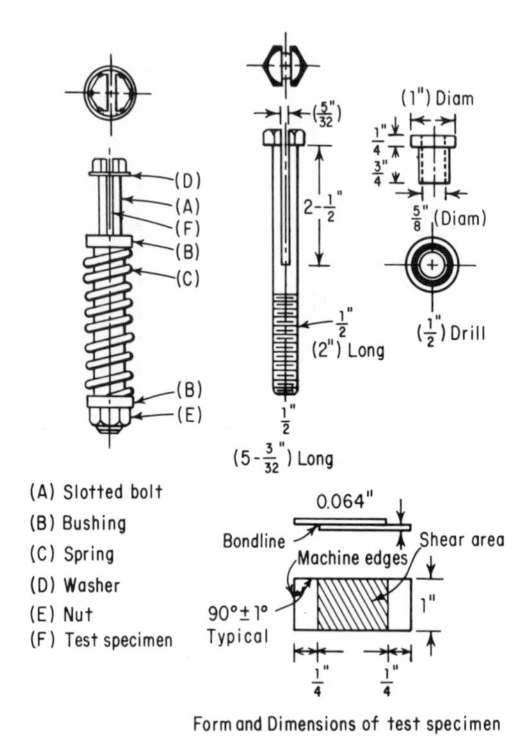

fig. 7-12 *Creep rupture test apparatus and specimen.*

creep of the metal, discount the test and redesign the specimen. The test would continue an established length of time or until the adhesive fails.

The report should include information pertinent to the adhesive and adherends, total observed actual deflection, duration of the test, test temperature, and mode of failure if complete failure occurs during the test.

ASTM D2294-62T defines a test for creep properties of adhesives when loaded in shear (metal-to-metal). This test procedure utilizes a spring-loaded apparatus and employs a universal tensile testing machine for applying pressure. The specimen is a standard lap shear specimen with scribe lines as described in the other creep tests. The specimen is placed into the special fixture and a tensile load applied with a tensile testing machine. When the desired load is reached, the load is locked in place by mechanical means which hold the compressed spring. The observations and reporting are similar to the test described in ASTM D2293-64T.

IMPACT TESTS

Specimens for impact tests usually consist of two blocks, made of wood or metal, bonded together. In testing, the lower block is held in a wire grip and the upper block is struck with a pendulum hammer travelling at a known velocity (usually 11 ft per sec) and in a direction parallel to the plane of the adhesive layer. The energy impact is reported in pounds per square inch.

Impact tests are not considered one of the more common tests for adhesives and are not widely used. The primary problem pertinent to impact tests is the accuracy of the reporting. With existing methods, the data may have considerable scatter due to the lack of preciseness of the test.

CLEAVAGE TESTS

The cleavage test is conducted by introducing a prying force at one end of a bonded specimen to split the bond apart. A suitable cleavage

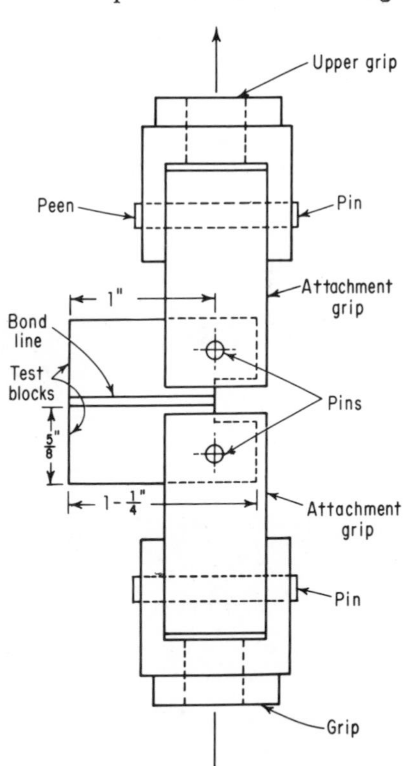

fig. 7-13 *Cleavage test set-up.*

test is described in ASTM Method D1062-51. Test specimen bonding surface is 1 in. square and ⅝ in. in thickness, with attachment holes as shown in Fig. 7-13. It is important that after bonding, all flash is removed from the edges of the specimen. Specimens are tested by means of special grips as shown in Fig. 7-13 with an applied load rate of 500 to 800 lb per min and the results reported in pounds-per-inch breaking load.

The report should include pertinent information concerning the adhesive and adherends, failing load, test temperature, rate of loading, and mode of failure.

FLEXURAL TESTS

A sandwich specimen (3.0 × 16.0 in.) is supported at or near the ends and loaded from the top center (one-point or two-point loading—see example of two-point loading in Fig. 7-14). By adjusting the length of the loading arms, the location of stress concentrations may be varied. This test procedure applies to shear as well as a bending moment on the sandwich. This test can produce failure in the sandwich by shearing the core, or by shearing the bond between the core and the facing sheets, or by direct compression or tension failure of the facings, or by localized wrinkling of the thin facings at the loading points. The test specimens must be designed so that the property sought is actually the one obtained. Long spans produce high face-sheet stresses and core or bond failures would be improbable. Short-span loading tends to produce core or bond failures provided the face sheets are relatively heavy. The stress concentrations at the load points are reduced by two-point loading, and for this

fig. 7-14 *Two-point-loading flexural test. (The AVCO Corp. Nashville, Tenn.)*

reason the two-point loading is more popular than one-point loading. If deflection readings are taken on a short specimen, it is possible to calculate an approximate value of the core shear modulus. If beveling stiffness of the sandwich is not known, it is possible to obtain the bonding stiffness by testing a long specimen.

The following formulae apply to the flexure test:

Shear stress

$$S = \frac{Ps}{(h+c)b}$$

where S = core shear stress
Ps = total load applied at two points of $A_s/4$ from each reaction
A_s = span length
h = total sandwich thickness
c = core thickness
b = sandwich width

The following formula is used for the facing stresses:

$$F = \frac{P_B A_B}{4F(h+c)b}$$

where F = facing stress
P_B = total load (two-point) located a distance of $A_B/4$ from each reaction
A_B = span length
F = facing thickness
h = total sandwich thickness
c = core thickness
b = sandwich width

The core shear modulus is calculated by the following formula:

$$G = \frac{P_s A_s C}{2W_s\,(h+c)^2 b\,\dfrac{(1 - 11\,P_s A_s{}^3)}{(768 W_s D_b)}}$$

where G = core shear modulus
P_s = total load, applied at two points located a distance of $A_s/4$ from each reaction
W_s = midspan deflection of sandwich P_s/W_s–slope of initial portion of the load-deflection curve
A_s = span length
C = core thickness
h = total sandwich thickness
b = sandwich width
D = flexural stiffness of sandwich; $D = \frac{E(h^3 - C^3)}{12L}$
E = flexural modulus of elasticity of facings
L = 0.91 for isotropic facings or 1.0 for most orthotropic facings

The report should include maximum facing and core shear stresses, load-deflection diagrams, core shear modulus, sandwich stiffness, test temperature, and description of the mode of failure. Flexural tests are defined in Federal Test Specification 175, ASTM D1184, and Military Standard 401A.

CORE DENSITY AND SPECIFIC GRAVITY

It is sometimes necessary to determine the density and specific gravity for a particular type of core, although reliable information in this particular area is usually available from the manufacturer. If a determination is necessary, the following procedure may be used.

The test specimens should be cut to a size and shape that can be conveniently measured. The test specimens are weighed and measured as accurately as possible and the data recorded. The following formulae are used to calculate the density in pounds per cubic foot:

$$\text{Density (lb/ft}^2) = \frac{\text{weight (lb)}}{\text{volume (ft}^2)} = 1728\ \frac{\text{weight (lb)}}{\text{volume (in.}^2)} =$$

$$3.81\ \frac{\text{weight (g)}}{\text{volume (in.}^2)} = 62.4\ \frac{\text{weight (g)}}{\text{volume (cm}^2)}$$

Specific gravity may be calculated as follows:

$$\text{Apparent specific gravity} = \frac{\text{density lb/ft}^2}{62.4}$$

This test is specified in MIL-STD-401A, "Sandwich Constructions and Core Materials: General Test Methods."

This concludes the basic mechanical tests used for adhesives. There are others, and new ones are being devised every day. Other tests for adhesives which fall into the environmental class will be found in Chap. 8.

TABLE 7.1 Effect of Loading Rate and Temperature on Aluminum/Aluminum Lap-Shear Strength (tensile lap-shear strength, psi)

Loading rate in/min	Epoxy phenolic								Modified Epoxy		Nitrile phenolic		Modified nitrile phenolic	
	10-mil		*12-mil*		*15-mil*		*20-mil*							
	73°F	300°F	73°F	300°F	73°F	300°F	73°F	300°F	73°F	300°F	73°F	300°F	73°F	300°F
0.05	2510	1990	2830	—	1990	—	2230	—	2730	—	4010	1220	3150	970
2.0	2350	—	3100	—	2170	—	2370	2420	2530	1740	4280	1220	3100	1110
200	2260	—	3510	—	2530	—	2500	2440	2700	1860	5930	—	—	1500
2000	4760	3530	3920	—	2960	—	2380	—	—	2550	—	3560	4300	2070

REFERENCES

Braune, R.: Effects of Rate of Loading on Shear Stress of Resin Adhesives, *Society of Plastic Engineers Journal,* May, 1960.

Donovan, C. F.: How to Test the Strength of Bonded Joints, *Adhesives Age,* March, 1966.

"Effects of Molecular Structure on Mechanical Properties of Structural Adhesives," Case Institute of Technology, Air Force Contract AF33(616)-7210, September, 1960.

Federal Test Method Standard 175, "Adhesives; Methods of Testing."

Hochberg, M. S.: Evaluating Adhesive Bonded Temperature-Resistant Sandwich Constructions, *Adhesives Age,* March, 1966.

Lewis, A. F., and W. B. Ramsey: Polymers and Adhesive Joint Strengths, *Adhesives Age,* February, 1966.

Lunsford, L. R.: "Adhesive Torsional Shear Test," General Dynamics/Fort Worth, ERR-FW-134, February, 1962, Fort Worth, Tex.

Military Standard 401, "Sandwich Construction and Core Materials: General Test Methods."

Military Specification MMM-A-132, "Adhesives, Heat Resistant, Airframe Structural, Metal to Metal."

Military Specification MIL-C-7438, "Core Material, Aluminum, for Sandwich Construction."

Military Specification MIL-C-8073, "Core Material, Plastic Honeycomb, Laminated Glass Fabric Base, for Aircraft Structural Applications."

Military Specification MIL-S-9041, "Sandwich Construction; Plastic Resin, Glass Fabric Base, Laminated Facings and Honeycomb Core for Aircraft Structural Applications."

Military Specification MIL-C-21275, "Core Material, Metallic, Heat-Resisting, for Structural Sandwich Construction."

"Product Dimension," Brookfield Engineering Laboratories, Stoughton, Mass.

"Structural Sandwich Construction; Wood; Adhesives," 1966 Edition, ASTM Volume 16, American Society for Testing and Materials, Philadelphia.

Tanner, W. C., and R. F. Wegman: "Variations of Adhesive Bond Strength with Rate of Load Application," Technical Report 3054, March, 1963, Picatinny Arsenal, Dover, N.J.

Terry, E. L.: "Static and Repeated Loading Characteristics of Joints in Beryllium Structures," The Brush Beryllium Company, March, 1965.

"Viscosity Measurement and Control," Brookfield Engineering Laboratories, Stoughton, Mass.

CHAPTER 8

Environmental Testing

The service life of a finished product must always be considered; therefore, after the mechanical and physical properties of an adhesive are determined, the environmental factor usually draws the second wave of concern. Environmental testing may be defined as the aggregate of surrounding things, conditions, or influences that are other than normal circumstances. Testing in these areas is necessary pertinent to adhesives to assuage fears of failure and to establish customer confidence in the fabricated hardware. The performance of an adhesive cannot be predicted with any great degree of accuracy in the expected service environment until it has been thoroughly tested to a simulated service condition. It may be desirable to subject the test specimen or assemblies to various stress loads simultaneous to the environmental exposure, thus testing the efficacy of the joint load-bearing abilities under adverse conditions.

The development and production of military hardware, aerospace products, and commercial aircraft pose severe environmental testing problems. If adhesives are to be employed, the service requirements must be known and are usually available from a design handbook or similar document; consequently, the necessary tests must be performed to ensure reliability for the service life of the product in question. It is not intended to suggest here that Federally financed programs are the only ones that require extensive environmental testing, but the reliability factors are usually high, the degree depending on the function of the part.

The most common environmental tests are elevated temperature, cryogenic, humidity exposure, salt spray, fluid immersion, static heat aging, and general weathering. The standard tests specified in military, Federal, and ASTM procedures are not considered adequate in many cases; thus, tests must be devised to simulate as nearly as possible the expected service conditions. To illustrate, MMM-A-132 requires that Type IV shear specimens be tested at 500°F after aging 192 hr at 500°F, and must retain 1000 psi in shear at 500°F after such exposure. Evaluation of an adhesive for use in the supersonic transports requires knowledge of the performance of the adhesive after exposure to elevated temperatures for many thousands of hours, thus the standard specifications do not represent more than a mere token guideline. Testing of adhesives for reentry space vehicles poses serious problems and requires special testing equipment. The bond lines in a reentry vehicle may reach 1400 to 1700°F for a few seconds, and the temperature rise is extremely rapid. Air-to-air and air-to-ground missiles are subjected to tremendous loads coupled with elevated temperatures when fired, thus standard tests are not adequate and a decision on this basis would still be sheer guesswork. The natural aging process applicable to missiles during extended shortage periods poses a major problem. Aging processes can be accelerated for testing purposes, but experience has cast some serious doubt that results are

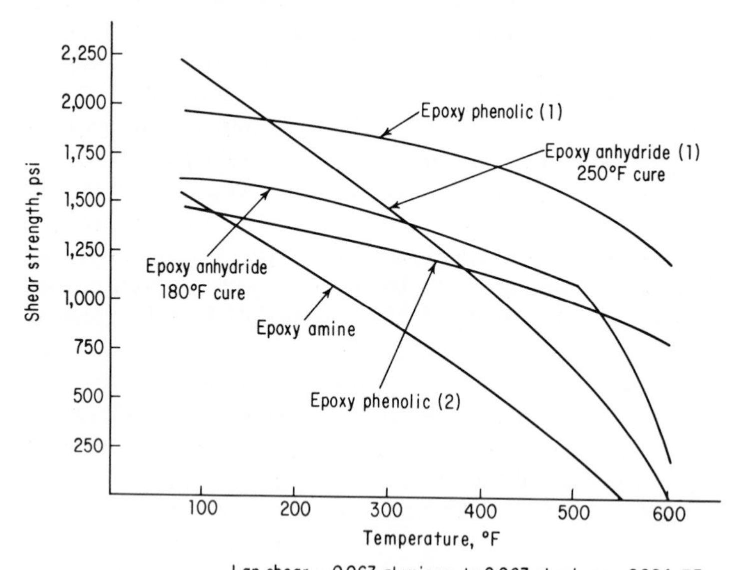

fig. 8-1 *Adhesives tested at elevated temperatures.*

indicative of the natural aging process of an assembly. Time becomes the critical element because it is difficult for most companies to carry out long-term studies due to tight delivery schedules and contract commitments.

The adhesive manufacturers perform a variety of environmental tests to qualify their respective products to military specifications (Fig. 8-1). Data obtained appear in their reports, which are available, and this information is considered useful. Present engineering requirements necessitate testing beyond the manufacturers' studies. Adhesive engineering personnel are skeptical of data from sources other than the tests they conduct themselves. This is normal and understandable because "a pig in a poke" could possibly spell disaster.

ELEVATED TEMPERATURE TESTS

Elevated temperature tests are performed on a variety of test-specimen configurations and undoubtedly the largest area of environmental testing encompasses these tests. They can be performed in conjunction with practically all mechanical tests defined in Chap. 7, the largest areas being tensile shear and peel. If elevated temperature tests do not exceed 800°F, they may be performed with test chambers available that are adaptable to practically all testing machines. An environmental test chamber consists of an insulated housing that encloses the jar extension mechanism. The elevated temperature tests specified in MMM-A-132 may be performed with this type equipment. When this type equipment is not available, electrical heating strips may be wrapped around lap shear tensile specimens and tests performed. The heat is monitored with thermocouples placed near the adhesive joint. Regardless of the heating approach, no less than 5 min or more than 10 should be required to raise the specimen to the desired temperature, and then it should be soaked at the desired temperature for a minimum of 10 min before testing. Thermocouples should be affixed near the center of the bond line to ensure accuracy and reproducibility of data, because a time reading versus chamber temperature would seldom be accurate. Alternately a prototype specimen may be used with a thermocouple embedded in the bond line and placed adjacent to the actual test specimen.

Care must be exercised when placing the test specimen in the holding fixture or grips to prevent stress loading as the temperature increases. Bond-line measurements should be precise and accurate as the bond-line thickness is an important factor when testing at elevated temperatures.

ASTM D2295-64T defines a test for determining strength properties of adhesives in shear by tension loading at elevated temperatures (metal-to-metal, tentative). This test utilizes a radiant heat source with a high-efficiency reflector and a standard tensile testing machine equipped with pin-type grips. Figure 8-2 illustrates the use of quartz lamps as a heat source that may be employed for test temperatures above the capabilities of a standard test chamber (chamber shown in photograph).

Elevated temperature test reports should contain information pertinent to the adhesive system and adherends, number of specimens tested, length of overlap, test temperature, temperature rise rate on specimen, soak time, and mode of failure.

STATIC HEAT AGING

Long-term elevated heat aging tests may be conducted by placing the specimen in an air circulating oven maintained at the desired temperature with thermocouples placed at selected locations to be assured of universal heating. The specimens should be spread out as much as possible, and not stacked one on top of the other.

A group of specimens should be processed in as much the same manner as possible, preferably autoclave cured. If a large number of specimens are to be tested, it may be impossible to fabricate the entire test group in one operation; consequently, it may be necessary to fabricate in multiple shots, causing slight variations in bond-line thickness or surface conditions. In the event this is the case, specimens should be fabricated and then thoroughly mixed before identification, thus preventing a variable that may be pertinent to one shot from creating an out-of-balance curve if the specimens are tested at various intervals. In some isolated cases it may be desirable to fabricate specimens with varying bond-line

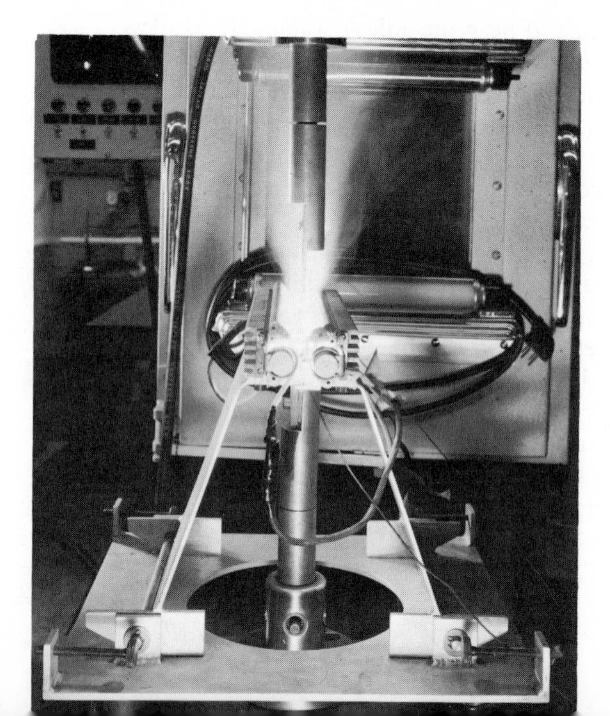

fig. 8-2 *Elevated-temperature testing utilizing a quartz lamp as a heat source.*

thicknesses or other process variables in order to observe the effects of heat aging on these variables, since these discrepancies may occur in production.

After fabrication of the specimens, bond-line and dimensional measurements are obtained and recorded, and control specimens tested and tabulated. Specimens are usually taken from the aging chamber at various intervals and tested to produce a degradation curve. The bond-line measurements should be taken again when the specimens are removed from the heat chamber, and compared to the original measurements and any changes recorded.

Test ovens should be equipped with a recording device that will produce a permanent record. If not, the temperatures should be monitored at regular intervals and any discrepancies recorded and reported to the cognizant individual. When the specimens are removed to the mechanical testing area, care should be exercised to prevent thermal shock. If the specimens are to be tested at elevated temperatures, a heated box of some kind is ideal for transfer. If they are to be tested at room or cryogenic temperatures, a slow reduction in temperature is desirable. It would be considered poor judgment to remove specimens from a hot oven and toss them onto a cold table because reproducibility of results may be difficult the next time around.

If an assembly is to operate in an elevated service environment that fluctuates, testing with varying temperatures may be necessary. The temperature fluctations and number of cycles are naturally dependent on the

fig. 8-3 *A missile structure with heating elements attached.*

design requirements or established service environment. If the temperature change is rapid, it may be necessary to utilize two or more ovens, stabilized at the desired temperatures and conveniently located for rapid transfer of the test specimens. If the temperature change is gradual, a cam driven temperature control device may be attached to an oven to produce the desired temperature changes.

It is necessary at times to place an entire assembly in a test fixture for heat aging utilizing electrical heating elements or infrared lamps since it would be impractical to place the test part in a chamber (Fig. 8-3). During the heating or aging the structure may be subjected to vibration, shear loads, bending, peel moment, or various other stress loads. Tests of this nature are quite complex and because there are no existing standards, the test must be devised to satisfy the need or requirement. Tests of this nature are often performed by specialist groups who make complex testing their profession.

HUMIDITY EXPOSURE TESTS

Tests termed standard humidity tests should be performed in a cam-controlled humidity cabinet specifically designed for this purpose. The chamber should have a device to record any deviation pertinent to heat or humidity.

Distilled or deionized water should be used to maintain humidity and the reservoir should be checked and replenished as often as necessary to maintain proper water levels. Drain holes in chambers should be checked for clearance at regular intervals, thus preventing water accumulation in the bottom of the chamber.

Specimens should be processed and fabricated under identical conditions for a test series, unless data pertinent to process techniques or bond-line deviations are desired. Precise bond-line measurements should be taken before subjecting the specimens to humidity tests and then taken again when the specimens are removed from the chamber. Specimens should be vertically suspended in the chamber by means of glass hooks, waxed strings or other nonmetallic materials. Small holes may be drilled in the ends of the specimens for attachment, and the specimens should be suspended in a manner that prevents their contact with each other or with any metallic object. They should also be located to prevent condensation from one specimen dripping onto another.

When specimens are removed from the chamber, they should be placed in moist polyethylene bags or their equivalent until they are tested. If possible, they should be tested immediately upon removal from the test chamber, because a delay may produce erroneous results due to joint recovery. When reporting results the mode of failure is important because this indicates where the degradation occurred, i.e., in the adhesive layer, in the primer system, at the interface, or degradation of the adherend. The mode of failure actually establishes the weak link in the system being tested.

MMM-A-132 specifies 30 days in a closed chamber maintained at 120± 5° and 95- to 100-percent relative humidity. This is the procedure used by the adhesive manufacturers in qualifying materials, but many times it is necessary for more severe tests to be used. An example applicable to this actually occurred a short time ago when a company was awarded a contract to fabricate airborne equipment racks. The adhesives were selected using results from humidity tests that conformed at that particular time to MIL-A-5090. When time came for final Air Force acceptance, the electronic equipment had been installed and the Air Force subjected the entire unit to the requirements of Mil-STD 202 which governs electronic equipment and is much more severe. Instead of one cycle at 140°F in 24 hr, they were subjected to two cycles from room temperature to 162°F every 24 hr. The racks failed and required a new adhesive selection. This is an example of the design responsibility that was mentioned earlier in the test. Everyone must be well informed and fully aware of all qualification and acceptance tests that the final product will encounter, regardless of how minute the assembly which a company or individual must produce. There are many sources of information pertinent to humidity tests and the effects of various cure cycles, pressure deviations, primers, effects of humidity after heat or room-temperature aging (or other environments), permeability of seals, effects on coatings and various substrates. Moisture coupled with heat is the prime cause of adhesive joint degradation, and often has an even more undesirable effect on the adherend. Table 8.1 shows the effect of humidity on various surfaces. The adhesive is an epoxy-nylon processed to manufacturers' recommendations, and this particular system was chosen because of its poor resistance to moisture exposure. Note that after 30 days' exposure, the plated surfaces failed adhesively, which indicates the moisture was penetrating the adhesive layer and oxidation was occurring at the interface.

Figure 8-4 illustrates the degradation versus exposure time, utilizing an epoxy-nylon adhesive with chronic acid anodized surfaces. The humidity exposure was performed in accordance with Military Test Standard 202B. Table 8.2 shows the effect of humidity exposure applicable to various adhesive types on aluminum adherends and tested per Military Specification MIL-E-5272.

One of the major objections to static humidity tests is that the joints are not stressed during exposure; consequently, results are not indicative of service environments. Efforts have been made to develop a test method that would place the specimen under stress during the accelerated hu-

fig. 8-4 *The degradation of an adhesive after moisture exposure.*

midity test (or other environment). One such test which has been proposed as an ASTM standard, utilizes a lap shear specimen and a sustained stress jig (Fig. 8-5). Specimens are fabricated, placed into the fixture, and subjected to the desired exposure medium.

The test report should include information pertinent to the adhesive and adherends, test conditions used to determine the strength of the con-

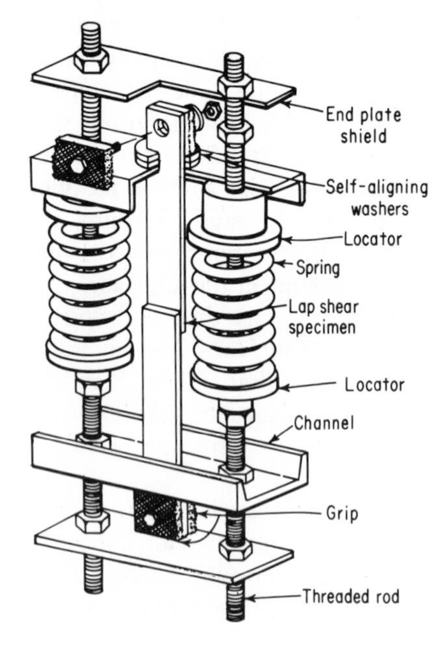

fig. 8-5 *Fixture for application of stress during Environmental exposure.*

trol specimens, description of the test environment, number of specimens tested, stress-load applied, length of time to failure, and mode of failure.

WATER-IMMERSION TESTS

Water-immersion tests are performed simply by immersing the specimen in distilled or deionized water. The specimens should be suspended so that they are not in contact with each other. The tests may be performed at room temperature or at elevated temperatures, and large Pyrex glass beakers may be used for small-scale laboratory testing. If it is desired to elevate the temperature, the beaker may be covered with a layer of aluminum foil or its equivalent and placed in an oven. The temperature is retained just under the boiling point of the water for the desired length of time. This is a rapid and very severe test, and when utilizing metallic adherends rapid oxidation usually occurs. The room-temperature immersion tests form a part of the qualification tests to MMM-A-132, and this data is reported in the manufacturers' data sheets. Water-immersion tests are effective in checking the effectiveness of seals, adhesion of primers, stability of conversion coatings, degradation of conductive adhesives, adhesive recovery, and surface-preparation techniques.

Figure 8-6 illustrates the surface preparation evaluation of some aluminum adherends with various surface treatments. Specimen No. 1

fig. 8-6 *Comparison of sodium dichromate-sulfuric acid etched specimens versus alodyne treated after water immersion and static heat age tests.*

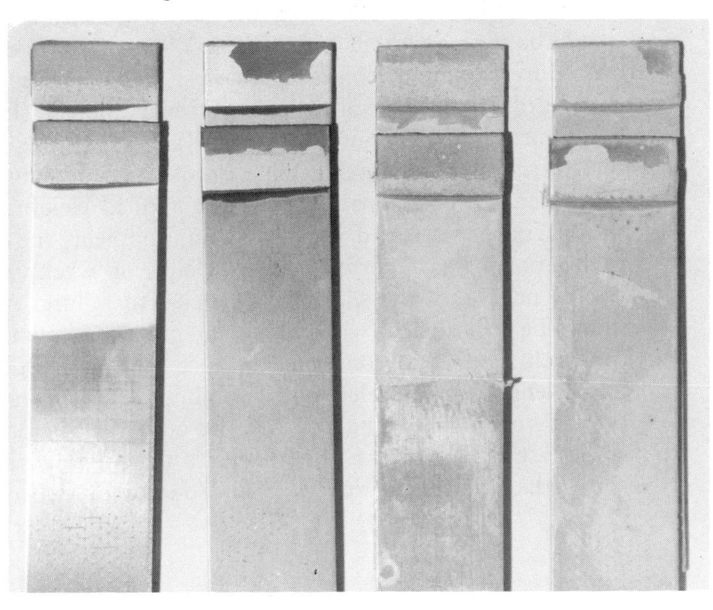

represents alodine 1200 treated aluminum after water immersion; No. 2 shows a sodium dichormate-sulfuric acid treated aluminum specimen after the same immersion. (Notice that No. 1 failed adhesively and No. 2 failed cohesively). Specimen No. 3 was alodine 1200 treated, heat aged, and then it failed adhesively. Number 4 was sodium dichromate-sulfuric acid treated and failed cohesively. They were immersed at the same time and under the same conditions. Numbers 2 and 4 were 40 to 50 percent stronger in shear than the alodine treated specimens (Nos. 1 and 3). The same adhesive and cure conditions were used.

Specimens may be water submerged for the desired period of time, some removed at intervals and tested. The remainder may be tested at various intervals as is deemed necessary or they may be dried in an oven, then tested for recovery. It is evident that this could be valuable information if an assembly is to be subjected to water or moisture for short periods of time, and then possibly a normal or elevated-temperature environment. Most adhesions have a high percentage of recovery after drying, thus the effect on the adherend becomes very important.

A tentative ASTM test method proposes a peel test for water immersion which may also be used for other fluid-immersion tests. High stress loads in the presence of water or water vapor may cause some adhesive to fail at less than 10 percent of the stress required to rupture a dry joint. This procedure is recommended to be used as a rapid screening test for assessing the durability of adhesive joints or to determine the effects of water on various surface preparations.

A standard T-peel specimen may be used for the test coupled with an apparatus as shown in Fig. 8-7. The successful use of this test method depends on preparing good-quality peel joints having as little variation in peel strength as possible over the entire test panel. Control specimens should be tested without water immersion, for comparative data. The test procedure for immersion is as follows: clamp one end of the unbonded portion of the peel specimen to a stationary object (Fig. 8-7) and to the other portion of the joint (or coupon) and attach the desired weight. The total force on the adhesive joint is equal to the sum of the weights of the water, the peel apparatus, and the applied weight. During the period the specimen is stressed in the test environment, the amount of the unpeeled portion is recorded hourly, daily, or weekly, depending on the rate of peel, until the specimen is peeled to failure. At least six readings should be recorded for each specimen and a minimum of three specimens for each test condition; the distance peeled (distance peeled equals initial length of unpeeled portion minus length of unpeeled portion at time t) versus time. The slope of the curve at any point is the peel rate at corresponding time t. The curve is then divided into six equal parts and if the peel rate varies less than a factor or two over all six parts of

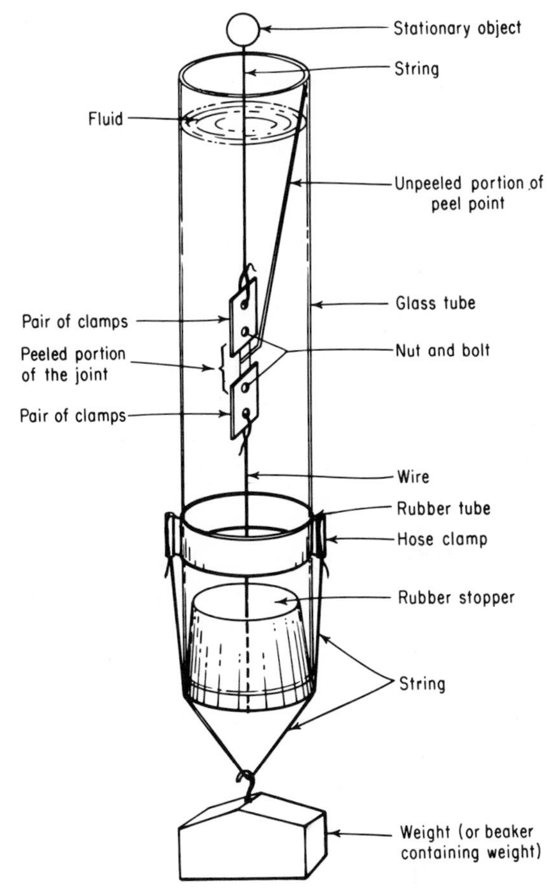

fig. 8-7 *Apparatus for measuring adhesive peel rates in water.*

the curve, then obtain an average peel rate (slope of curve) for the specimen. In order to assess durability, compare the peel rates under different levels of stress and water temperature by plotting a log of peel rate versus stress.

The report should include information pertinent to the adhesive and adherends, a description of test environment, angle of peel, number of specimens tested, glue-line thickness (by all means), average peel rate, load applied, and mode of failure.

SALT-SPRAY TESTS

Salt-spray tests are usually conducted in commercially available equipment which is available in a variety of designs and engineered to control items such as spray velocity, temperature, and salt content of the solution. The same precautions are taken in the preparation and suspension of specimens as specified under humidity tests. Specimens for salt-spray exposure are sometimes racked in a nonmetallic rack and the specimens (especially lap shear) are placed in the rack from a 45 to 60° angle and then placed in the chamber with the panel parallel to the principle direction of flow of salt spray through the chamber.

The sodium chloride should be pure within 2 percent and the sodium chloride content of the spray may vary from 5 to 20 percent, usually depending upon the design requirements. Salt-spray tests form a part of the qualification requirements of MMM-A-132 and tests specified in Federal Test Standard 151, MIL-E-5272 and Federal Test Standard 202.

Salt-spray tests are important in the area of surface preparation. Figure 8-8 illustrates the effect of salt spray exposure on adherends. The specimens in the photo are magnesium with various conversion coatings. When adherend surface degradation occurs, it becomes difficult to evaluate the adhesive system, as was suggested by statements made in Chap. 5 concerning surface preparation. If the surface preparation is not adequate for the service environment, the entire design is worthless, and in the case of magnesium, several coatings that offer excellent protection against environmental conditions are poor coatings to utilize with adhesives.

A salt-spray test report should include information pertinent to adherends and the adhesive system, exposure time, salt content of the solution, test values after exposure (if any), visible conditions of the adherends, and mode of failure.

fig. 8-8 *Magnesium specimens which have been subjected to a salt spray environment.*

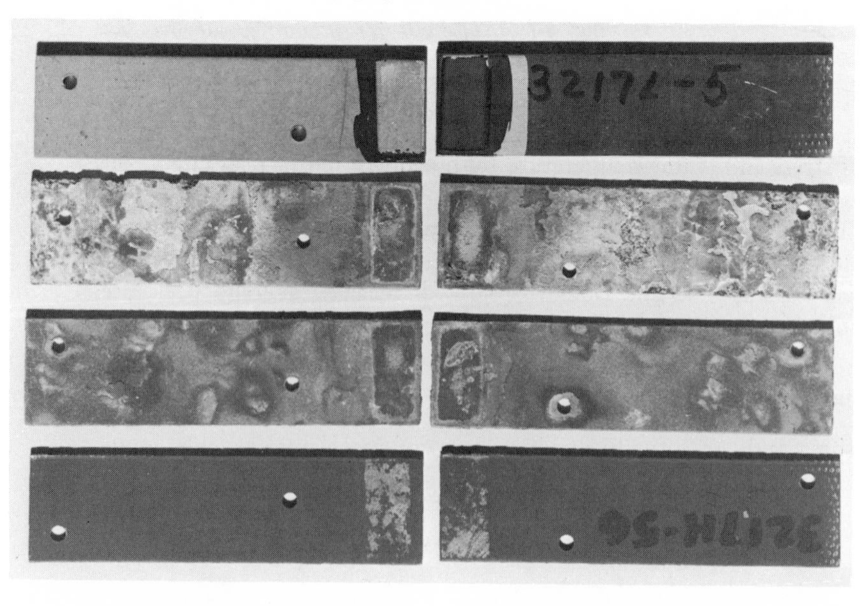

FLUID-IMMERSION TESTS

Fluid-immersion tests may be immersion in any fluid that the adhesive system may be exposed to in service. The standard test fluids for qualification to MMM-A-132 are listed as follows:

Fluids	Specifications	Immersion
Water	Distilled or not containing more than 200 parts per million of total solids	30 days ± 2 hr
Jet engine fuel JP-4	MIL-J-5624	7 days ± 2 hr
Anti-icing fluid	MIL-F-5566	7 days ± 2 hr
Hydraulic oil	MIL-H-5606	7 days ± 2 hr (For Types I and II adhesives)
Standard test fluids	TT-S-735	7 days ± 2 hr

These tests are usually performed at room temperature (70 to 80°F) unless the adhesive is expected to see service under more severe conditions. For example, adhesives used to bond heat exchangers and cooling devices may be subjected to hot fluids such as heated ethylene glycol. This application for adhesives poses serious problems and requires extensive testing under simulated service conditions to establish a high confidence level. Tests of this nature usually utilize the actual assembly if at all possible, after the candidate adhesives have been selected by a screening process, i.e., the prime candidates selected by hot immersion tests utilizing lap shear specimens and service performance validated later by proof testing.

There are other types of fluid immersion tests. For example, small pieces of the cured adhesive may be subjected to fluids to obtain the swell factor. Swelling occurs with many adhesives when subjected to a particular fluid, especially if the fluid is heated. This brings out the importance of bond-line measurements before and after immersion when utilizing conventional lap shear or peel specimens. It must also be taken into consideration that the joints may be loaded during service due to fluid pressurization, shift of load-applicable fuel tanks, etc.

The report should include information concerning the adhesive and adherends, immersion time, type of fluid, temperature during immersion, type of specimen, any change in bond-line thickness, color changes in the adhesive or primer system, length of immersion time (hours, days, months), strength levels after mechanical testing, and mode of failure.

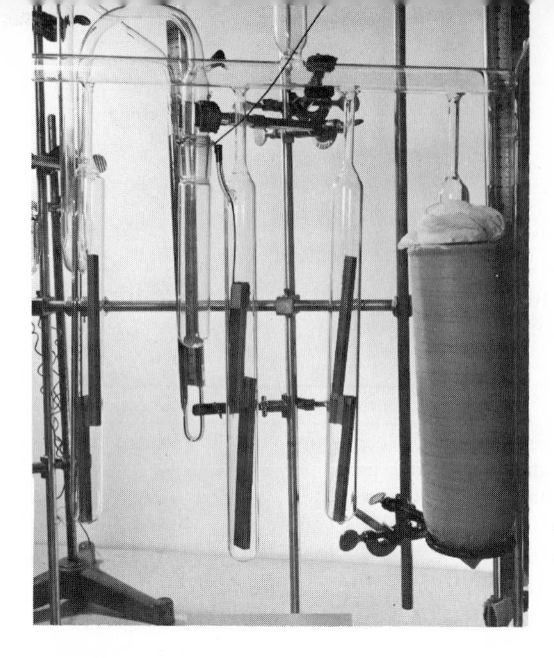

fig. 8-9 *Vacuum test utilizing lap shear specimens conducted at —300°F.*

VACUUM TESTS

Space-age adhesives have propelled the aerospace industry into a new era of testing which includes radiation exposure, LOX compatibility, and effects of a vacuum environment, to mention a few. Adhesives systems, approved for usage in a vacuum, that possess desirable mechanical properties can be listed on a small index card with ample space left for doodling.

There are several current vacuum tests that are used. Figure 8-9 illustrates a vacuum test at —300°F and 10^{-6} mm of mercury. The specimens are lap shear which are exposed the desired length of time and temperature. They are then mechanically tested and compared to control specimens.

Another approach is the weight loss under vacuum technique. A portion of cured adhesive ($0.5 \times 0.5 \times 0.125$-in. slug) is weighed on accurate pharmacist scales or an analytical balance. The specimen is then subjected to the necessary vacuum and desired temperature, removed from the chamber, reweighed, and weight loss calculated. Visual observations may also be valuable during the test because if the specimen is outgassing, the test tube will appear hazy or milk colored. A better description would possibly be Los Angeles on a smoggy day, but fortunately (or unfortunately) not everyone has witnessed the latter comparison.

THERMAL SHOCK TESTS

Thermal shock tests provide necessary data for adhesive selection for assemblies or details that are subjected to rapid temperature changes.

When these rapid changes are present in an assembly, heavy stresses are placed on the bonded joint because of the expansion and contraction of the adherends, and they change within the adhesive system. Test specimens of various configurations are utilized, the more common being the simple lap shear and flatwise tensile. The ideal specimen is the part in question if facilities are available to properly conduct the test. It may be necessary to provide facilities if it is considered absolutely necessary to proof test a critical large assembly.

A test procedure that would fall into a standard category is in the temperature range from +185°F to —40°F. This necessitates two chambers stabilized at these temperatures and located as closely together as possible for quick and convenient transfer of test specimens. The specimens should be placed in the hot chamber, stabilized at the desired temperature, and maintained for a minimum of 15 min, then transferred to the cold chamber, allowed to reach chamber temperature, and held for 15 min. This is one cycle, and the number of cycles should be a minimum of three. The temperatures, soak time, and number of cycles would naturally depend on the design requirements. A missile may change from a low temperature environment to extreme elevated temperatures in a matter of seconds, thus a vast amount of internal stresses are placed on the bond line. Figure 8-10 illustrates the effects of thermal shock on a cork ablative coating bonded to a simulated missile structure. The adhesive in this example is an epoxy-pyromellitic dianhydride (PMDA) which cures to a very rigid state, consequently there was little, if any, movement at the interface, possibly accounting for the numerous cracks at the outer surface. Figure 8-11 shows an identical test structure bonded with an epoxy-phenolic adhesive system which continued to outgas during thermal aging and thermal shock tests. This problem was eventually solved by utilizing a slightly flexible modified-epoxy system that performed well under these test conditions.

When performing thermal shock tests it is advantageous to use chambers that are equipped with glass windows, thus the test specimens may be observed during the study. The specimens should be equipped with thermocouples and the temperature recorded with an adequate recording device. The visual inspection should be continuous, and changes in color, appearance of cracks, or other changes considered abnormal should be recorded. If mechanical test specimens are thermal shocked the specimens should be mechanically tested at various temperatures and compared to original control specimens that have also been tested at the same temperature levels to determine the effects on the adhesive at elevated or low temperatures. If miniature or microscopic cracks are present in the adhesive layer, they may not adversely affect the test values at room temperature but greatly affect test results at cryogenic or

fig. 8-10 *The effects of thermal shock on a cork ablative coating bonded to an aluminum structure.*

fig. 8-11 *Cork ablative bonded with an epoxy-phenolic after environmental tests.*

elevated temperatures. This is especially true with most adhesives when the cohesive properties are affected, but if interfacial degradation occurs test values at various temperatures would possibly be more consistent from a percentage standpoint when equated to the controls. If a large or disturbing difference did exist, it would be an exception rather than a rule.

Thermal shock tests are also valuable in evaluating sealants, ablative coatings, paints, surface coatings, conductive adhesive, and plated surfaces.

The test results are reported in actual test values (psi or linear inch pounds) if mechanical test specimens are used, or results may be repeated in percentage bond-strength change using the following formula:

$$R = 100 \frac{Y - X}{X}$$

where R = percentage change in strength
X = initial strength of control specimens
Y = strength of tested specimens after exposure

The report should include information pertinent to the adhesives and adherends, number and type of specimens tested, cycle temperatures, time required for stabilization from one temperature to another, amount of soak time at stabilized temperature, any visible change in appearance and the cycle in which the change occurred, complete information concerning mechanical tests, and mode of failure.

CRYOGENIC TESTS

The word cryogenic is defined as low-temperature research; cryo means cold or frost-like. The term low-temperature testing has essentially the same meaning but is not always used synonomously with cryogenic testing applicable to adhesives. MMM-A-132, "Adhesives, Heat Resistant, Metal to Metal, "refers to low-temperature tests performed at –67°F, but testing and research below that temperature are usually referred to as cryogenic tests. The reason –67°F is considered a standard test is difficult to determine, but it is possibly because in earlier adhesive tests, this temperature could be obtained by the use of dry ice and acetone. Standard environmental refrigerated test chambers will yield temperatures in this range but testing below that temperature poses more problems. Low-temperature tests (–67°F) are required as acceptance tests by most companies and in many cases, are considered routine tests associated with production parts. The tests may be performed with lap shear, peel (T and climbing drum) or flatwise tensile specimens. Strength properties pertinent to low temperature tests should be determined after a 10-min soak and the bonded area should reach the desired temperature in a minimum of 10 min. Tests of this nature should always be performed strictly in accordance with a standard procedure, especially during the evaluation of an adhesive system. This enables us to compare the data obtained to information received from the manufacturer and other sources. Low-temperature tests using standard specimens and procedures are conducted in the same manner as tests described in Chap. 7, with the exception of the test environment.

A vast amount of research has been performed with adhesives in the cryogenic area in the last five years. The aerospace firms are highly concerned about the performance of adhesives when subjected to subzero temperatures, because many systems become brittle and lose desirable mechanical properties when exposed to these conditions. Elaborate test equipment has been designed by agencies such as the National Aeronautic and Space Administration for cryogenic testing of adhesives in conjunction with the space programs. One such test setup has a simulated rocket with the bonded insulation installed and the fuel cavity filled with LOX. The test is then monitored from concrete bunkers via several television cameras located at strategic points. It must be kept in mind, however, that this type of test setup is expensive, and most companies are compelled to devise more economical tests which many times result in a "witch's dream."

The simple lap shear is used extensively for cryogenic testing because of simplicity, and much can be learned about the compatibility of adhesives with subzero conditions utilizing this type of specimen and a cryostat as shown in Fig. 8-12. The cryostat is mounted between the grips of a standard tensile testing machine. Pin-type grips are recommended to prevent preloading and the specimen must be located to coincide with the direction of applied load through the center line of the grip assembly. The specimen is tested with a load rate of 1200 to 1400 lb per min.

The report should include information pertinent to the adhesives and adherends, failing load, load rate, temperature during test, length of overlap, and mode of failure.

NORMAL WEATHERING

Many times specimens or assemblies are placed outside, i.e., on the roof of buildings, exposed to salt air near the beach, etc. Tests of this

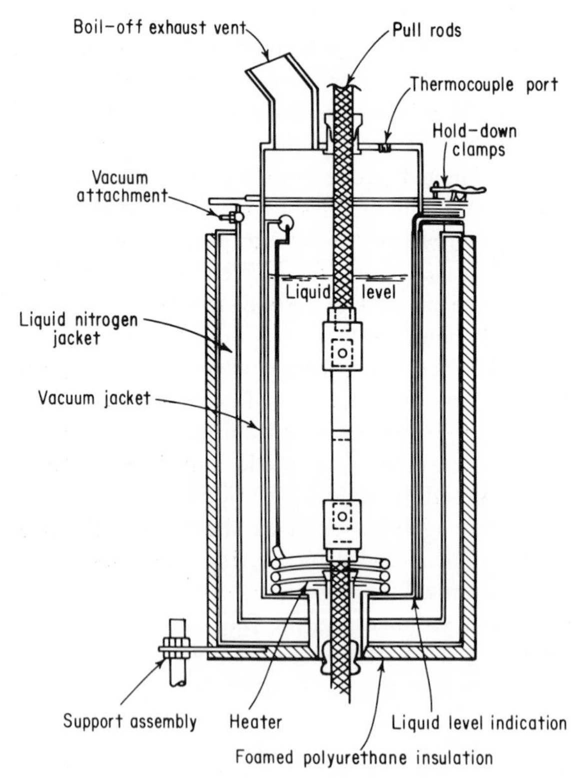

fig. 8-12 *A cryostat and accessories for cryogenic testing of adhesives.*

nature are performed to assess the effects of weathering on the assembly or specimens. Testing adhesives is sometimes so thorough that specimens are exposed to climates in different parts of the world to evaluate their response to various geographical areas.

SUNSHINE TESTS

Sunshine tests may be a requirement for the fabrication of military aeronautical equipment which may include adhesive-bonded assemblies. The equipment is mounted within a test chamber in a simulated operating position and subjected to radiant energy at the rate of 100 to 200 watts per sq ft. Of the total energy, 50 to 60 percent shall be in wavelengths above 7800 A and 4 to 6 percent in wavelengths below 3800 Å. The test-chamber temperature is maintained between 110 and 115°F during the course of the test. The unit is visually inspected or destruct tested at the completion of the specified test period.

RAIN TESTS

If rain tests are necessary they are accomplished in a chamber and the specimen or assembly mounted in a simulated installation position. The rain test chamber temperature should be maintained between 70 and 80°F. A simulated rainfall of approximately 4 in. per hr is recommended, and is produced by means of water-spray nozzles of such design that water is emitted in the form of small droplets rather than a mist. The temperature of the water should be maintained between 50 and 70°F. The rainfall should be dispersed uniformly over the test area. The duration of the test is normally 2 hr, but may be longer to conform with design requirements. This type of test is usually performed on an assembly rather than a mechanical test specimen.

Provided that test specimens are used on weathering tests, they would be tested in the specified manner, and all information accurately reported by the testing function. If the tests are performed on adhesive bonded parts or assemblies, they should be inspected for corrosion or any undesirable effect that would prevent the assembly from meeting the service requirements. Damage that would hamper the operation during service life would provide reason to consider the equipment as having failed to meet the test to which it was subjected. The assembly may also be tested destructively to obtain the necessary data.

FUNGUS GROWTH

Fungus-growth tests are performed to determine the effects of this growth that might be encountered, for example, in service in the tropics.

These tests are performed in a special chamber having external connections, and the test specimens are sprayed with a suspension of mixed spores. Specimens may be tested mechanically, analyzed by a chemist, or visually inspected. If deterioration or fungus growth of the components would prevent a piece of equipment from meeting these service requirements, the adhesive or substrate would be unacceptable.

Several of the more common environmental tests have been covered in this chapter. There are many more and new ones are being devised as new applications of adhesives are found. Adhesives may be subjected to altitude tests, pressure tests under various conditions, explosion-proof tests, exposure to ozone and radiation under various loading conditions, and tests under simulated space conditions to perfect space-bonding techniques. An ugly specter has already reared its head, i.e., the ability to bond in outer space.

No attempts have been made in this test to develop an adhesive design chart, i.e., a particular adhesive system for a specified adherend. The reason is obvious: it would only lead to misuse and confusion. Seldom, if ever, would two bonding requirements be the same; therefore, testing is essential to ascertain the correct adhesive for a specific application.

TABLE 8.1

Material	Control psi	Type failure	10 days moisture	Type failure	Percent degradation	psi 30 days	Type failure	Percent degradation
Aluminum no etch	3051	Cohesive	774	Adhesive	74.3	484	Adhesive	83.6
Aluminum gold plate	4623	Cohesive	2209	40% Adhesive	50.1	843	Adhesive	81.7
Aluminum silver plate	2116	90% Adhesive	1368	Adhesive	35.3	176	Adhesive	91.2
Aluminum C. acid anodize	6118	Material Failed	4291	80% Cohesive	29.6	2256	80% Cohesive	63.2
Aluminum cadmium plate	2148	70% Adhesive	522	Adhesive	75.7	321	Adhesive	85.1
Aluminum sodium dichromate	5900	Cohesive	3941	Cohesive	33.1	2867	Cohesive	51.4
Aluminum spot-weld etch	5400	Cohesive	3869	Cohesive	28.4	2274	80% Cohesive	57.9
Steel no etch	4733	Cohesive	2581	60% Cohesive	45.6	847	Adhesive	82.1
Steel prebond 700	4205	Cohesive	3284	60% Cohesive	21.2	1722	60% Adhesive	59.0

TABLE 8.2

	Control	30 days	60 days
Epoxy	3000	3000	2800
Vinyl phenolic	4000	3800	3600
Nitrile phenolic	5500	5000	4800
Polyurethane	2300	2000	1600
Polysulfide epoxy	2500	2500	2400
Epoxy phenolic	3200	2800	2500
Epoxy nylon	6000	4200	2300

* Adherends: 2024–T3 aluminum

REFERENCES

DeLollis, N. J.: "High Temperature Testing of Re-entry Bonds," The Scandia Corporation, May, 1960.

Dunn, M. B., M. D. Musgrove, and O. T. Ritchie: "Applications of Current Materials to Hot Airframe Design," SAE National Aeronautic Meeting, Los Angeles, Calif., October, 1959.

Hartman, A.: "The Influence of Low Temperature on 'Redux' Bonded Lap Joints," Ciba Technical Service Department.

Hochburg, M. S.: Evaluating Adhesive Bonded Temperature Resistant Sandwich Construction, *Adhesives Age,* March, 1966.

Kausen, R. C.: Adhesives for High and Low Temperatures, *Materials in Design Engineering,* August, 1964.

Long, R. L.: "Cryogenic Adhesive Application," presented to Structural Bonding Symposium, Marshall Space Flight Center, Huntsville, Ala., March, 1966.

Military Specification MIL-E-5272, "Environmental Testing, Aeronautical and Associated Equipment, General Specifications for."

Military Specifications MMM-A-132, "Adhesives, Heat Resistant, Airframe Structural, Metal to Metal."

Pascuzzi, B., and J. R. Hill: Structural Adhesives for Cryogenic Applications, *Adhesives Age,* March, 1965.

Volume 16, American Society for Testing Materials (ASTM), 1966.

Wegman, R. F., and W. C. Tanner: Strength of Epoxy Adhesives where Stressed to Failure in Milliseconds, *Adhesives Age,* September, 1965.

CHAPTER 9

Quality Control of Adhesive Joints

Quality is described as a property or attribute that denotes usefulness; control relates to exercizing restraint, to holding in check, or to curbing. When these two terms are combined, the true meaning is obvious. Quality of a bonded joint is related to the functional adequacy of the assembly, but it is well to remember that perfection is the enemy of adequacy. Perfection, when referring to quality control or reliability of adhesive-bonded structures, is not a choice word; it may even create a negative attitude, because people realize that bonded joints can never achieve a level even closely approaching perfection. One of the best known quality cliches, "Quality cannot be inspected into a product," receives full recognition in the field of adhesive bonding, but quality-control personnel must not accept the attitude that an adhesive-bonded joint is a chaotic failure that is going to accidentally happen somewhere.

There are certain functions that can be performed to prevent substandard parts from reaching a customer, but the present quality controls applicable to adhesives are often totally and completely inadequate. One of the important services that quality-control activities perform is the compilation of defect statistics which provide information to all responsible bonding personnel. It is not implied here that quality control is faltering in the performance of its responsibilities as related to the definition of those responsibilities, by company and customer documents. The quality-control function is an absolute necessity.

The greatest problem is that quality control (QC) must deal with the human element, more so in adhesives than in other fabrication techniques, and at times this appears to be an absolute impossibility. When one looks beyond the three common problem areas (raw materials, design, and manufacturing processes) to a fourth dimension, i.e., general operations, an extremely high percent of defects is traceable to poor workmanship. It would be ironic indeed, when searching for ways to improve quality, to overlook this large picture that comes into focus when the causes of quality breakdowns are analyzed. When the results of several motivation and training programs are analyzed, comparably high gains in quality are the profits received from programs aimed at improving employee performance. What does this mean? It means that perhaps we have not taken a broad view of what constitutes our real opportunity for quality advancement in adhesive bonding. In all probability, too much emphasis is placed on paperwork and "crash correction programs," but far too little attention is paid to personnel. Vice Admiral Paul Stroop advanced this theory in an article in *Industrial Quality Control Magazine* when he wrote, "In spite of automation and other technical advances made in the last decade, a major element in achieving our quality-reliability goals is to obtain the full contribution of all the people involved in the process. The industrial world has become so complex that it is difficult for the individual to recognize his personal responsibility for the end product. Every device available should be employed to focus attention on the contribution that should be made by each employee—no matter what his particular job may be. He must be able to identify his own well-being with the success of the product. He must be persuaded to couple his own reputation to that of his employer. He must be made to feel morally obligated to do his best on his job for the good of his community and his country."

The advancing technology makes the people-factor in quality control even more critical and complex. It demands increased performance from the human being. The dependence on quality shifts from the employees' dexterity and skill to their judgment, knowledge, and desires. Management can call a crash meeting (first sign of trouble) to talk about preventing scrap or MRB action, but if long-range programs have overlooked the human factor then someone is looking down the wrong end of the telescope. The typical question is "Why are our profits low in the bonding shop?" The next step is to appoint a committee to study the problem for three months and compile a report on current problems. If the report is ever finished, it will not create the excitement of an old battered copy of the Kinsey Report (which had nothing to do with manufacturing hardware) and most likely would be lost in a maze of red tape. Most people close to adhesive bonding have written so many reports of this nature that a standard form is always available; just fill in the date and release in three months. The reply from upstairs: "The next time I want an idiot to do something, I'll do it myself."

Trained inspectors in the art of adhesive bonding are required for satisfactory performance. An inspector, or quality-control technician, cannot be expected to perform an adequate inspection job if he (or she) is not fully aware of what the objectives of the search should be. An inspector cannot do an intelligent job of plotting a quality curve placed in his work area unless he has appreciation for what he is doing. If he does not understand the problems that faulty or sub-standard bond lines may produce, a supreme effort could not be expected of him in looking for it with an x-ray or sophisticated ultrasonic apparatus.

The next requirement is clearly detailed specifications to work with, supported by check-off sheets. It must be determined during the manufacturing process at what points inspection is necessary, since constant surveillance would be impractical no matter how desirable. Not only should an inspection sign-off sheet accompany the assembly, but a sheet for the technician to sign after each operation should be available. Many times defective parts are discovered but there are no existing records as to who performed the job, so the day shift places the blame on the swing shift and vice versa. If personnel are required to sign off operations, both inspectors and technicians, they will naturally be more careful simply because this is an element of human nature.

There are many times when conflicts between quality control and manufacturing occur. Quality control is accused of hampering production, and this is natural since manufacturing personnel must produce. It must always be kept in mind, however, that quality is everyone's responsibility, not only to the parent company, but to the customer as well. Future contracts depend upon the performance of the finished product, regardless of the type of assembly or parts being manufactured. The ideal solution that many companies have adopted is placing a clause in the general specification that requires the manufacturing personnel or technicians to be qualified under a specific training course, and the issuance of a certificate to that effect. The floor inspector may then be sure that qualified people are performing the operation.

The prime military document applicable to inspection requirements is MIL-A-9067. The quality control of adhesive bonding is the qualification of adhesives for use, in-process control, and final inspection.

Adhesives to be used in production bonding of parts should be sampled and tested within a period of time not exceeding 24 hr prior to use. This test determines if the adhesive has surpassed its shelf life, and test standards of this type are usually specified in general adhesive-bonding specifications. Adhesives should never be used by the technician unless he has documented proof that the adhesive has just been received from the vendor. Acceptance tests must be performed before usage. All operations during the bonding process must be monitored and accurate records maintained.

Process control may be defined as the continual surveillance of all the steps associated with adhesive bonding. The range of coverage is extensive and control starts when the substrates and adhesives arrive and is not finished until the assembly is complete (Fig. 9-1).

VENDOR CONTROL

The first control applicable to adhesive bonding usually starts with the adhesive manufacturer. Most companies have established procurement specifications which the manufacturers have agreed to conform to. There

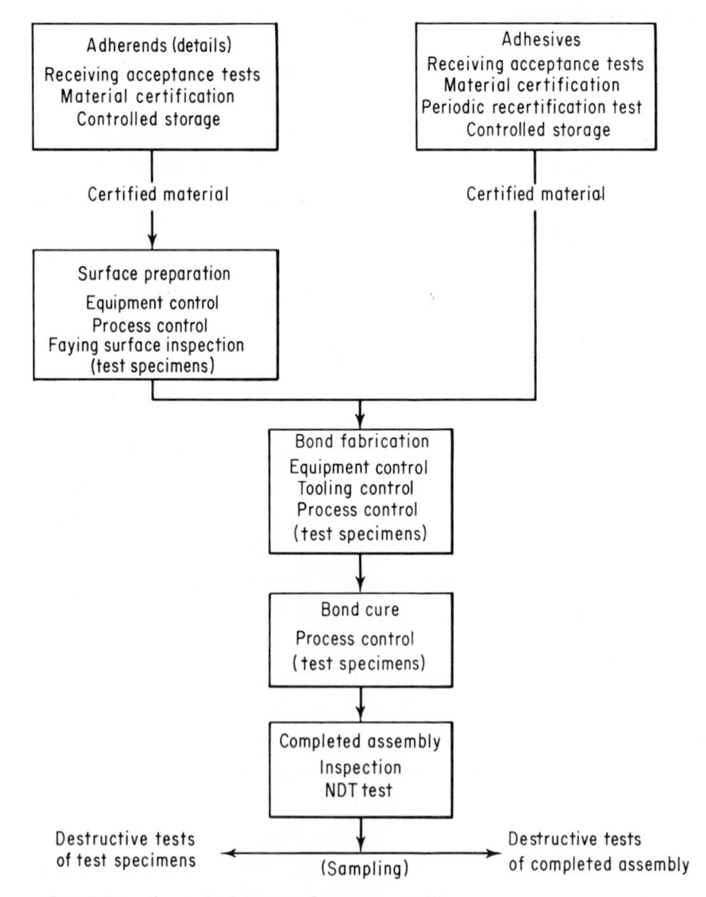

fig. 9-1 *A typical flow chart of quality control system for adhesive bonding.*

will usually exist a qualified products list, generated by the cognizant company, which usually lists more than one source of procurement. This is insurance to the company in the event an adhesive manufacturer cannot deliver the required volume.

The adhesive manufacturer will perform tests specified by the purchaser to ensure conformance to the procurement specifications. When the purchaser receives the adhesive material it should be accompanied by the following information, and records should be generated at this point that will be available for a minimum of two years, depending on the service of the assembly.

1. The test data received from the adhesive manufacturer, with warranty, if any.
2. Purchaser's lot number.
3. Date of manufacture.
4. Date shipped from vendor and date received by purchaser.
5. Expiration date of manufacturer's shelf life.
6. Date of acceptance sampling, if any, and results of those tests.
7. Scheduled reinspection dates during storage.
8. Date and time into or out of refrigerator. (Companies are now sending small samples of films which do not necessitate opening the roll before use. This permits acceptance tests but prevents air and moisture from getting to the adhesive during storage).

If the adhesive was shipped in a cold condition, one must be positive the preservation was adequate and it may be necessary or desirable to weigh the package.

COMPANY QUALIFICATION TESTS

These tests usually consist of requirements beyond those placed on the adhesive manufacturer. This responsibility usually rests with the quality-control laboratory, which has the responsibility to release adhesives for production use. When the adhesive is qualified and released for production, the floor inspectors may keep a close check on the time expiration date.

PRODUCTION RESPONSIBILITY ON MATERIAL CONTROL

1. When the adhesives reach the production area, qualification data should be checked and the floor inspector notified before storage.
2. Production should maintain an "adhesive inventory chart" showing all adhesives in storage and in current use in the shop. The chart should show the type of material, lot and batch number, date, and results of the quality-test lab report. This should be posted near the storage area

allowing the shop, quality, and process personnel to insure a first-step-in, first-out basis.

3. Store and handle materials as prescribed by the applicable specifications. Time in storage is affected if adhesive is left out of the freezer too long or in/out too many times. This should be specified in the pertinent specification. For unknown reasons, it is often left out.

4. Liquid adhesives should be mixed or thinned according to the applicable specification (by qualified personnel) and should be accomplished in an isolated area. It is a good idea to have certain people to do all mixing. If an error occurs, the accusing finger may be pointed at the proper individual. This will also eliminate the need for the floor inspector being present each time a batch is mixed. It is not a usual practice, but a safety precaution would be to have a check sheet showing the individual who did the mixing, the batch number, time of mixing, the ratio of adhesive to hardener; this document should be verified and stamped by the floor inspector.

RESPONSIBILITY OF THE ASSIGNED FLOOR INSPECTOR

1. The floor inspector should maintain his own file with laboratory test reports on the adhesives in work.

2. Qualification and acceptance reports on other raw materials.

3. When notification from production arrives that a new batch of material is in the house, he should verify lot or batch number and make sure the proper information is recorded on the production inventory chart.

4. Verify and stamp the adhesive inventory chart when all information is available.

5. Survey tagging of liquid adhesives.

6. Before use, verify that the adhesives are qualified and that all materials are stored, mixed, and handled as prescribed by the applicable specification.

INSPECTION OF OTHER INCOMING MATERIALS

Sheet materials should be checked for thickness, alloy, cladding, T-condition, and general appearance. Many incoming specifications require tensile sheets on incoming substrates, especially plastic materials.

Honeycomb should be checked for thickness if procured already cut to size. It should be checked for burrs, cuts, crushed cells, node separation, regular hexagonal shape of cells, and general appearance as to corrosion and distortion. It may be checked for strength as required by a specification, or require a manufacturer's affidavit.

IN-PROCESS INSPECTION

The floor inspector should make routine checks on all tooling such as presses, autoclaves, recording equipment, jigs, weighing devices, pressure-regulating equipment, and any other equipment associated with the operations to be sure they conform to satisfactory performance. This type of equipment may require a certification stamp (especially scales and measuring devices). Checks should be made to be sure these stamps are current. Records should be maintained in this area by the inspector, recording all tooling checks and any corrections that may be needed. The discrepancies may be brought to the attention of production personnel and other cognizant people.

The inspector must observe the prefit before the bonding operation, and check for excessive gaps, bending, burrs, and general appearance. After prefit, the assembly should be stamped or identified to be sure the same group of details go into the final assembly.

When the details are transported to the cleaning area the inspector has several functions that are very important. He must be sure the cleaning tanks have been titrated and meet the requirement of the pertinent process specification. The tanks should have a process chart nearby with all the necessary data available. The temperature of the tanks should also be checked for specification conformance. It is the responsibility of the inspector to be sure the process coupons are processed with the assembly they are to accompany. If results of these process coupons are recorded and plotted on a graph, the danger signals of tank concentration are easily recognized as the test values will begin to slowly decrease.

After the parts are cleaned the inspector should be sure they are rinsed well and wrapped in a manner to prevent contamination before reaching the layup area. Brown kraft paper or polyethylene bagging is usually recommended. The time-in/time-out of the particular solution should be recorded. The time-out will provide a record to ensure the lead time between cleaning and priming or bonding is not exceeded.

If a primer is being utilized on any of the details, the primer area should be checked for humidity and temperature. The operator should handle the parts with clean, white gloves; primer tables should be clean, and storage racks checked. If the primer requires a heat cure, the time and temperature should be maintained and stamped off. The primer should be checked for qualification, if any, and the lot and batch number recorded on the process sheet. The cured primer should be observed for color, checked for complete cure, and thickness measurements taken in various areas. Primers applied too heavily create many production problems. The primed areas should be checked for dust, blisters, blush, fingerprints, and general appearance.

The inspector should be sure the details are grouped and tagged so as to be first in, first out of the layup area. Before the layup begins, the tools or fixtures should be checked for cleanliness, making sure all resin from previous operations has been removed. A spot of cured resin will leave an indentation, especially on honeycomb assemblies with thin face sheets. The operators should be observed as to handling techniques, including gloved hands, etc. The application of mold release or use of solvents should be monitored. The adhesive used for the layup should be checked again at this point for specification conformance.

When the assembly is ready for cure, the time should be recorded again and signed off. Cure cycles, temperatures, cool-down rates, and assembly removed time should be recorded and kept as permanent records.

The assembly should not be cleaned up before quality control has inspected it. Production has a habit of cleaning parts up to improve appearance before submitting them to inspection, such as solvent cleaning, removing flash, etc. The inspector should inspect the part before the flash is removed. The flash reveals many things, such as cure, pressure, and, in the case of composite films, it would reveal if the film had been reversed.

The final inspection by the floor inspector would include a visual inspection and possibly coin or mallet tapping. The part would be inspected for dimensional conformance, warpage, and sometimes a weight check. The process coupon would be tested either by the floor inspector or routed to the quality-control laboratory. When these results are tabulated, the part is ready for the final buy-off, unless a non-destructive testing requirement is specified.

PRODUCTION RESPONSIBILITY TO QUALITY

Production should be sure all personnel working on the bonding programs are qualified, except where unqualified personnel would be permissible, such as tool cleanup, production movements, etc.

Production supervisors should post charts on all equipment such as presses, autoclaves, and any other pressure equipment, specifying required gauge settings to be sure the technicians are aware of the proper settings to obtain the required heat and pressure (especially pressure). They should request periodic checks on all equipment that does not have a requirement to be checked.

Production should make sure all parts are properly identified and should keep their own records as to number of parts, production flow, and tooling discrepancies. Production supervision should also watch the sign-off sheets to determine if recurring problems single out a particular individual who may need assistance or added instruction. When new jobs come into the shop, production should be sure the shop personnel are aware of any changes that might create problems.

One of the things missing in most bonding shops is a complete book of company specifications. Every shop should have an area with the necessary documentation that could be used not only by the floor inspector, but by shop personnel as well.

QUALITY CONTROL AND PRODUCTION RESPONSIBILITY ON A NEW ASSEMBLY

After the bonding of the first production assembly utilizing a new or modified tool, the assembly is usually tested destructively. This function is usually performed by quality control as a tooling check, the information being related to materials and process engineering, tool engineering, and production. A production part will be destroyed at specified points thereafter to keep a check on the tooling. If nondestructive testing methods are available, they should be used before the destructive tests. Reliability in these techniques can be established, thus reducing the amount of destructive-test assemblies. At present, an average figure for destructive-tested parts is approximately 2 percent of total production, which can be very expensive where large assemblies are concerned.

When a new assembly, new tooling, and possibly a new adhesive reach the production stage, a coordinated effort is required by all functions. The inspector and production personnel should first check the paperwork to be positive that all specifications, route sheets, sign-off sheets, adhesive qualification sheets, etc., are in order. Many times this is not the case because someone "upstairs" wanted parts "yesterday." The assembly should be fabricated to the applicable specification, but every precaution should be taken to produce a part indicative of future production. It should go even beyond that with added thermocouples attached to the tool in an effort to detect hot or cold spots. Many times, on a new production assembly, quality control is not aware of the more likely problem areas. For this reason the floor inspector must observe every operation, thus preparing for a quality paperwork modification on later assemblies. The quality-control personnel can play an important role in producing reliable bonded assemblies.

REFERENCES

Bodner, M. J.: "Structural Adhesive Bonding," John Wiley & Sons, New York, 1966.

Cagle, C. V.: "Adhesive Training Manual," NASA, Huntsville, Ala., 1965.

Epstein, G.: "Adhesive Bonding of Metals," Reinhold Publishing Company, New York, 1954.

Federal Aviation Agency: Bonding Inspection, *Quality Control Digest*, U.S. Government Printing Office, Washington, D.C., 1960.

Fitzgerald, E. B.: "Quality–Whose Responsibility?" presented at the 19th annual Technical Conference, American Society of Quality Control, Los Angeles, May, 1965.

"Handbook of Adhesives," American Cyanamid Company, Havre de Grace, Md.

Military Specification MIL-A-9067, "Adhesive Bonding and Inspection, Requirements for."

Sharpe, L. H.: Assembling With Adhesives, *Machine Design*, August, 1966.

Sharpe, L. H.: Designing With Adhesives, *Machine Design*, August, 1966.

CHAPTER 10

Nondestructive Testing

The pros and cons of nondestructive testing (NDT) of adhesive bonding have been the topic of many heated discussions in recent years. There are still a few skeptics who are not convinced that nondestructive testing equipment is necessary to determine the quality of a bonded joint. They still rely on the trusted half-dollar. An experienced NDT technician would not argue with the validity of the coin-tapping technique, because a void or complete discontinuity is easily detected by the coin-tapping or mallet-tapping method. Many times a very poor joint can be detected in this manner because the sound characteristics and mechanical properties of many adhesives have a distinct relationship. The primary problem with the tapping method is that some substandard joints, which are short of optimum expectations, cannot be detected with this type of inspection. For example, with a particular adhesive system, one expects 5000 psi in shear, but the actual joint is 3500 psi. This substandard joint is of great concern because there is a monumental amount of data indicating that the substandard joint deteriorates much faster under environmental conditions than does the sound, solid joint.

The major problem in selling ultrasonics to "unbelievers" can be summed up with two words: misunderstanding or stubbornness, the latter associated with the individual who has closed his mental mechanisms to the subject because his opinions have been formed. He believes that NDT just isn't practical.

Nondestructive testing is either expected to do the entire job, or deemed worthless. Naturally, neither is true. When used properly, NDT is a valuable aid in determining the quality of a bonded joint. Many excellent instruments are in inactive storage because someone fabricated six specimens, "yanked" them apart and declared, "It's no darn good!" This type of attitude could be compared to the story of the man who journeyed to the Grand Canyon, motored up to the edge, left the engine running, jumped out and peeked over the rim. He dashed back and told his wife, "Let's go; it's just a big gully full of rocks." One quick look at NDT is not sufficient to determine its merits. It requires research, money, training, and practice. Regardless of the type of equipment available, it can be no better than the operator. Manufacturers of ultrasonic equipment have recognized the need for training and offer training courses in this field. The Society of Nondestructive Testing has performed an excellent service in educating people in this area.

The theory of ultrasonics is quite complex; the end use is very simple once enough experience has been acquired to understand the use and limitations pertinent to a piece of testing equipment.

Companies in the aerospace industry that have pursued NDT are in a much better position for a bright future in adhesive bonding than those companies which have not. Nondestructive testing is needed for these reasons: (1) poor joints may be detected at an early production stage, thus eliminating cost or failure; (2) as the confidence level grows higher, the amount of destructive test units are reduced. One destruct test unit saved may pay for the equipment. One manufacturer reports destructive test units reduced from every twenty-fifth unit to every seventy-fifth unit, and they expect to reduce test units still more. (3) Cost savings in the reduction of final inspection time; and (4) shop personnel will be more careful in assembly if convinced that discrepancies will be detected. Techniques that will be discussed later have actually produced fingerprints on panels that could be presented to the Security Office and the guilty person identified. (5) The stress engineer may be informed concerning reduced quality of a particular joint, and thus take adequate measures. (6) NDT serves as a tooling check. If tooling pressure is reduced due to a tooling defect, it can be detected and corrected before many inferior parts are manufactured. (7) NDT elevates the company's confidence in adhesive bonding and, even more important, NDT elevates the customer's confidence in the product.

The progressive company is always looking to the future. Nondestructive testing of adhesive bonds is realistic and in the near future will be mandatory by absolute necessity for all aerospace firms.

The nondestructive testing of adhesives can be classified into three attitudes: (1) do nothing, (2) calculated risk, and (3) need to know. The frightening thing about these classifications is that the majority of people associated with adhesive bonding belong in the first two categories.

The largest and almost universal complaint registered by NDT personnel is that they are called into the picture too late. To be effective, a program of NDT should start in the design phase. But the opposite usually occurs, i.e., NDT people are notified that NDT is desired two weeks prior to production. This brings about unsatisfactory results as the responsible people are forced to deliver a snap judgement because there is not enough time, and possibly not enough funding, to do an adequate job. The NDT effort should be coordinated between the design and materials engineering activity because when it is passed to quality control it is usually too late. There exists a gap here that is hard to span; consequently, the NDT effort is juggled and sometimes dropped into the act. To put it simply and bluntly, NDT very often becomes an exercise in frustration. NDT should be planned just as carefully as other functions, starting at the design concept, or even before, i.e., when the proposal is written. There is very little evidence that contracts have been lost due to proposed NDT and added cost for a program; but there is considerable evidence that an NDT effort is valuable in obtaining contracts. The designer should design with NDT in mind because the product of a gamble is failure.

Proponents of NDT are not without blemish. Many times they fail to point out or define problem areas to the proper authorities. NDT programs are unsatisfactory at times because they fail to prevent failures from occurring, and the blame then falls "you know where." NDT personnel assigned to adhesive bonding programs should be familiar with adhesive properties and the prime causes of failure. One cannot prescribe a solution unless it is thoroughly understood what causes bonding failures.

The NDT field is not in need of new methods and techniques but the present methods need more research and improvements. There is available a good kit of tools but no single one is a universal tool. The best tool or device must be utilized to do a particular testing job; many times the technique cannot be adapted to the job, but, rather, the job is geared to the technique. It must be kept in mind that an NDT technique of any type is futile by itself because certain process variables cannot be detected. The proper approach for NDT of adhesives is as follows:

1. Have a program plan.
2. Define problem areas.
3. Define acceptance levels.
4. Select methods that may be applied.
5. Fabricate simulated standards.
6. Screen with selected methods.

7. Destruct test sample specimens.
8. Analyze data and select technique.
9. Test actual assembly and apply information obtained from laboratory tests.
10. Destruct test actual assembly.
11. Generate specification for implementing technique, clearly defining applications and limitations.

The present techniques which are defined in this chapter fall into the following categories:

1. Sound
2. Thermal
3. Photographic
4. Strain-sensitive
5. Visual
6. Mechanical

The most widely used of these for adhesives is sound, primarily ultrasonic, although x-ray techniques rate a close second.

THE PULSE-ECHO TECHNIQUE

Man has possessed the knowledge since ancient times that sound projected in any given direction will reflect from distant objects. In the language of the layman, as well as the scientist, this phenomenon is known as an echo. Around this rudimentary observation, man has added a package of facts, theories, and observations of basic laws which constitute the physics of sound. The origin of ultrasonics (sound above the audible frequency range) can be traced back to a number of early experiments. Perhaps the beginning was the discovery which resulted in the generation of high-frequency sounds. The Curie brothers, experimenting in the late 1870s, observed that when mechanical pressure was applied to certain natural crystals, an electrical instability occurred within the crystal and the surfaces become oppositely charged. They further observed that when the pressure on the crystal gave way to tension, the field in the crystal was reversed. Thus, by applying alternate loads of tension and compression an alternating electrical current was produced. Later experiments showed that the effect was reversible and that high-frequency electrical energy could be converted to high-frequency mechanical energy (sound) by the use of these crystals. Electromechanical energy conversion has more recently acquired the name piezoelectricity, named after the technical term, "piezoelectric effect."

One of the present methods for applying ultrasonics to the inspection of adhesives is based for the most part on what is known as the pulse-echo technique. This technique usually employs a piezoelectric crystal which acts as both an ultrasonic generator and receiver. This is accomplished by first energizing the crystal with a very-high-voltage, short-duration pulse of direct current. The current causes the crystal to change dimensions rapidly and continue to vibrate at its own natural frequency until the energy is dissipated. The vibrating crystal will impart a series of high and low pressure vibrations to whatever medium it contacts. These regions of pressure variations constitute a sound wave of a frequency equal to the number of pressure variations per second.

Between the pulses of energy transmitted, the crystal vibrations dampen sufficiently to "listen" for reflected energy (echoes). When a reflected pulse reaches the crystal, its energy is imparted to the crystal which in turn converts the mechanical energy to electrical energy of the same frequency. The electrical energy is amplified by means of a receiver circuit tuned to the natural frequency of the crystal. Amplified electrical signals, representing sonic pulses, are displayed on a cathode ray tube in terms of an amplitude-time relationship (Fig. 10-1). The pulse-echo technique (one probe) is capable of detecting voids or discontinuities as follows:

1. Voids between adherends in adhesive-bonded laminate-type panels
2. Voids between the adhesive-to-skin interfaces in honeycomb sandwich assemblies

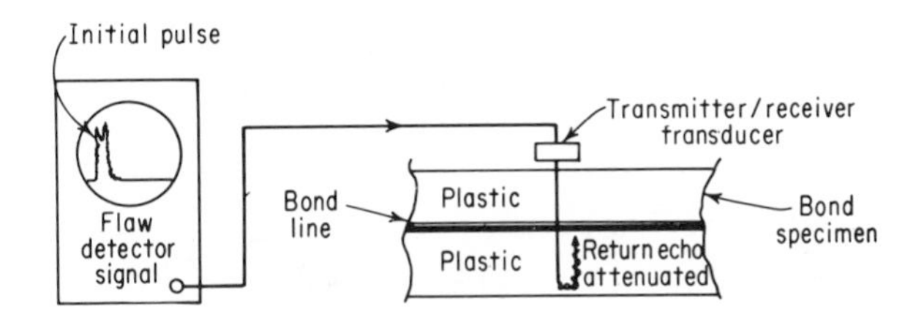

Illustrates bonded condition between highly attenuating materials (plastic) where return echo becomes attenuated before completion of round trip

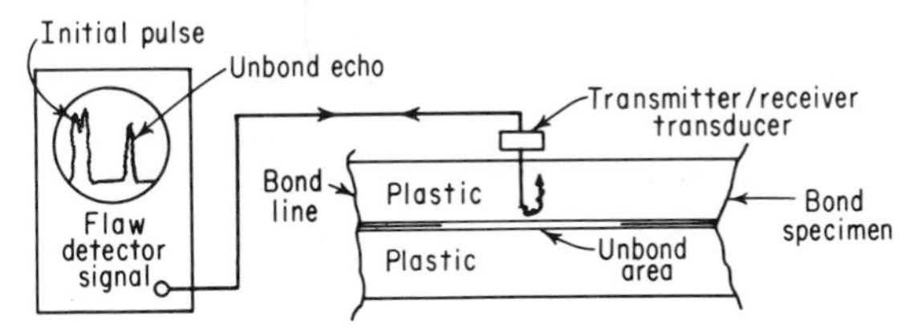

Unbonded condition illustrated by the presence of the reflection from the plastic/void interface

fig. 10-1 *The pulse-echo technique. (AVCO Corporation, Lowell, Mass.)*

3. Voids between adhesive-to-core interfaces in honeycomb sandwich assemblies
4. Excessive porosity in bonded joints in any type construction that is accessible
5. Delaminations in various glass-type laminates

This type of pulse-echo (one transducer acting as transmitter/receiver) testing is sometimes performed by evaluating energy reflection from discontinuities from the back surface of the part in question or from a reflector plate placed in back of the part. Multiple reflections from the unbond interface are frequently evaluated when the face sheets or laminates are relatively thin (metal or plastic). The length of the ultrasonic pattern along the base line is evaluated as well as the signal amplitude. One of the primary problem areas of this type of pulse-echo method is realized in testing plastics due to the signal loss experienced in many high-attenuating plastic adherends. The one-crystal pulse-echo system may be automated but the manual method is more common.

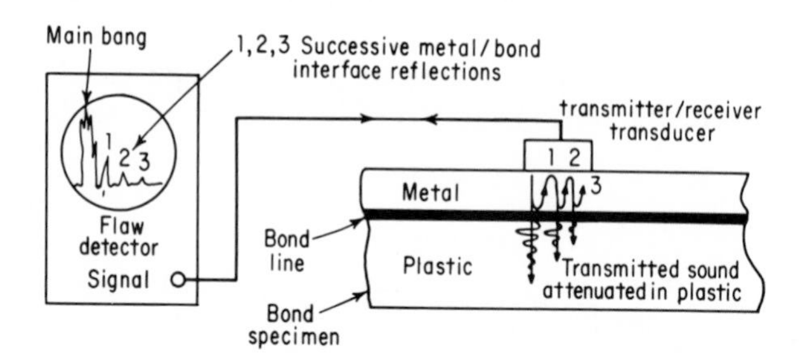

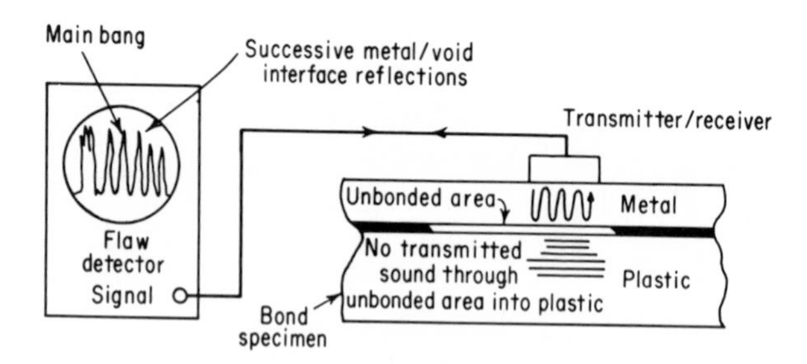

fig. 10-2 *Pulse-echo ringing technique. (AVCO Corporation, Lowell, Mass.)*

The pulse-echo ringing technique (Fig. 10-2) is a common method used for metal adherends which are not highly attentuating to the ultrasonic signal. Voids in the bond line which are near or at the adhesive interface can be readily detected. This method is predicated on the absorption or loss of energy into the adhesive material, or conversely, a buildup of reverberation and possible resonance over the unbonded areas. This technique differs from most single transducer techniques because resolution is of little importance and long pulse lengths are used to induce standing waves and interference patterns. Tests may be performed by either contact or immersion methods. To produce standards the transducer is placed over a known unbonded area of a standard test specimen and the sensitivity and pulse lengths adjusted until the desired pattern appears on the scope. The sweep length is then adjusted until the secondary pattern of relatively lower amplitude appears to the right of the major pattern. The transducer is then placed on a well-bonded area and a vast reduction of amplitude of the overall pattern will be observed. The use of reject control or a method of introducing nonlinearity into the receiver amplifier sometimes greatly emphasizes this effect; pulse tuning is then adjusted for the best pattern. This pattern is

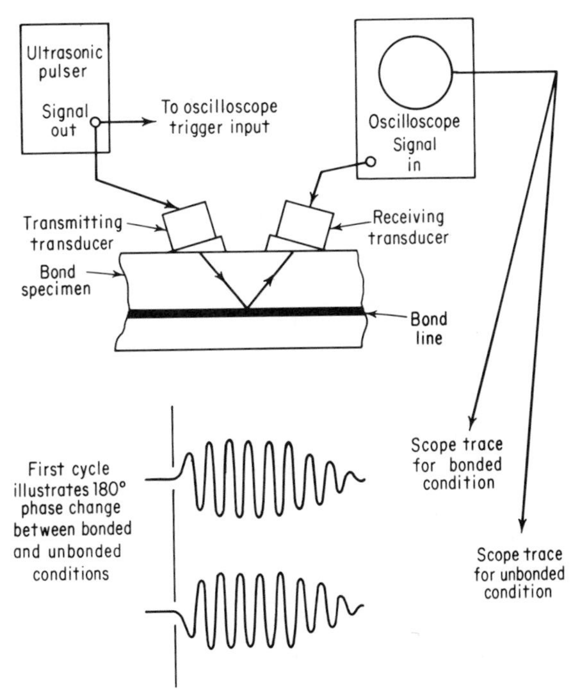

fig. 10-3 *The phase inversion technique. (AVCO Corporation, Lowell, Mass.)*

generally considered to be the most easily interpreted visually. For automatic recordings, the same general pattern is obtained; however, the gate may be placed at any point to monitor the amplitude of a pulse according to its sensitivity to the bond/unbond condition.

The lower frequencies are generally more effective applicable to the thicker face sheets or more absorptive adhesive materials (1.0 to 5.0 Mc). In some test cases it may be desirable to purposely choose a resonant or near-resonant frequency in order to obtain sufficient information/amplitude. It may be desirable to experiment with several frequencies coupled with optimization of the instrument controls to produce the most drastic variation in signal pattern for a change from fully bonded to completely unbonded conditions. In many instances the transducer may be driven at frequencies other than its natural frequency, with improved results.

The phase inversion technique may be employed successfully when the difference in acoustic impedance between the adherends is widespread and the adherend with lower impedance is accessible. The phase of the signal reflected from a bond to an unbond is 180° (Fig. 10-3). Obviously, this technique has limited application, but current research utilizing this method may produce a system with more widespread capabilities.

The through-transmission technique of inspecting adhesive bonds (Fig. 10-4) is a popular approach and considered most applicable to

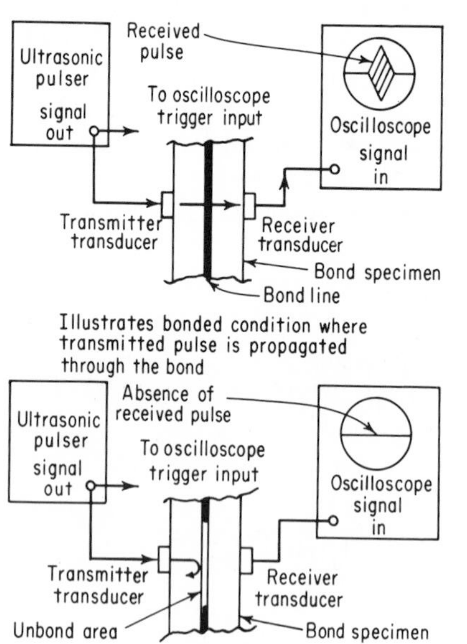

fig. 10-4 *The through-transmission technique.*

materials of high attenuations. Discontinuities or unbonds approaching the size of the probe are easily detected and possibly areas of relative high porosity. This method is considered less complex as compared to the majority of pulse-echo methods because only one signal must be evaluated. Transmitted energy is not affected as much as reflected energy by variables such as part alignment, surface irregularities, detail geometry, and other inherent discontinuities. Penetration of high-attenuating materials is two to three times as great with through-transmission as by pulse echo. However, through-transmission inspection is usually limited to applications (see pitch-and-catch method) where one has physical access to both sides of the part, and the water immersion may at times be undesirable or impractical for use with adhesive-bonded details.

There are three general methods for signal read-out as follows:

1. A-scan, in which the amplitude of the signal or displaced spot on the cathode ray tube would indicate the cross-section reflecting the size of the unbonded area. The distance from the adherend face signal to the discrepant signal as compared to the distance from the adherend face signal to the material back reflections, approximates the debond location.
2. B-scan, which presents a cross-sectional view of the unbond in relation to the top and bottom face signals.
3. C-scan, which presents a paper recording of the part with a visual view of the unbonded area. This is the most popular because the recording can be retained as a permanent record of the assembly.

The through-transmission reflector-plate technique is similar to the technique just described except that a single transducer is used and the sound energy passes through the water path, the part under test, reflects from a reflector plate, and returns to the transducer. The advantages of this method are: (1) on materials of relatively low attenuation, losses are multiplied by a factor of two or more; consequently greater signal changes are available for recording; (2) on thin sections, gating on the reflector surface renders a better recording.

The reflector plate must be flat and the part under test must be maintained at a constant distance from the plate during test, thus reducing loss differentials due to a varying water path, angles of incidence other than 90°, and scatter due to surface imperfections of the substrates.

The through-transmission "pitch-and-catch" method utilizes two transducers that operate from the same side of the adherend. This method is useful where pulse-echo single-transducer techniques lack resolution or sufficient power for the reflector plate method or if through-transmission (opposed crystals) is not feasible due to part geometry. The two transducers are placed in close proximity to each other and at a slight angle. This angle is best determined by results obtained from a standard test specimen. It may be necessary to place a baffle plate between the transducers to eliminate direct surface reflection and "cross talk." The signals obtained are monitored as other through-transmission signals.

The latest technique in through-transmission coupling to eliminate the water immersion utilizes the "wheel," having fluid-filled rubber wheels with crystals mounted inside the wheel (Fig. 10-5).

FOKKER BOND TESTER

Operational theory

The Fokker Bond Tester is an ultrasonic resonant frequency device which measures the effect of variation in loading upon the resonant frequency and amplitude of the transducer. The resonant frequency and amplitude readings of an unbonded sheet are determined first, after which the transducer is coupled to a bonded joint, and thus, an indirect measure of the strength of the bonded joint is obtained. Use of the available instruments by numerous investigators (references 1 through 11) has shown the Fokker Bond Tester (Fig. 10-6) to give consistently the most accurate quantitative predictions of bond strength. These studies show that void areas could be determined accurately and that weak bonds resulting from pressure deviations during cure can be detected.

The transducer of any ultrasonic device is of prime importance because it is the contact medium between the part and some type of read-

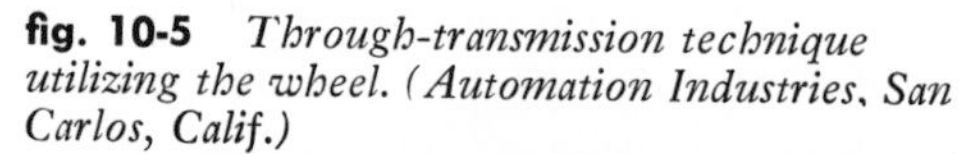

fig. 10-5 *Through-transmission technique utilizing the wheel. (Automation Industries, San Carlos, Calif.)*

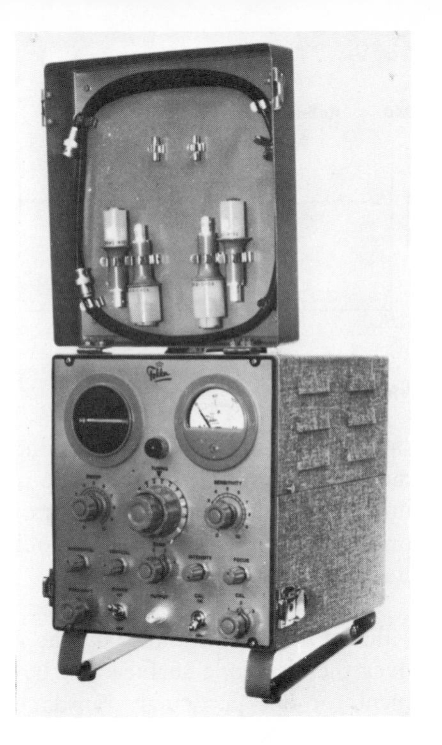

fig. 10-6 *The Fokker bond tester.*

out device. The transducer for the Fokker Bond Tester (Fig. 10-7) is manufactured from barium titanite, and has a cylindrical shape. It is baked in an oven at high temperatures and then polarized in a powerful magnetic field in the preferred manner. The advantages of barium titanite over the natural piezoelectric crystals (Rochelle salt, tourmaline, quartz, etc.) are (1) low sensitivity to humidity and small temperature variations; (2) stable piezoelectric output up to approximately 160°F; and (3) the profile and position of the electrodes may be determined before manufacture; thus the vibration made can be controlled.

After polarizing, attachment grooves are cut into the transducer, and

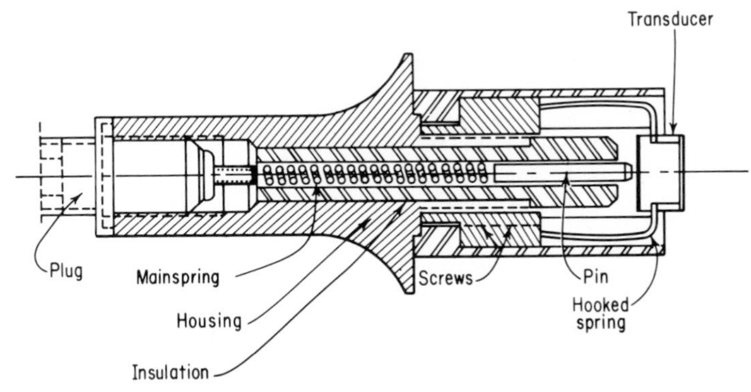

fig. 10-7 *Fokker transducer.*

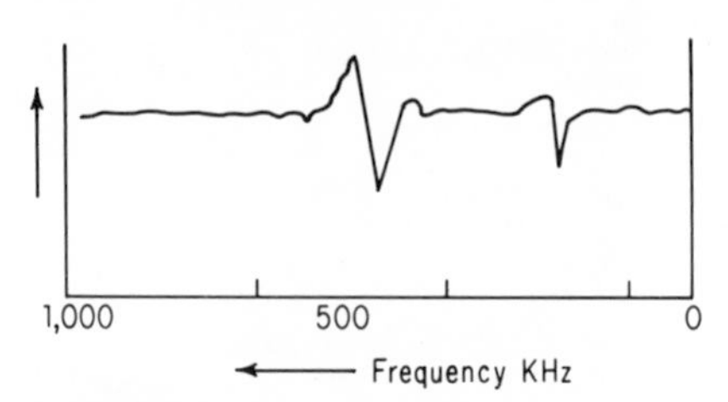

fig. 10-8 *Resonance spectrum of a transducer.*

since barium titanite is not conductive, the upper and lower surfaces are nickel or chromium plated. The platings also serve as a protective measure for the contact area of the transducer. If this plating becomes cracked or worn in any manner that would impair effectiveness, readings would be erroneous.

When a frequency of up to approximately 10,000 vibrations per sec is induced to the transducer by the means of a signal generator, the voltage over the electrode surface can be measured. The result of this measurement for one particular transducer is represented by Fig. 10-8. At the position *a* and *b* on the graph, a sudden change in voltage occurs across the transducer. This is because the transducer joins the vibration in the frequency supplied to it. This vibration is called the natural resonant frequency of the transducer and is determined by the mass of the crystal.

In Fig. 10-8, the frequency of the peak *a* is determined by the diameter of the transducer. The vibration mode of the transducer will generally be in a radial direction with this frequency. The frequency of peak *b* is governed by the transducer thickness and provides a vibration mode in an axial direction (Fig. 10-9). The amplitude of the required vibration is then determined by the diameter-thickness ratio of the transducer. The large transducers have low resonant frequency and the small transducers are usually very high.

The various transducers employed by the bond tester are coded as follows: The first and second figures indicate the diameter, and the third

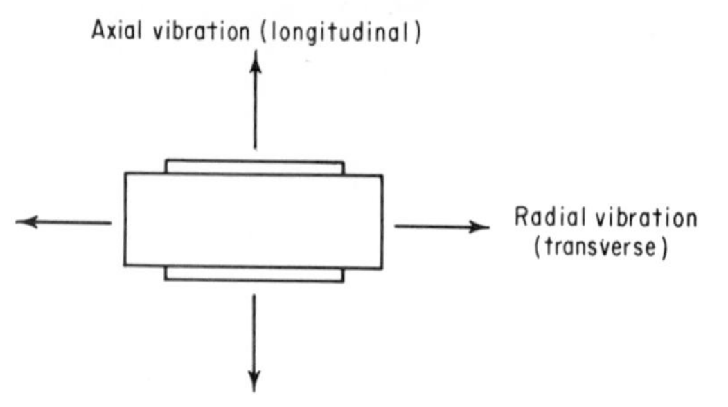

fig. 10-9 *Transducer vibration modes.*

and fourth figures denote the thickness (3814 is a ⅜-in. diameter, ¼-in. thick crystal).

The transducer is suspended to allow it to vibrate freely; the suspension clamps are also the grounding mode. The positive supply of alternating current is supplied to the transducer via a connector plug and a spring-loaded silver-plated pin that rests snugly against the transducer (see Fig. 10-7). The entire probe assembly is covered with a Teflon case to prevent erratic indications.

The transducer resonance method is the effect of variation in load on the natural resonant frequency and impedance of the transducer, which is measured and reflected on a cathode ray tube and a microammeter. The load is imposed by coupling a certain mass to the crystal. If the mass is large, then the resonant frequency and the impedance will change. This system operates on low frequencies and is applied primarily to bonded structures. The prime disadvantage is the thickness limitations pertinent to penetration. When testing aluminum, testing becomes more difficult when ½-in. total joint thickness is exceeded.

The influences acting on the transducer must be made visible, and the correct frequency must be applied to the crystal. This is accomplished as shown in Fig. 10-10. Note the following functions in the diagram: (*a*) the power supply, (*b*) the measurement circuit, (*c*) the calibration circuit, and (*d*) the visual units.

The power supply is based on a 110- to 120-volt, 50- to 60-cycle source. The main supply, which is electronically stabilized in *a*, provides a constant voltage within a range of ± 10 percent to clipper *b* with a sinusoidal ac voltage, and 50 to 60 cycles per sec; the output of this unit is a square-wave voltage which is used for blanking the sweep-back of the cathode ray tube (*v*).

The square-wave voltage is also fed into the integrator circuit *c*. This unit delivers a triangular voltage to the modulator *D* and the horizontal deflection amplifier *H-H*. The input to the modulator can be controlled by a continuous sweep attenuator.

The triangular voltage and the adjustable dc voltage (frequency shift) are mixed in an amplifier, and a linear output current is fed into the control coil of the oscillator inductance. With the adjustable capacitor (tuning), this inductance forms the resonance circuit of the oscillator *E*. The circuit is designed in a manner to allow the instantaneous value of the frequency shift to follow a linear flow with the horizontal deflection.

The oscillator output is supplied to the probe and mixer *K* in the calibration circuit, and the output from the probe passes through detector *M*. The output of the calibration circuit can also be supplied to the detector via the switch marked "calibration."

When passing the detector, the signal is fed into the vertical deflection

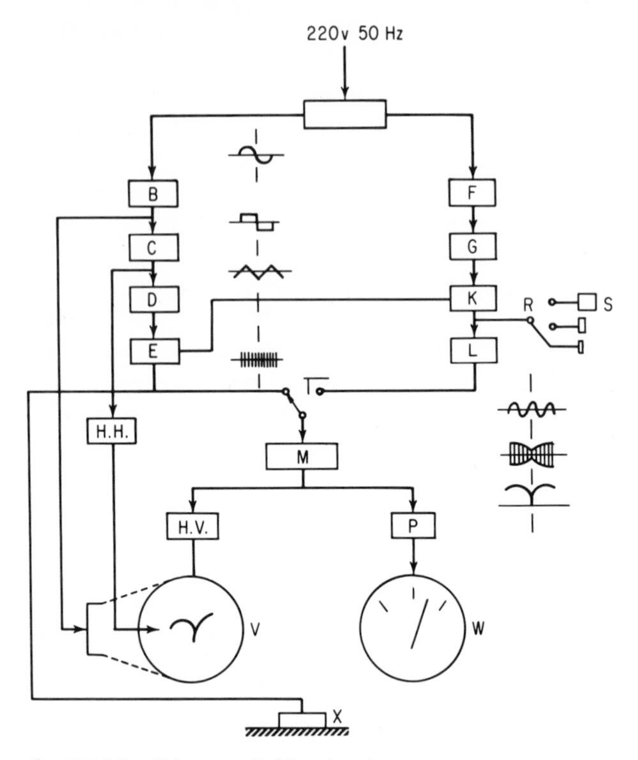

fig. 10-10 *Diagram, Fokker bond tester.*

A, power supply; *B*, limitor; *C*, integrator; *D*, modulator; *E*, oscillator; *F*, calibration oscillator; *HH*, horizontal deflection amplifier; *HV*, vertical deflection amplifier; *K*, mixer; *L*, biased amplifier; *M*, detector; *P*, vacuum tube voltmeter; *R*, calibration range switch; *S*, calibration crystals; *T*, calibration on/off switch; *W*, cathode ray tube (A-scale); *X*, probe.

amplifier (*HV*) and into the vacuum tube voltmeter *P*. This meter indicates the amount of induced damping of the crystal vibration on the *B* scale. The shift of the resonant frequency is represented on the cathode ray tube (*A* scale).

The calibration circuit is used to determine the amount of probe resonance peak shift. This circuit is comprised of a 10,000-vibrations-per-sec reference generator (*F*). The signal of this amplifier is fed into a nonlinear amplifier *G*. The output signal (including the frequency of the 10,000 vibrations per sec up to the sixth harmonic) is mixed with the output of oscillator (*E*). The output of the mixer (*K*) passes one of the three calibration crystals (*S*) and is fed into a biased amplifier (*L*) which passes the input signal only when it exceeds a certain level. The calibra-

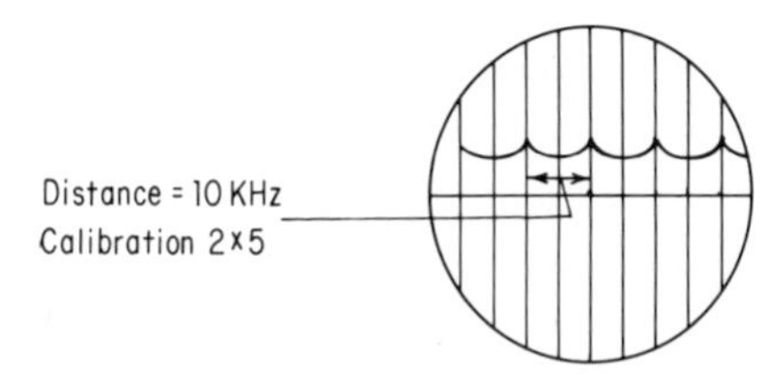

fig. 10-11 *Calibration distance.*

tion crystal acts as a sensitive parallel resonance filter. The output of the biased amplifier (L) passes through switch (T) to detector (M).

In the position "calibration on," a number of calibration peaks appear on the cathode ray tube. The distance between two peaks amounts to 10,000 vibrations per sec (Fig. 10-11).

Knob F operates the condenser and thus the fine adjustment of the center frequency in conjunction with the bond selector (J). Figure 10-12 represents the frequency which coincides with the position of these knobs. For each adjustment, the related resonance pattern is given. A small portion of this resonance pattern is shown on screen A with calibration switch (P) in the on position. The visible part of this image may be adjusted with the sweep G which controls the frequency sweep of the carrier.

The voltage difference across the transducer (equals amplitude height of the image) may be amplified by adjusting the sensitivity (E). The height of the amplitude coincides with the indication on the ammeter (B).

Image adjustment

The image on the cathode ray rube (A) may be adjusted with the horizontal (H), vertical (I), intensity (K), and focus (L).

The horizontal knob (H) is utilized to adjust the image in the horizontal direction. The position of the image may be adjusted in the vertical direction with knob (I). The intensity (K) is used to adjust the clarity of the image, and focus (L) is used to produce a sharp image.

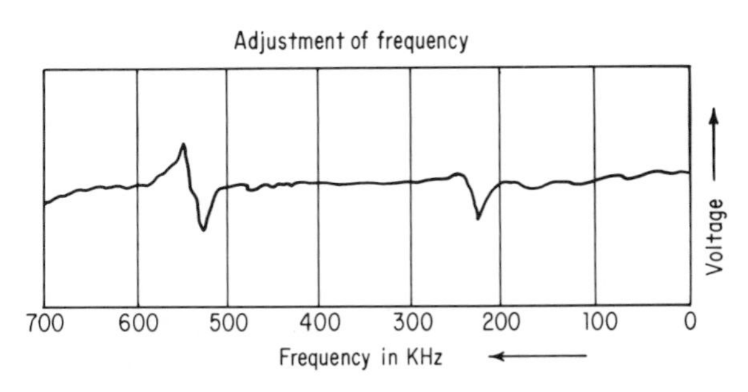

fig. 10-12 *Adjustment of frequency.*

Uses and limitations

The Fokker Bond Tester measures the cohesive properties of a bonded joint. It will only measure adhesion as related to cohesion. It will not detect discrepancies due to improper surface preparation of adherends or the presence of foreign matter such as grease or release agents at the interface of a bonded joint. These discrepancies must be controlled by close adherence to process controls.

This instrument is capable of detecting discrepancies in the adhesive layer of a bonded assembly which is caused by: (1) reduced strength caused by thick or porous bond lines, which are normally caused by pressure deviations due to poor tooling, laxation of tooling procedures, or dimensional mismatch of the adherends; (2) reduced strength caused by exposure of the adhesive to weathering provided the degradation occurs in the cohesive layer; (3) complete discontinuity or voids between the adherends or adhesive layer; (4) an undercured joint provided there is a distinct change in elastic modulus or density from the undercured to the fully cured state.

Operating principle

The cohesive layer of an adhesive is closely related to its elastic properties. These elastic properties are, in turn, related to the ultrasonic (acoustical) properties of the adhesive layer. The instrument makes a comparison between the ultrasonic properties of an unbonded face sheet and a bonded joint. This is accomplished by first placing the transducer on an unbonded face sheet which would be considered to have the cohesive strength of zero. The transducer is then coupled to a bonded joint, thus inducing loads into the adhesive layer. The change in loading from the unbonded sheet to the bonded joint causes a resonant frequency shift and a change of amplitude which is related to the dynamic mechanical properties of the joint, and thus is an indirect measure of the strength of the joint.

Since the Fokker Bond Tester measures primarily the cohesive strength of the adhesive layer, two basic problems are evident.

1. The adhesive properties of the adhesive layer at the interface are not measured but can cause premature failure during the destructive test. Therefore, it is of great importance that the adhesion of the adhesive layer is greater than the cohesive strength. With adequate surface preparation, a majority of adhesives possess this property.

2. The yield strength of the adherend is not measured by the bond tester, but influences the destructive test results of a lap shear tensile joint (honeycomb not applicable). Therefore, specimens with constant yield strength and thickness of adherends should be used to prepare a

correlation curve. The quality curve should be used to correlate specimens of varying overlap strength. (Quality curves are the result of T over L curves defined in Chap. 4.)

Correlation curve

A correlation curve should be established for a particular adherend and adhesive system. This is accomplished by preparing test specimens simulating as nearly as possible the joint to be tested. The test specimens are of varying strength. They are first tested nondestructively and then destructively. Fokker bond-quality units are then plotted versus actual test values and the correlation curve is developed (Fig. 10-13). The test specimens may be prepared by placing shims or spacers between the adherends to create pressure deviations on the joint, at the same time keeping platen or vacuum pressures and heat constant. Lap shear tensile specimens are prepared for metal-to-metal parts (MM-A-132) and flatwise tensile specimens for sandwich-type construction. Other techniques may be used to produce various bond-strength levels, but care should be exercised to induce only one variable to produce a correlation curve. Teflon should never be used to produce complete voids in honeycomb panels for two basic reasons: (1) this type of discrepancy never occurs in production and (2) sufficient adherence may occur during the bond cycle to give good joint indications. If complete voids are desired, Mylar or metallic shims are recommended. The thickness of shims

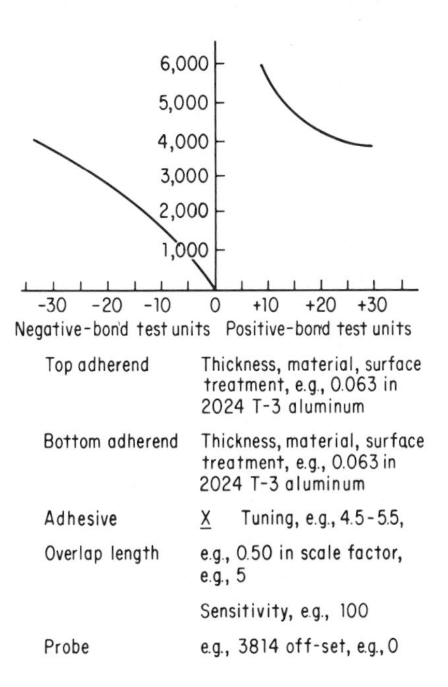

fig. 10-13 *Correlation curve for metal-to-metal joints.*

depends on the quality bond desired and the reaction of a particular adhesive system to pressure deviations.

The correlation curve attempts to represent the problems foreseen in most production operations. It has already been stated that the prime processes which affect adhesion, i.e., surface preparation and heat, can be controlled by careful processing and rigid controls, although it is recognized that pressure is a factor in adhesion, especially the structural film tapes. The cohesive properties are affected by heat, but are usually more dependent on pressure.

The area of importance then becomes the substandard joints caused by pressure deviations which are related to poor or faulty tooling, loss of pressure due to bag leaks, or dimensional mismatch. Poor joints of this type can be produced via the shim method.

The ideal system would be one with better adhesive than cohesive properties. The epoxy-phenolics possess this property. They will normally fail from the maximum expected (approximately 3200 psi in shear) to 100 psi in shear in a cohesive manner. The nitrile-phenolics usually have better cohesive properties than adhesive. Starting at the maximum expected (5500 psi) they normally fail cohesively to approximately 4000 psi and then revert to an adhesive failure to zero psi. The epoxy-nylons usually fail cohesively from the maximum (7000 psi) to approximately 3500 psi where they begin an adhesive mode of failure, but at a low level (approximately 1000 psi) they may revert back to a cohesive failure. If adhesion is ruled completely out of the Fokker testing theory, then it would be impossible to obtain a curve from the latter two systems. Instead of saying the cohesive properties are measured, it would be more correct to state, "Cohesive properties are measured as related to adhesion." This relationship is more evident in adhesive systems that are highly dependent on pressures.

In preparing correlation curves, it is evident with a particular system that the bond-line thickness, ultrasonic response, and mechanical test values correlate. This leads one to believe that bond-line thickness is being measured, which is false in many instances because many adhesives are not dependent on bond-line thickness for strength even though the thin bond line is preferable due to better resistance to severe environmental conditions and fatigue.

After experience has been gained in preparing correlation curves, the procedure is less time-consuming. A thorough knowledge of the adhesive system to be utilized is an important factor, consequently the number of specimens necessary to produce a curve depends on the experience of the operator. The confidence level may be determined by a statistical analysis, but correlation curves are unnecessary if the assembly in question is being tested only for a complete void.

Readings normally should be taken from the oscilloscope (A scale) for metal-to-metal and from the microammeter (B scale) for sandwich structures. When developing curves, readings should be retained from both A and B scales until it can be determined which readings are most reliable. In testing composite structures the A scale may indicate a void, or substandard area, but the B scale may be used to determine the layer in which the discrepancy exists.

If large-diameter probes are to be used for metal-to-metal joints, lap shear tensile specimens large enough to accommodate the probe must be used. Fokker readings should be recorded and then the specimen sawed to 0.5-in. overlap before destruct testing (saw one side only). If quality curves (*T* and *L*) exist for this particular substrate and adhesive, corrected computations may be obtained, eliminating the sawing operation.

Calibration

The Fokker Bond Tester should be calibrated by the operator as specified below each time the instrument is used, and frequent calibration checks should be made during testing procedure.

Metal-to-Metal Calibration

PRIMARY CALIBRATION

a. With power on and calibration off, allow the instrument to warm up for a minimum of 45 min prior to use. Insufficient warm-up time will cause drift (Fig. 10-14 for location of these and all other controls).

b. Set the sweep, tuning, and sensitivity controls to zero. The positions of other controls are not important at this time. The microammeter indicator must drop to zero if the instrument is working properly.

c. Using the horizontal and vertical controls, create a horizontal linear image approximately centered and extending the minimum possible length across the oscilloscope (A scale). Both ends of the horizontal image should be within the face of the scope (A scale). Adjust intensity and focus as required to obtain a sharp image or wave form.

d. Temporarily center the sweep and sensitivity controls to approximately the number 5 position.

e. Connect the proper probe for the joint to be tested to the coaxial cable and plug into the Fokker Bond Tester or into applicable attachment if limit indicator is being utilized.

f. Set the band and the calibration controls for the particular probe as specified in Table 10.1.

g. Approximately center the frequency control within its travel and adjust the tuning control within the range shown in Table 10.1.

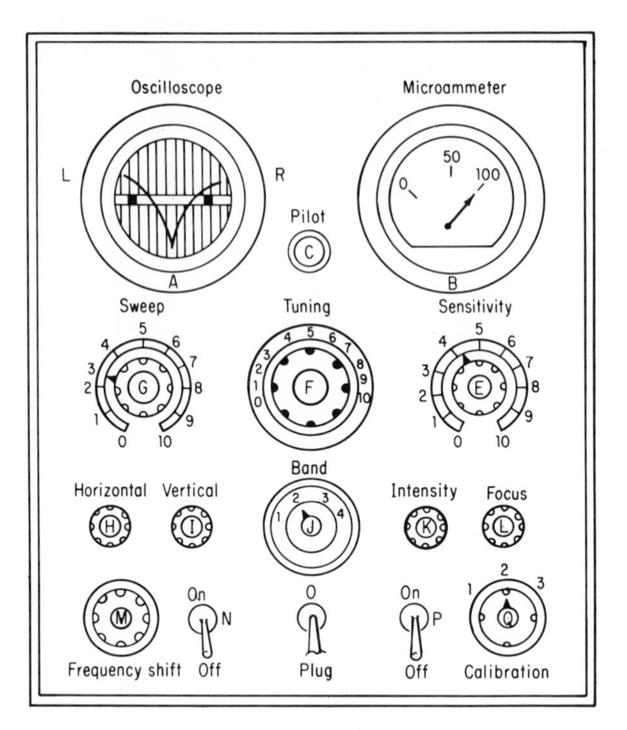

fig. 10-14 *Fokker control panel.*

A, oscilloscope; *B*, microammeter; *C*, light; *E*, sensitivity control; *F*, tuning control; *G*, sweep control; *H*, horizontal control; *I*, vertical control; *J*, band selector; *K*, intensity control; *L*, focus control; *M*, frequency shift control; *N*, power switch; *O*, plug; *P*, calibration switch; *Q*, calibration selector.

h. Adjust the fine tuning knob until a wave form with one or more sharp "dips" appears on the oscilloscope (A scale). If no dip appears, check transducer for adequate contact. The instrument is not operating properly if the dip cannot be obtained within the tuning range specified in Table 10.1. This may sometimes be accomplished by moving the frequency-shift control. If not, do not use instrument. The image may be adjusted vertically by using the vertical control.

i. Move the calibration switch to the "on" position, thus creating a ripple of peaks across the oscilloscope (A scale) as shown in Fig. 10-15. The wavelength between the ripple peaks is ten units, and may be adjusted horizontally with the sweep control to any desired number of A-scale dimensions. Set the distance between peaks to the proper scale factor (Fokker instrument units per oscilloscope scale dimension) as specified in Table 10.1.

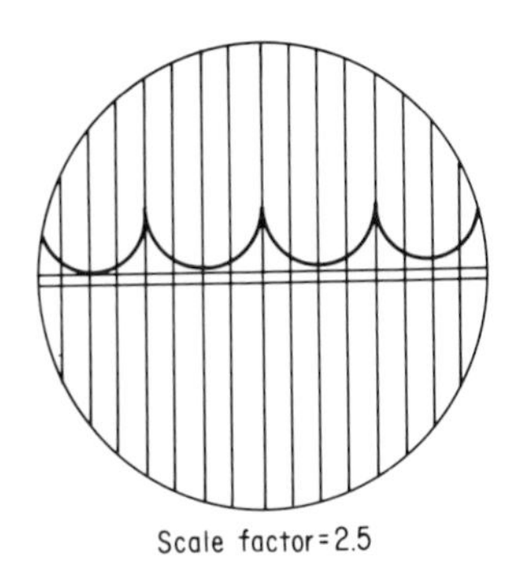

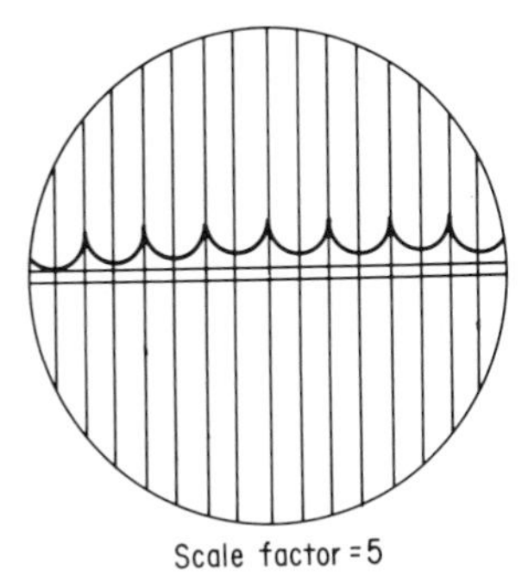

fig. 10-15 *Calibration ripple peaks.*

j. Adjust the scale factor by using the frequency-shift control to adjust the linear distance between peaks to the desired value. It may be necessary to alternately adjust the frequency shift and the sweep control several times to obtain a stable wave form. Vertical adjustments may be made with the vertical control. This scale-factor calibration must be extremely precise to enable readings to correlate with the correlation curves. After this adjustment, turn the calibration switch off and place probe on calibration specimen.

SECONDARY CALIBRATION

a. The probe shall be calibrated to the particular bond using the proper calibration specimen as specified in the detail requirements. The recommended specimen configuration should be used whenever possible because it eliminates secondary ultrasonic resonance. If this is impossible, the probes may be calibrated to an unbonded portion or similar face sheet of the joint to be tested. This calibration shall be performed as specified below each time the top face adherend to be tested is changed or the calibration of the instrument has changed.

b. Apply a small amount of mineral oil, cellulose gel, or other contact medium to the surface to which the probe is attached.

c. With probe on calibration specimen, use the frequency-shift control to adjust the largest dip on the A scale to the center or zero (Fig. 10-16) unless a different offset is desired. Do not readjust the tuning unless it is absolutely necessary, in which case the instrument must be rechecked for scale factor.

d. With the probe on the calibration specimen, set B scale (using *E*) to read 100 plus or minus 1. The largest dip on scale A must remain centered. If necessary, the vertical or horizontal location of the dip may be adjusted using only the vertical or the frequency-shift controls. The instrument is now calibrated for use.

Calibration for Honeycomb and Sandwich Structures

Same as metal-to-metal with the exception of the 1510 probe which will be specified below. The honeycomb probes shall be calibrated to the top face sheets or fascimile of the sandwich to be tested. Readings are

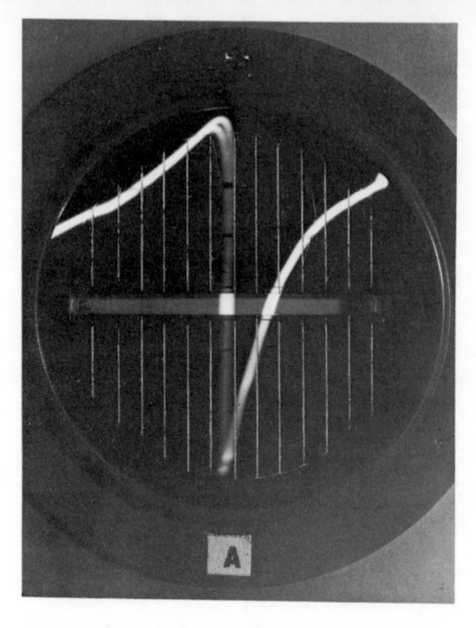

fig. 10-16 *Center calibration of image. (Hughes Aircraft Company.)*

taken from the B scale (mA) and must be taken from both sides of the panel with the exception of the 1510 probe. (A sample curve for a honeycomb sandwich structure is shown in Fig. 10-17).

Operation of the 1510 Probe

a. Adjust to permanent settings in Table 10.1.
b. Position the probe on a single sheet or sample specimen as the top face sheet of the part to be tested.

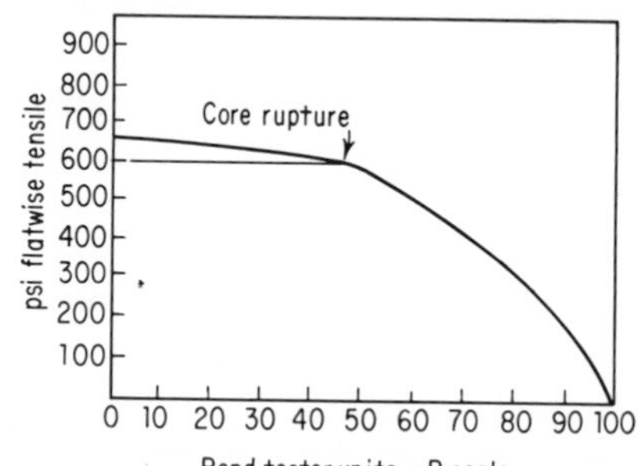

Top adherend	0.016 2024 T-3 aluminum
Bottom adherend	0.020 2024 T-3 aluminum
Adhesive	X
Core	X
Probe	3414
Scale factor	2.5
Off-set	0
Sensitivity	100

fig. 10-17 *Honeycomb sandwich correlation curve.*

c. Set sensitivity at 0.5 and tuning knob at 1.5. Shift dip to center using frequency-shift knob.

d. Adjust sensitivity control to indicate 100 on scale B. Transfer probe to part and test.

The 1510 probe is the most versatile of the probes utilized by the tester. Two or more bond lines may be tested from one side only with this probe; therefore, it is usually preferred for composite structures. The low frequency of this probe also adds to its merits.

The calibration procedures given here are considered standards. If these standards do not produce desirable results, settings may be designed to fit a specific application. With the variety of probes available, several hundred combinations could be attempted and many times one proves profitable.

Testing Metal-to-Metal, Metal-to-Plastic, Etc.

Place the probe in contact with the assembly to be tested, coupling the transducer with a suitable contact medium such as glycerin, mineral oil, cellulose gum, 3-in-1 oil, etc. (Cellulose gum is recommended where a contamination problem exists due to subsequent bonding operations, or when testing plastic substrates that would absorb oils. The cellulose may be removed with a cheesecloth moistened with water.)

Readings are then taken from A scale as shown in Figs. 10-16, 10-21, 10-22 or B-scale readings as indicated on the microammeter.

Sandwich testing

A standard correlation curve for a honeycomb sandwich structure is shown in Fig. 10-17. Probes utilized for testing sandwich structures are usually loaded in the axial direction (3414, 1214, 1414) and are influenced as shown in Fig. 10-18. The transducer is coupled to the top face sheet in the same manner as metal-to-metal, metal-to-plastic, etc. The peak is then centered on the A scope and the microammeter is set on 100. (It may be advantageous at times to set the mA pointer on 50.) The probe is then placed on a sandwich panel that is typical of maximum expectation. The small mass of the bonded core will not usually cause a resonant shift, but it will dampen the vibrations of the transducer, thus resulting in a decrease in voltage and a drop in amplitude. The amount of damping is

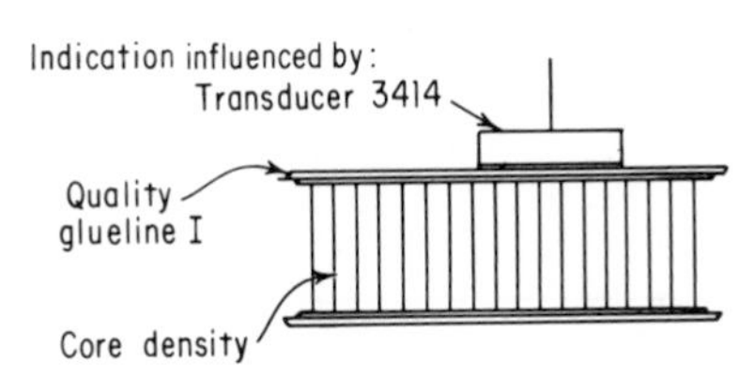

fig. 10-18 *Factors influencing standard honeycomb curves.*

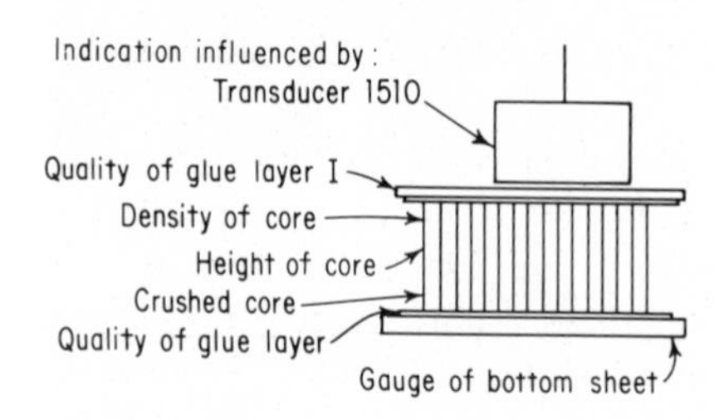

fig. 10-19 *Factors influencing the 1510 probe.*

determined by the quality of the bonded core. When using the standard probes, both sides of the panel must be inspected. When using the 1510 probe (Fig. 10-19), only one side must be tested and the visual indications are influenced by:

a. Quality of the top glue line
b. Density of the core
c. Height of the core
d. Crushed core
e. Quality of the lower glue line
f. Thickness of the bottom sheet

Multiple glue lines are tested with the 1510 probe and curves are produced as shown in Fig. 10-20. Curves of this nature may become quite complex, but may be simplified for production use by limit indicators. The following combinations in quality of glue-line quality can be assumed as related to Fig. 10-20.

a. Glue layer I void. Quality of glue layer II cannot be established.
b. Layer I varying. Layer II void. Indication as for a single-bonded joint.
c. Layer I poor. Layer II poor. The indications for these combinations are difficult to interpret. In certain cases, the indication on the B scale can be of some assistance.
d. Layer I poor. Layer II ideal. The indications are as for a thin top sheet with heavy gauge sheet bonded to it. The shift for quality is indicated by curve *D*.
e. Layer I ideal. Layer II poor. Indication as for single bonded structure (point *E*).

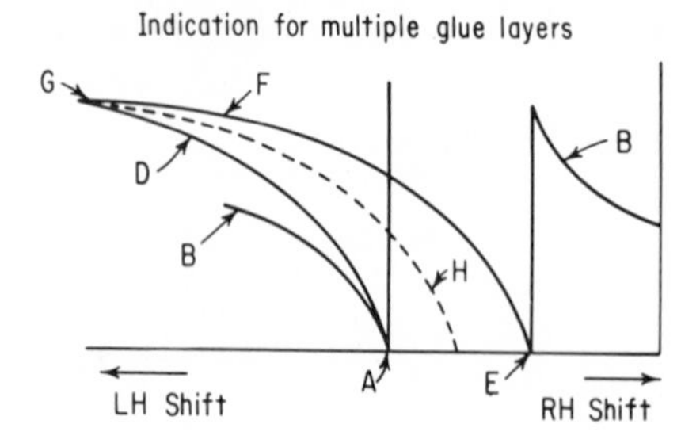

fig. 10-20 *Correlation curve for multiple bond line structures.*

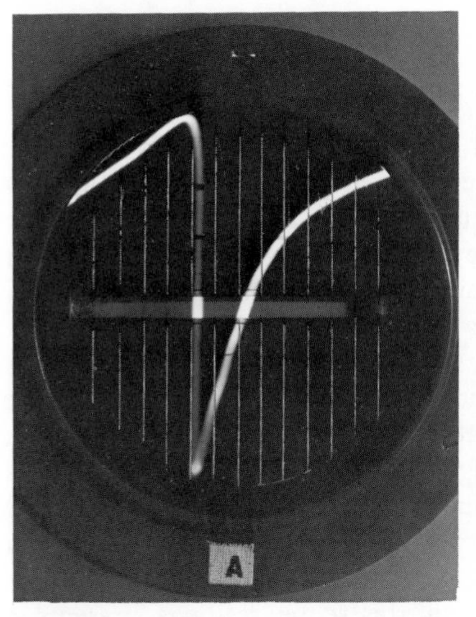

fig. 10-21 *Minus or negative 10 quality units.*

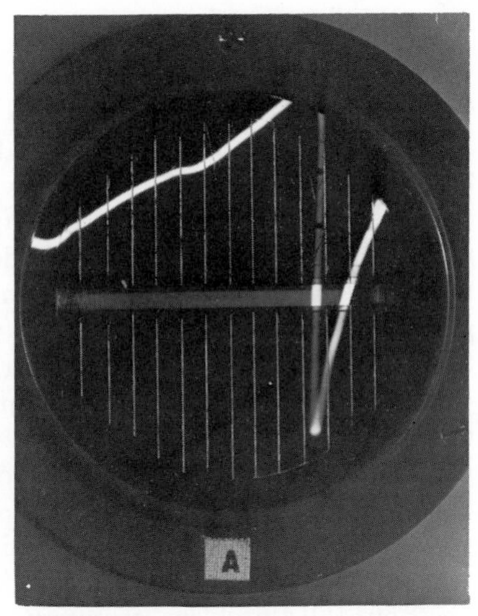

fig. 10-22 *Plus or positive 17 quality units.*

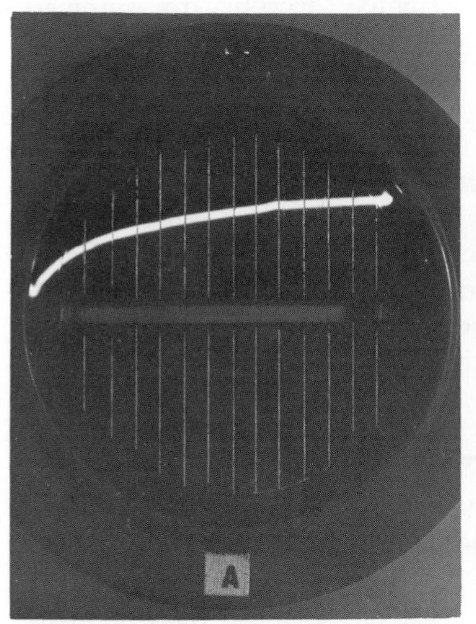

fig. 10-23 *Image in transition.*

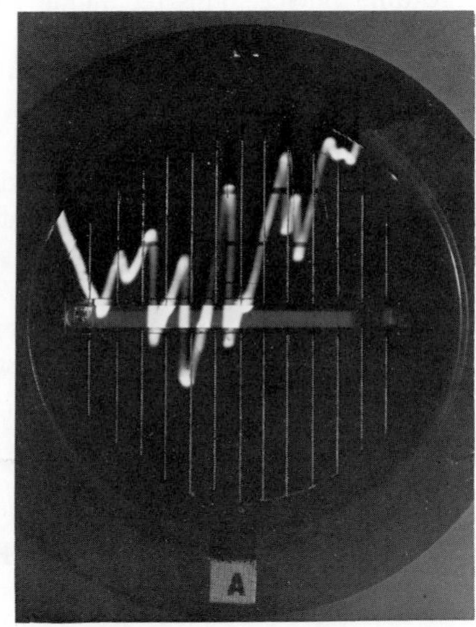

fig. 10-24 *Overloaded probe condition.*

f. Layer I ideal. Layer II varying. The curve *F* from *E* to *G* represents the course of the quality.

g. Layer I ideal. Layer II ideal. Indication as for thin top sheet to which a heavy gauge sheet is bonded with 100 percent quality (point *G*).

A scale interpretation

Problems have been encountered in the A scale (CRT) read-out. Figure 10-21 is a minus or negative 10 quality units. Figure 10-22 is interpreted as plus or positive 17 quality units. Figure 10-23 indicates the peak with the largest amplitude is to the left of the scope or in transition and would usually be considered approximately 50 percent of maximum expectation as applied to a standard curve. This is often confused with an infinite image which would yield mechanical test values of maximum expectations. The infinite image does not present the slope to the left that is evident in Fig. 10-23, and the line is very ragged. Obviously, if these two scope readings are not separated, a correlation is impossible. Figure 10-24 indicates an overloaded probe condition, and when interpreted as minus 10 quality units, may be misleading. When this condition occurs, a larger probe should be utilized for the joint in question. The problem of interpreting the scope readings presents a stronger argument for utilizing the limit indicators which are capable of accurately defining scope images, and transmitting this information to a recording device. This technique is absolutely necessary for automated equipment.

Correlation of peel tests

Studies have produced very little useful data in this area because the strength is revealed only at one point during the peel test. The Bond Tester gives an average indication of the area where the probe is in contact; thus test results show a wide scatter range. The peel test is rather sensitive in the area of adhesion, and since this is not the performance factor of the Bond Tester, the accuracy of the results may be questioned. The peel test is influenced by the stiffness of the top face sheet, which also creates problems.

Detection of bond degradation due to environmental factors

Mild success has been experienced in the detection of reduced bond strength due to environmental exposure. Two prime problems exist that must be considered as follows:

1. Lap shear specimens are usually used for tests, and the degradation first occurs near the edges of the joint where the large percent of the load is carried during a mechanical test. Bond Tester readings are taken near the center of the overlap; thus the readings are not indicative of the edges of the joint where the deterioration is concentrated.

2. When degradation occurs at the interface, which is common with many adhesives, it is difficult to detect until a nearly complete layer of oxidation is present on the adherend interface.

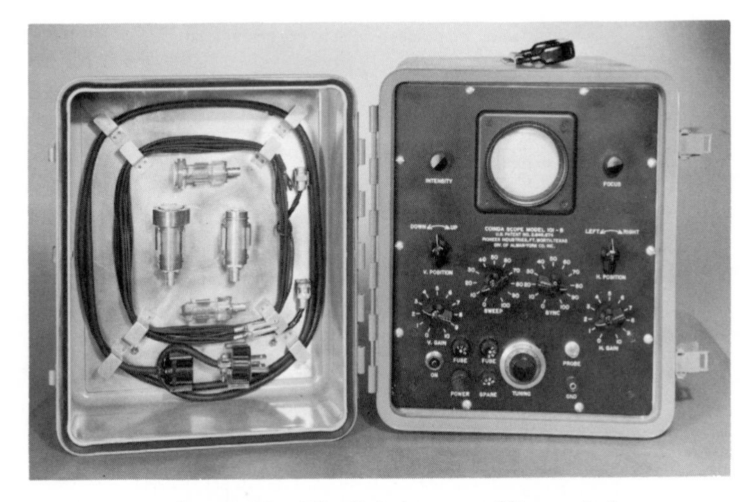

fig. 10-25 *The Coinda-scope. (Pioneer Industries, Fort Worth, Tex.)*

The Coinda-Scope

The Coinda-Scope is a resonant frequency testing device that utilizes a barium titanate crystal transducer (Fig. 10-25). The instrument is portable and requires a liquid contact between the transducer and the part being tested. It is adaptable to metal-to-metal, plastic-to-plastic, or combinations (honeycomb and metal-to-metal). Voids and strength levels are easily detected to some degree, depending on the adhesive system and adherends. As other resonant frequency devices, it is limited by thickness penetration and also requires grounding, especially with plastic substrates.

Voltage and frequency of the transducer are reflected on a cathode ray tube, reduced to a single pattern, and interrupted as shown in Fig. 10–26. This device compares an unbonded standard specimen to a good bond. A standard panel of the same adherends and adhesives as the part in question is fabricated with a known void, and the instrument is tuned to obtain the large V peak (pip) with the highest amplitude with the probe placed on the void area. A sound, well developed joint will reflect a somewhat flat pattern when the transducer is coupled utilizing the same frequency (Fig. 10-26). The Coinda-Scope has been faced with the same problems as other test instruments in the resonant frequency area, i.e., limited research. It is the firm belief by many knowledgeable persons associated with adhesive bonding that if one-tenth of the money exhausted on outlandish and expensive contraptions had been invested in resonant frequency research, nondestructive testing of adhesive bonds would unquestionably be far ahead of today's state of the art. The problem

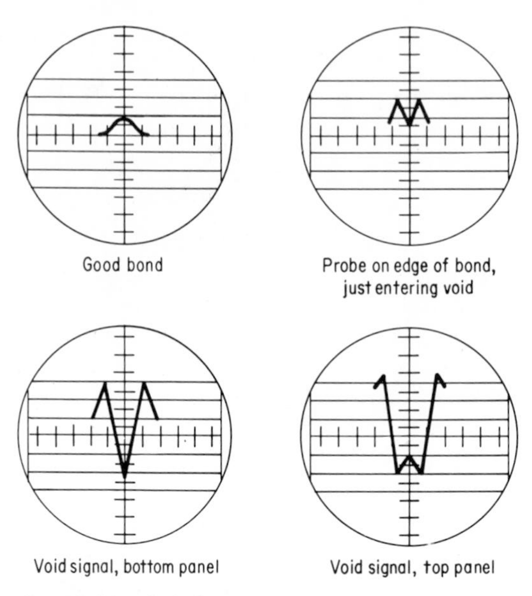

fig. 10-26 *Coinda-scope wave form.*

has been centered in vain attempts to use or modify elaborate existing equipment to adhesives, equipment that was designed before the need for NDT of adhesive bonds was known.

Sonic test system

The sonic test system shown in Fig. 10-27 was developed by North American Aviation. This is a dual module portable instrument for nondestructively testing composite structures. The system features the capa-

fig. 10-27 *The sonic test system. (North American Aviation.)*

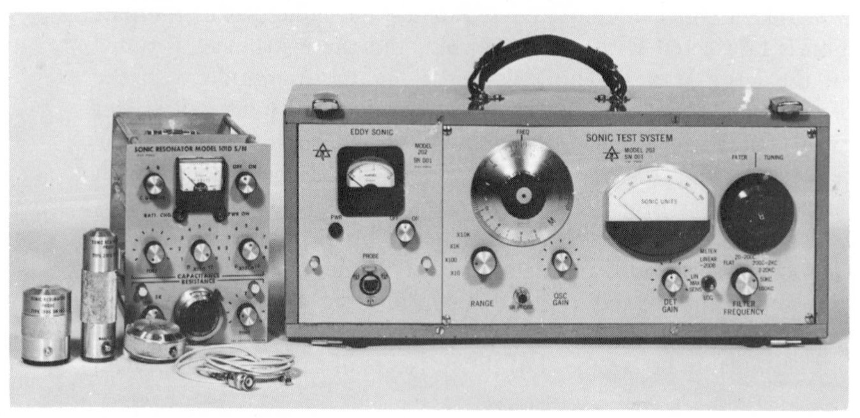

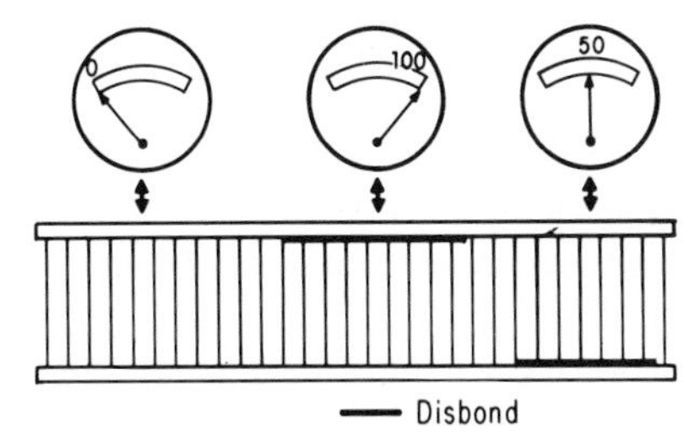

fig. 10-28 *Sonic instrument response for unbonded honeycomb panel. (North American Aviation.)*

bility to detect defects located anywhere throughout the depth of a composite while requiring access to only one surface. This feature is vital when testing in-service or completed aerospace vehicles, such as commercial jet airliners, where access is very frequently limited to only the outer mold line surface of the airframe. Typical defects detectable include face sheet disbonds, core crushing, and core fracturing. A typical response to defects in honeycomb is shown in Fig. 10-28.

The sonic resonator plug-in module for the sonic test system is particularly oriented for testing bonded nonmetallic composites. However, it can also be employed for testing metallic or combination metallic-nonmetallic composites. The sonic resonator operates on the basic principle of an interferometer, where the instrument generates and monitors the characteristics of a standing wave in the material under test. Defects will change the vibrational properties of the test material, and thus change the instrument response. Several sonic resonator probes are shown in Fig. 10-29. To accomplish the test, a light film of liquid must be applied to the surface of the test material. If the liquid film poses a material contamination problem, a protective, strippable plastic coating can be placed (usually sprayed) on the test-material surface prior to couplant film application.

fig. 10-29 *Probes for sonic resonator. (North American Aviation.)*

The eddy-sonic plug-in module for the sonic test system is particularly oriented for testing all-metallic or combination metallic-nonmetallic composite materials. The eddy-sonic test method operates on the principle that acoustical vibrations are inherently associated with the flow of eddy currents. This instrument both induces eddy currents in the metal portions and analyzes the resulting acoustical radiation from the test material. Changes in acoustical radiation, which have been named eddy-sound, are used as a criterion for defect detection. The eddy-sonic test method offers the special advantage of not requiring a liquid couplant. Because of this, testing speeds are greatly increased and the test operations are simplified.

The eddy-sonic method, like the sonic resonator, has the capability of detecting disbonds or core damage while anywhere throughout the depth of a composite material with access to only one surface. Figure 10-30 shows a conventional eddy-sonic probe which contains a meter displaying the instrument read-out. The built-in meter significantly helps the operator to scan the test material more thoroughly, accurately, and with greater speed because he doesn't have to keep looking back and forth between probe and instrument. A typical instrument response is shown in Fig. 10-31. Typical applications for the sonic system are shown in Table 10.2. There are no claims by the developers and users of this instrument that bond strength can be predicted, but the possibility cannot be ruled out because of limited research utilizing this system, plus the fact that current research is constantly improving the capabilities of the equipment.

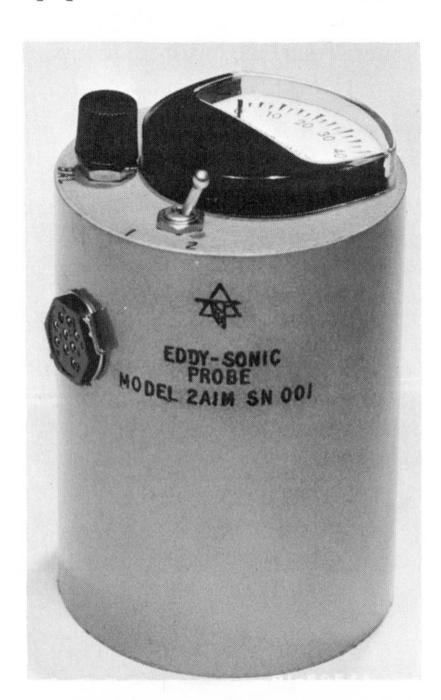

fig. 10-30 *The eddy-sonic probe. (North American Aviation.)*

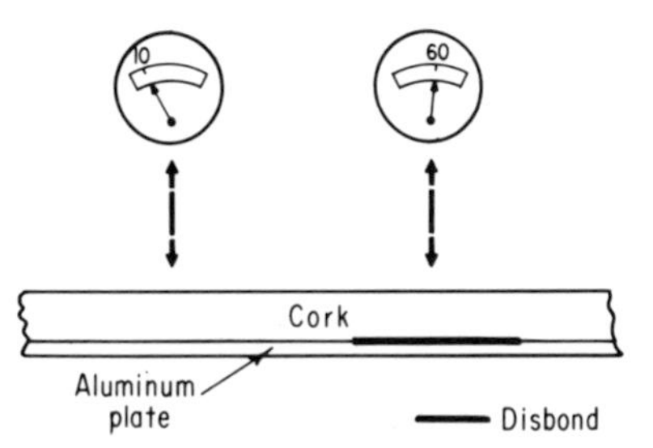

fig. 10-31 *Instrument response of sonic system for aluminum-cork bond. (North American Aviation.)*

Microwave testing

Microwave techniques may possibly hold the answer to many difficult test problems in adhesive bonding that utilize plastic or other nonmetallic adherends. This is an area of great concern since a majority of the present techniques do not perform satisfactorily (including resonant frequency) on many nonmetallic substrates, especially when these substrates are incorporated into composite structures. Test problems related to plastic composite bonded structures which do not respond to any type of nondestructive testing technique previously described in this chapter may possibly be tested by the microwave technique. These methods are being evaluated by several companies and the results reported are promising.

The microwave frequency range is from approximately 300 to 1,000,000 Mc per second. If expressed in gigacycles (1Gc is equal to 1000 Mc), the range is 0.3 Gc to 1000 Gc, or wavelengths of 2 to 0.3 mm. They are similar to light waves since they travel in straight lines; thus they can be equated to the optical laws. They are naturally invisible, and consequently they operate on the theory of reflections, i.e., they are reflected from objects that block their passage, which are usually metallic in composition (this may be metallic-filled adhesives). Consequently, for this reason, the microwave theory is worthless in the detection of voids in adhesive-bonded metallic structures. A system used for nonmetallic honeycomb sandwich structures is shown in Fig. 10-32. This method utilizes the principle of scattering as a mechanism for void detection, with the void acting as the scattering mode. The result is a sharp decrease in a signal received by two horns placed side by side and spaced symmetrically about the void. The sandwich panel is bonded to a metallic structure with a breakaway (epoxy) adhesive. The voids tend to scatter the microwave transmission pattern and the signal losses are observed on on oscilloscope or microammeter.

Microwave testing offers the advantage of rapidity; thus it may be satisfactorily adapted to production-line use. The equipment may be portable and have the ability to penetrate thick substrates, and the cost is considered nominal. The major limiting factor is the restricted area of

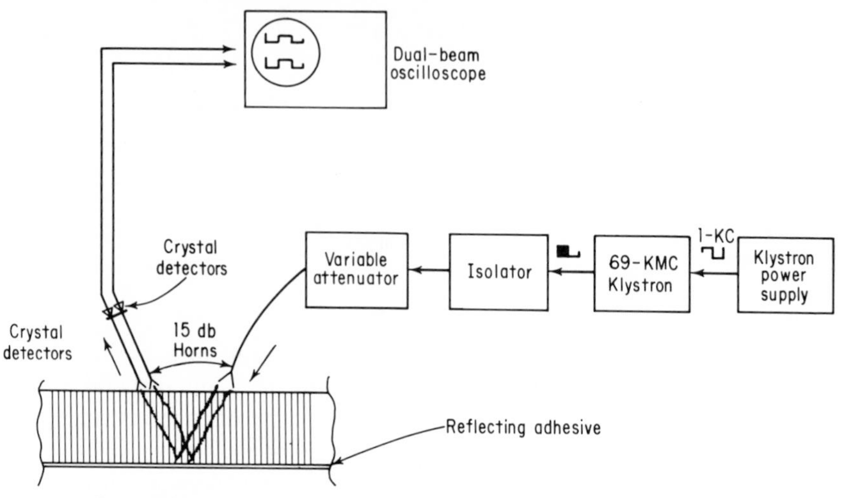

fig. 10-32 *Microwave void-detection system for honeycomb structures. (AVCO Corporation.)*

usage applicable to adhesives, which possibly stems, in part, from the limited research that has been accomplished.

Radiography (x-rays)

The radiographic technique utilizes the sciences dealing with x-rays or rays generated by radioactive substances. This method is widely used for the detection of voids and irregularities in adhesive-bonded joints, especially honeycomb structures. Experienced x-ray technicians are able to obtain a considerable amount of useful information pertinent to bond quality from x-ray film. This field has expanded into an enormous industry which has produced systems ranging from the basic method (Fig. 10-33) to complex systems equipped with television image displays. The x-ray techniques offer the advantage of rapid inspection coupled with good penetration, but have the following disadvantages:

1. Skilled technicians are essential.
2. The method is expensive as compared to other techniques.
3. It is difficult to determine the bond line in which a void occurs in a composite or multiple-bond-line structure.

X-rays are usually produced with equipment rated at 15 Kvp (low energy) to 31 Mev (high energy). Radiation beams scan the bonded assembly and rays are absorbed according to the density of the substrates. Consequently, an area that is unbonded will absorb fewer rays than the solid, well-developed joints and the detector (plate or film) will display a stronger reaction under the void as compared to the well-

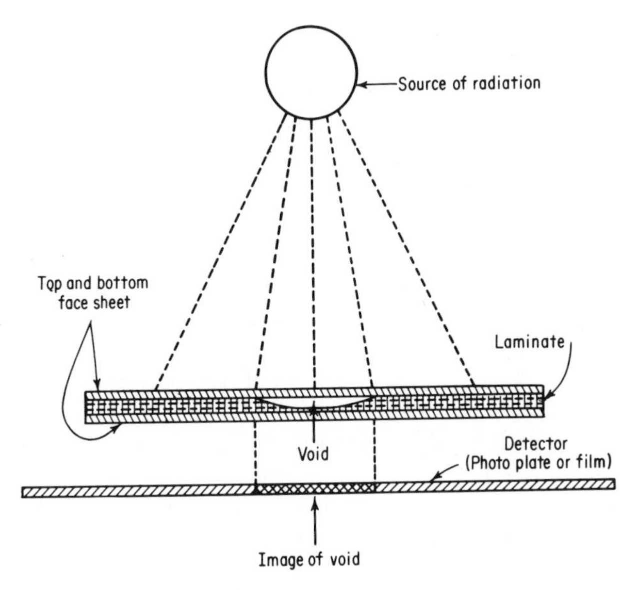

fig. 10-33 *Basic x-ray system.*

bonded areas. The visual details or information obtained from the x-ray depends on the orientation of the beam, the thickness of the part being tested, and the type of film employed. X-rays have a place in NDT of adhesive bonds but will only detect unbonds and areas of extremely high porosity. The x-ray techniques, regardless of the sophistication, leave many unanswered questions concerning the reliability of a bonded joint, and the degree of confidence depends on a thorough and complete knowledge of the physical and mechanical characteristics of the adhesive system being tested, with less emphasis on the adherends.

Birefringent coating inspection

This unusual method of testing shows promise for rapidly testing large honeycomb structures, and may be adapted to many test areas if other modes of implementation are discovered. Studied are currently in progress at Marshall Space Flight Center, Huntsville, Ala. to develop better inspection methods utilizing this technique. Voids can be detected by this technique, which employs a transparent plastic film. A honeycomb sandwich panel is stressed with air pressure, consequently the defective areas bulge slightly outward. These deflections carry through the plastic surface film and are visible when illuminated and viewed (or photographed) through a polarized filter.

This technique in the present state of development offers the following advantages:

1. It is inexpensive as compared to elaborate test equipment.
2. It is adaptable to extremely large structures.
3. It does not require highly trained inspectors.
4. It could possibly detect substandard areas that are not complete voids because stresses would cause a slight deflection when pressurized.

In the present state of development the following limitations are apparent:

1. Honeycomb core material must be perforated and the sandwich panel edges must be sealed to develop internal pressure.
2. These coatings are sensitive to extreme environments during cure.
3. Relatively high pressures are necessary for heavy face sheets.

Birefringence results from deformation of optically isotropic materials, i.e., materials which transmit light at the same velocity in all directions and which exhibit a single value for a refractive index. These materials become anisotropic or birefringent when deformed by pressure because two optical axes are formed at any selected point of observation. These axes are orthogonal at every point and the refractive index of one axis is not necessarily equal to that of another. It has been established that the difference between refractive indexes is proportional to the difference between principal stresses at the point considered, and the optical axes coincide with the principal stress directions. If the indexes differ at a point, a portion of the incident light amplitude emerges behind its complimentary component. The term "relative retardation" is used to define this phenomenon, and a polariscope is used to measure this quality in terms of a selected wavelength of light. A constant of proportionality between relative retardation and principal stress difference can be determined by a simple calibration.

To inspect a panel via the birefringence technique the optically isotropic material is brushed or sprayed onto the adherends and cured. The structure is then pressurized with air and surface strains are transmitted to the coating, which then becomes anisotropic, or birefringent. The birefringence is detected by a reflective polariscope since the observed light is reflected from the adherend surface. (If the surface of the adherend is not reflective, a reflector coating is applied prior to the indicator coating.) A simple polariscope consists of a simple sheet of polarized film placed directly on the coating surface. (A commercial polariscope is also available.) The stressed areas (voids) are displayed by an array of colors. The coating may be slightly heated (150 to 160°F) by the utilization of a hair-dryer-type heat gun and easily peeled from the panel.

Liquid-crystal inspection

One of the latest techniques, which is currently under study for void detection in sandwich structures, employs liquid crystals. These crystals have been described as compounds which exhibit in the liquid state optical properties similar to crystalline solids. They are termed cholesteric because their molecular structure is characteristic of a large number of compounds that contain cholesterol, although cholesterol alone does not have a liquid-crystal phase. The use of these compounds for NDT is based on the thermal characteristics of the adherends, the adhesive, and the ability of the cholesteric substance to produce a variety of colors when its thermal environment is changed. When voids are present in a bonded panel the thermal properties are reduced, and when liquid crystals are present on the panel surface these voids are visible if heat is applied to the substrate.

Liquid crystal offers certain advantages as follows:

1. The test is simple. If satisfactory results are not obtained on the first attempt, the part may be heated again.

2. The color response is rapid; color changes may be observed in less than 1 sec.

3. It is reproducible, i.e., the same color may be present at the same temperature with a given mixture.

4. It eliminates expensive equipment, although the crystals are quite expensive at this stage of development.

At the present stage of development, certain limitations are evident:

1. There is no evidence that anything but complete discontinuities can be detected.

2. Materials with poor heat-transfer properties are difficult to test, e.g., aluminum face-sheet honeycomb assemblies have been unsatisfactory, due to the rapid heat dissipation.

3. Maximum skin thickness that can be successfully tested is approximately 0.060 in.

4. A panel must be accessible from both sides for a complete evaluation.

The crystals must be applied on a dark background since the colors arise because of reflection rather than absorption of light. A suitable background is prepared by applying a water-soluble black paint to the adherend to which the crystals are to be applied. The crystal mixture may be brushed or sprayed on the panel and approximately 1 g will cover 1 sq ft with a film thickness of approximately 10M, which is sufficient to produce the desired color indications. If less than 10M are present on the panel, colors will appear in an irregular pattern quite different in appearance from the colors obtained with a real temperature gradient. When the crystals have remained on the adherend for a long period of time, the colors appear dull but may be intensified by warming and lightly brushing the surface.

The crystals may be heated by heat lamps, or their equivalent, to slightly above the color-transition range, and then cooled with a stream of cool air. Rapid heating and cooling of the surface results in the maximum sensitivity to bond discrepancies. If a part is free of defects, colors are uniform and the surface cools consistently, changing from one color to another. If an unbonded area is present it will be evident because of the slower transmission of heat through the unbonded area. The colors in this area will lag in transition, thus presenting a quite clear picture of the void area. If a permanent record is desired, a color photograph is recommended. After inspection, the liquid crystals and black paint are easily removed with hot water. The applications and limitations of this technique have not been clearly established, as research is still being performed in this area of nondestructive testing.

The Porta-Shear

The Porta-Shear (Fig. 10-34) is a portable pneumatic unit that is used to determine the strength of a small plug that is removed from the part in question at room or elevated temperatures. This test could not be classified as a true nondestructive test because a small portion of the assembly is removed, but the technique certainly has its merits. The most important information received from this test is the surface condition of the adherends relative to surface preparation. This test method has also been utilized with mild success in conjunction with the Fokker Bond Tester to equate pounds per square inch in shear to bond-quality readings, thus eliminating the slow process of mechanical testing and calculations.

The instrument utilizes a controlled-depth hollow cutter to machine a circular recess through the outer adherend and into the adhesive material, allowing a $\frac{1}{4}$-in. diameter button remaining bonded to the interface of the adherend "plug." The channel or circular groove cut around the plug is approximately $\frac{1}{16}$ in. wide, thus allowing attachment of the Porta-Shear grip to the bonded button or plug. A small-resistance heater may be used which seats against the face of the adherend button

fig. 10-34 *The Porta-shear. (Pioneer Industries.)*

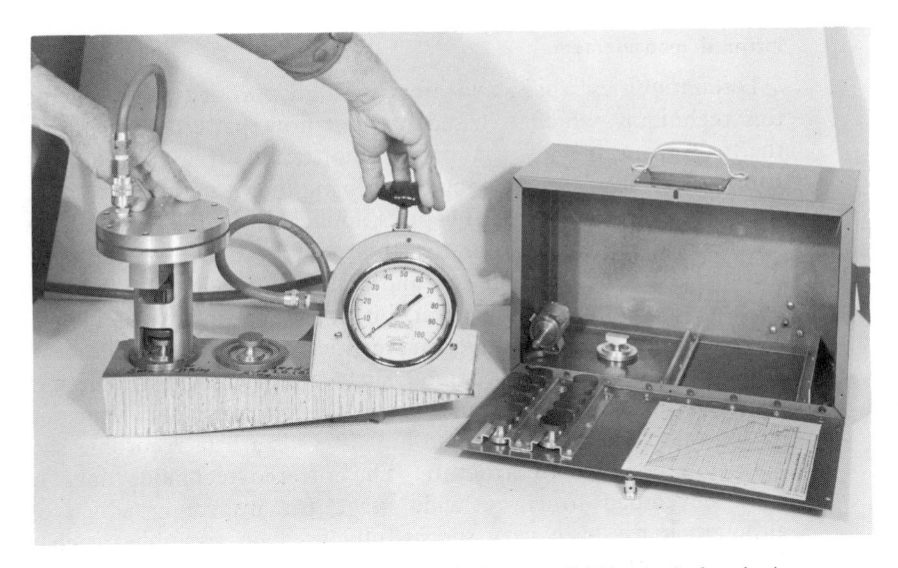

fig. 10-35 *The Porta-pull (Pioneer Industries.)*

if elevated-temperature testing is desired. The pneumatic load is applied to the button through the shear tip of the Porta-Shear head. The force required to shear the button is recorded and converted to shear strength in pounds per square inch via a cross-reference sheet. It may be desirable not to remove the button if the load meets a predetermined standard, or the load may be applied and then released without removing the plug. If the plug is removed, it may be bonded back in place or the hole in the part potted with a room-temperature-setting adhesive.

The Porta-Pull (Fig. 10-35) is a similar device that is utilized on honeycomb or sandwich structures. This device consists of an air regulator, air gauge, hole saw, air-operated pulling device, test pads 1 sq in. in diameter, and a chart to convert test values to pounds per square inch.

The test pads are bonded to the area to be tested and the hole saw used to cut through the face-sheet material. A predetermined value may be selected, and if a bonded joint fails before the value is reached, the joint is considered substandard. If it is desirable to remove the plug, the hole may be potted or a small repair may be made on the test area.

The major objection to this technique (and Porta-Shear) is the established fact that a plug removed at a selected point on a bonded panel may not be indicative of other or even adjacent areas on the same panel. The Porta-Pull and Porta-Shear are marketed by Pioneer Industries, Division of the Almar-York Company, Fort Worth, Tex.

Infrared measurements

Discontinuities, voids, and areas of high porosity can be detected by this technique, which is based on heat-flow patterns presented by the thermal properties of the part being tested. This method has been approached from several angles utilizing both static heat sources and scanning heat techniques. Theories applicable to this method are based on the rate of heat flow or the ability of heat to flow through a solid material. If a bonded joint is sound and well developed, heat can usually be transferred through the joint, but if a void or area of high porosity exists, thermal conduction characteristics are reduced. The read-out is usually accomplished by a remote surface-temperature measurement with an infrared radiometer. Figure 10-36 is a schematic of a scan device for testing sandwich structures, and Fig. 10-37 is a scan recording showing a void in a honeycomb structure. The infrared technique must be more fully developed to enjoy wide usage for inspecting adhesive bonds, although it has been very successfully used on assemblies such as large fiberglass rocket-motor cases. It has the advantages of noncontact with the test piece, nonalignment problems, and rapid substrate coverage.

"Coin" or "mallet" tapping

A sensitive human ear can distinguish frequencies from approximately 10 to 20,000 cycles per sec and the sound variable between a void and some type of bond falls within this area when tapped gently with a coin, mallet or other metallic object. When a panel is tapped lightly with an object, the sound emitted from a well-bonded area is quite different from that from an area containing a void.

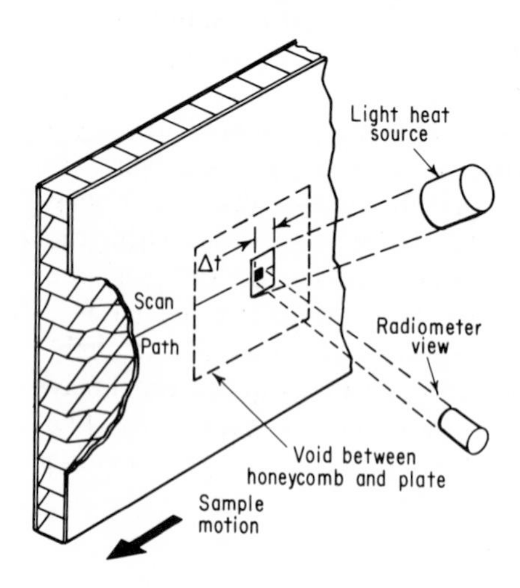

fig. 10-36 *Infrared test system applied to honeycomb structures.*

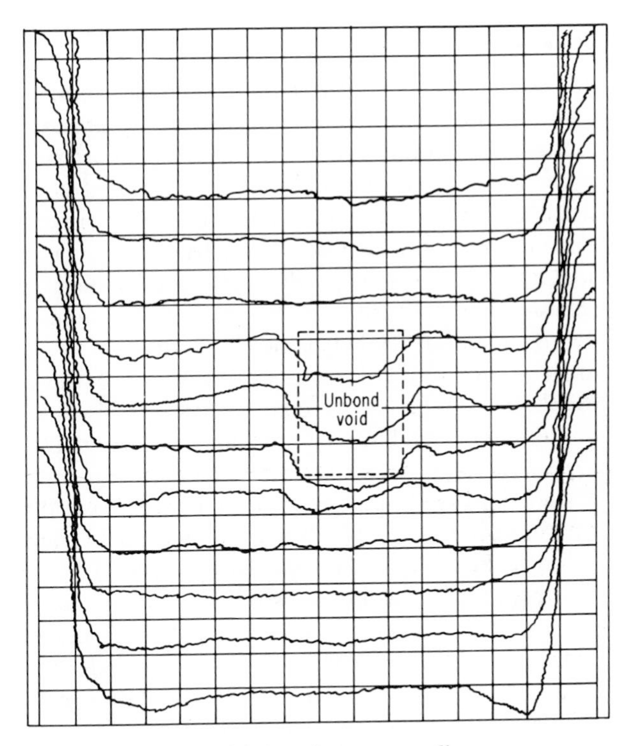

fig. 10-37 *Typical infrared scan recording.*

Studies in this area are underway that utilize electronically controlled tap devices, having a controlled tap force and tap dwell. It has been estimated that a "good mechanical tapper" equipped with a good coin has a dwell time of approximately 54 msec, and emits a sound wave at that dwell on a good bond of approximately 20,000 cycles per sec. The electronic tapping device uses a shorter dwell time and the sound is picked up by a microphone and relayed to an oscilloscope. It is reported that good, sound joints show a fairly consistent frequency of 25,000 cycles per sec, while poor bonds show a decreasing frequency as the waves die away. This approach seems feasible because of the relationship between sound characteristics and mechanical strength of adhesive joints.

The major problems encountered with the tapping techniques are that (1) the areas which are joined in a substandard manner cannot be detected, (2) composite structures, especially if they employ plastics, are almost impossible to test, and (3) if the manual technique is used, the technician's ear may be out of calibration, or someone may accidentally switch coins or mallets with him.

Visual inspection

This method, coupled with the coin-tapping technique, is the oldest form of inspecting bonding panels or assemblies.

Visual inspection consists of a visual examination of the assembly for discrepancies. The flash would be examined and the part inspected for wrinkles or visible voids. An experienced operator may determine to some degree the quality of a part, but visual inspection alone is not adequate. The color of the flash may indicate to some extent if the adhesive was properly cured. An overcured joint (aged) can be detected by this method, provided there is a substantial change in adhesive color; or, if a composite film was reversed during the assembly of a sandwich panel, this could be detected by careful visual examination. Should the polyethylene backing be left intact during assembly utilizing a film-type tape, this could also be detected. Too much emphasis cannot be placed, however, on examination of the flash because a good flash has been interpreted as an adequate bond and brought disaster. Examine the photo in Fig. 10-38. The photo represents a bonded missile hinge which is a primary joint in the structure. Note *the flash* around the outer periphery of the hinge. This assembly was passed by inspection by the visual technique, but detected during a nondestructive test. This is proof that high confidence in the visual-inspection technique may bring many problems. The cost saved in a discrepant bond detection, cited by this one example, could possibly pay for several completely equipped non-destructive test laboratories because this was a test unit and could have cost the company an entire account.

fig. 10-38 *Typical unbonded condition that may be bypassed by visual inspection because of satisfactory flash.*

TABLE 10.1 Probe Selection Guide

Probe number	General application	Top face sheet	Bottom sheet gauge	Band "J"	Tuning "F"	Calibration "Q"	Scale factor
3814	Metal-to-metal	Maximum 0.063 in.	Minimum 0.020 in.	3	4.5–5.5	2	5
1414	Metal-to-metal	Maximum 0.020 in.	Minimum 0.002 in.	3	5.5–6.5	2	2.5
3412	Metal-to-metal	Maximum 0.200 in.	Minimum 0.063 in.	2	6–8	1	2.5
3414	Face-to-core	Maximum 0.040 in.	Minimum core density 2 lbs/ft^3. Maximum cell 5/16 in.	3	7–8.5	2	2.5
1214	Face-to-core	Maximum 0.064 in.	Minimum core density 2 lb/ft^3. Maximum cell ¼ in.	3	6–8	2	2.5
1218	Face-to-core	Maximum 0.020 in.	Minimum core density 1.5 lb/ft^3 Maximum cell 1½ in.	4	5–6	3	2.5
1012	Face-to-core (brazed joints also)	Maximum 0.200 in.	Minimum core density 7 lb/ft^3.	2	5–6.5	Not applicable	Established on application
1234	Metal-to-metal	Maximum 0.150 in.	Minimum 0.025 in.	3	5–7	2	2.5
1510	Face-to-core	Maximum 0.200 in.	Minimum core density 2.0 lb/ft^3.	1	6	Not applicable	Established on application
	Metal-to metal	Maximum 0.030 in.	Minimum 0.090 in.	1	6	Not applicable	

TABLE 10.2 Typical Applications for Sonic Test System

Composite Type	*Description*
Adhesive-bonded honeycomb	Metallic face sheets-metallic core
Adhesive-bonded honeycomb	Metallic face sheets-nonmetallic core
Adhesive-bonded honeycomb	Nonmetallic face sheet-metallic core
Adhesive-bonded honeycomb	Combinations of above
Adhesive-bonded honeycomb	Double-layered sandwiches
Diffusion-bonded honeycomb	Titanium and titanium alloy
Brazed honeycomb	Stainless steel face sheets and core
Adhesive-bonded laminars	Glass filament-rubber-propellant
Adhesive-bonded laminars	Cork-steel-rubber-propellant
Adhesive-bonded laminars	Cork-aluminum plate
Adhesive-bonded laminars	Multiple-layer glass filament cloth
Adhesive-bonded laminars	Metal-to-metal layers
Adhesive-bonded laminars	Plywood
Adhesive-bonded laminars	Plastic bonded to metal sheet
Adhesive-bonded corrugate (Raypan)	Glass filament face sheets and corrugated core
Diffusion-bonded corrugate (Jaffee Metal)	Titanium face sheets and corrugated core
Adhesive-bonded combination	Nonmetallic corrugate (Raypan)-to-nonmetallic laminar

REFERENCES

Arnold, J. S., and C. T. Vincent: "Development of Nondestructive Tests for Structural Adhesive Bonds," WADC Technical Report 54-231, April, 1959.

Baldanza, N. T.: "A Review of Nondestructive Testing for Plastics: Methods and Applications," Picatinny Arsenal, Dover, N.J., August, 1965.

Bayer, R. G., and T. S. Burke: Application of the Ultrasonic Resonance Technique to Inspection of Minature Soldered and Welded Junctions, *Materials Evaluation*, February, 1967.

Botsco, R. J.: Nondestructive Testing of Composite Structures with the Sonic Resonator, *Materials Evaluation*, November, 1966.

Cagle, C. V.: "Adhesive Bonding Training Manual," NASA, Marshall Space Flight Center, 1965.

———: "Procedure for Ultrasonic Testing of Adhesive Bonds Utilizing the Fokker Bond Tester," MEL Technical Report MD-39-65, Hayes International Corp., December, 1965.

Clemens, R. E.: "Adhesive Bonded Honeycomb Integrity-Fokker Bond Tester," Norair Division of Northrop Corporation, 61-273, November, 1961.

———: "Evaluation of Fokker Bond Tester System for Nondestructive Testing of FM-47 Adhesive Bonded Honeycomb and Metal-to-Metal Structures," presented at Los Angeles Society of Nondestructive Testing Meeting, February, 1962.

Cribbs, R. W.: "A Method for Correlating Defects Detected by Nondestructive

Tests with Systems Performance," presented at the Spring Convention of The Society of Nondestructive Testing, March, 1963.

Dick, P.: "Nondestructive Test Equipment for Predicting and Preventing Failure," 1964.

Gonzales, H. M., and C. V. Cagle: "Nondestructive Testing of Adhesive Bonded Joints," ASTM STP-360, 1964.

Hand, W.: Testing Reinforced Plastics with Ultrasonics, *Plastics Technology*, February, 1962.

Hastings, C. H.: "Reducing Failures through Quantitative Nondestructive Inspection Criteria," American Society of Mechanical Engineers, March, 1964.

———: "Summary of Nondestructive Testing Philosophy," presented at Air Force Systems Conference on NDT, El Segundo, Calif., 1967.

Hertz, J.: "Investigation of Bond Deterioration by use of the Fokker Bond Tester," General Dynamics A-ERR-AN-682, December, 1964.

Hitt, W. C., and J. B. Ramsey: Ultrasonic Inspection and Evaluation of Plastic Materials, *Ultrasonics,* March, 1963.

———: Personal correspondence, 1966.

Hughes, E. T., and E. B. Burstein: The Evaluation of Bond Quality in Honeycomb Panels Using Surface Wave Techniques, *Nondestructive Testing Journal,* December, 1959.

Kramer, J. M., A. F. Nuzzo, and G. Epstein: Large Plastic Moldings O.K.'d by Ultrasonic Inspection, *Materials in Design Engineering,* February, 1961.

LaRoe, T. A., and E. T. Hughes: Ultrasonics Solve Testing Problems, *Metal Progress,* June, 1961.

Liquid Crystals Find Flaws via Color in Nondestructive Testing, *Materials Evaluation,* September, 1965.

LoPilato, S. A., and S. W. Carter: Unbond Detection Using Ultrasonic Phase Analysis, *Adhesives Age,* December, 1966.

McGonnagle, W.: "Nondestructive Testing Materials Evaluation," December, 1964.

———, and F. Park: Nondestructive Testing, *International Science and Technology,* July, 1964.

McKown, R. D.: "Learning the Inside through Ultrasonics," Automation Industries, 1964.

McMaster, R. C.: Why Nondestructive Testing Is Needed, *Metal Progress,* February, 1961.

Maley, D. R.: "Two Thermal Nondestructive Testing Techniques," Automation Industries, Boulder, Colo., 1966.

Merhib, C. P.: The Nondestructive Testing Information Analysis Center, *Materials Evaluation,* 1966.

Miller, N. B., and V. H. Boruff: Adhesive Bonds Tested Ultrasonically, *Adhesives Age,* June, 1963.

———: "Evaluation of Ultrasonic Test Devices for Inspection of Adhesive Bonds,"

Navy Bureau of Naval Weapons Contract 59-6266C Summary Report, Martin-Marietta Corp., Baltimore, Md., December, 1962.

Morgan, J. B.: Ultrasonic Testing Procedures, *Nondestructive Testing Journal*, February, 1959.

Ramsey, J. B.: Ultrasonic Techniques for Plastic Inspection, *British Plastics*, February, 1964.

Rockawitz, M., and L. J. McGuire: A Microwave Technique for the Detection of Voids in Honeycomb Ablative Materials, *Materials Evaluation*, February, 1966.

Sabourin, L.: "Nondestructive Testing of Bonded Structures with Liquid Crystals," presented at Conference on Adhesive Bonding, NASA, Huntsville, Ala., March, 1966.

St. Clair, J. C.: An Infrared Method of Rocket Motor Inspection, *Materials Evaluation*, August, 1966.

Schliekelmann, R. J.: "Course, Fokker Bond Tester," Fokker Corporation 479, May, 1965.

———: "Fokker Bond Tester Operation," Report R-217B, Fokker Corporation, Amsterdam, 1958.

———: Nondestructive Testing of Adhesive Bonded Metal Structures, Parts 1 and 2, *Adhesives Age*, May and June, 1964.

———: "Quality Control of Adhesive Bonded Structures," Fokker Corporation, Amsterdam, 1963.

Siedel, R. A.: Factors to Consider in Selecting Ultrasonic Test Equipment, *Metal Progress*, August, 1961.

Smith, D. F., and C. V. Cagle: Quality Control of Adhesive Joints, in "Quality Control Adhesives," John Wiley & Sons, New York, 1967.

———: Ultrasonic Testing of Adhesive Bonds Using the Fokker Bond Tester, *Materials Evaluation*, July, 1966.

Ultrasound Seeks Out Inclusions, *Iron Age*, April, 1965.

Walker, H. M.: "Inspection of Honeycomb Structures Using Bi-Refringent Plastic Coatings," delivered at Structural Bonding Symposium, NASA, Huntsville, Ala., March 16, 1966.

Woodmansee, W E.: Cholesteric Liquid Crystals and Their Application to Thermal Nondestructive Testing, *Materials Evaluation*, October, 1966.

CHAPTER 11

Adhesives Literature and Information Files

The large growth of technical adhesives literature and data in recent years has fostered the need for developing faster and more efficient adhesive information storage and retrieval systems.

John T. Milek has recently reported on a computer system to store and retrieve information on all phases and aspects of adhesive bonding and materials. Such a computer-prepared reference file of adhesives information is of great value to adhesives and materials research and development laboratory scientists and engineers for the preparation of technical reports, technical proposals, state-of-the-art surveys, and solutions to bonding problems.

Frank Swanson also has discussed the formidable problems of coping with the large mass of adhesives design data and information facing the adhesives engineer. The basic problem resolves itself to one of obtaining a maximum amount of useful information at a minimum cost and in a minimum time. A McBee punched card (8 × 11) system for accomplishing this is described by this author.

When appraising the extent of adhesives information and literature to be stored one finds the adhesives can be grouped or classified into the following general areas:

1. Adhesive materials
2. Bonding methods

3. Surface preparation
4. Joints and joint configuration
5. Adhesion theories
6. Adhesion and related properties
7. Adhesive formulations (including chemistry and polymerization)
8. Applications
9. Substrate materials (adherends)
10. Environmental effects and parameters
11. Testing and evaluation
12. Specifications and standards
13. Patents and miscellaneous literature (symposia, conferences, etc.)

It is important to recognize first that information or data on adhesives technology are contained in a wide range of formats: books, periodicals, reports (internal and external to companies), patents, indexes, abstracts, microfiche, microfilm, vendor data sheets and catalogs, specification documents (military; NASA; USA Standards Institute, formerly ASA; ASTM; Federal; Commercial), symposia, proceedings, handbooks, pamphlets, booklets, reprints, and preprints. Second, the information or data must be properly and completely identified with regard to author (personal and corporate), publisher or source, title, journal or book publisher, data, pagination, volume number, accession number (for reports) in order to store and retrieve the information as well as to reference the source of the information. A typical literature citation is shown in Fig. 11-1.

ESTABLISHING AN INFORMATION SYSTEM

In order to organize an adhesives file system, one must initially establish certain guidelines or a philosophy:

1. Will the file be concerned with current and future information and literature?

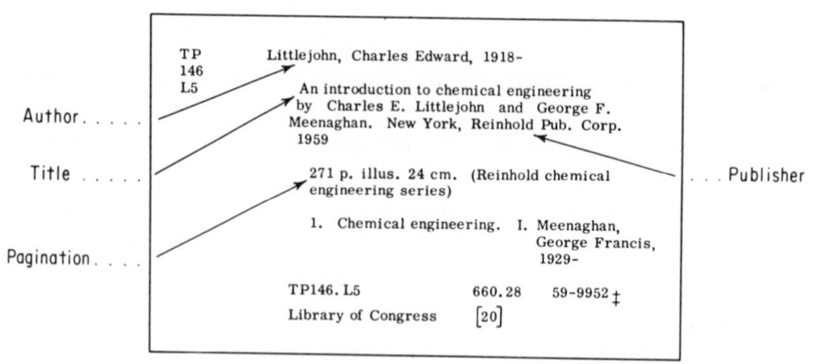
TP
146
L5

Littlejohn, Charles Edward, 1918-

An introduction to chemical engineering by Charles E. Littlejohn and George F. Meenaghan. New York, Reinhold Pub. Corp. 1959

271 p. illus. 24 cm. (Reinhold chemical engineering series)

1. Chemical engineering. I. Meenaghan, George Francis, 1929-

TP146. L5 660.28 59-9952 ‡

Library of Congress [20]

fig. 11-1 *Typical literature reference and citation information.*

2. Will retrospective and searching of the older literature be necessary or required?

For the latter, a number of indexing and abstracting services are available in most libraries: *Engineering Index, Applied Science and Technology, Metals Review of Literature, Chemical Abstracts, Defense Documentation Center Technical Abstract Bulletin, National Aeronautics and Space Administration STAR.*

Company librarians have literature resources to indicate which periodicals are available for scanning and reviewing the adhesives (e.g., *Adhesives Age*) and related literature fields, e.g., fastener technology (e.g., *Assembly Engineering, Fasteners*).

In the following sections of this chapter are a brief discussion of the sources, types, and availability of data, information, and literature in the above 13 general categories. The extent to which information and literature will be acquired, stored, or indexed will vary from company to company and library to library, depending on the funding, manpower, space, and equipment available. The assistance of technical adhesives specialists can greatly enhance such an information system and is strongly recommended by the author.

Swanson recommended a selective adhesive-evaluation system to cope with the mass of adhesive design and test information existing in the literature. The goals of his system are the following:

1. It should be selective in that it provides information specific to the user's area of interest.
2. It should provide a maximum amount of information at a minimum cost.
3. It should provide the basic information necessary in establishing subsequent extensive test programs that may arise for large and complex applications.
4. It should provide design information useful in solving small straightforward bonding applications that arise.
5. The information generated should be easily retrieved for application on adhesive problems.

The adhesive evaluation system used at the Honeywell Ordnance Division consists of a three-step program: The first step consists of the actual test program, which includes adhesive selection from various manufacturers, initial screening tests to narrow the choice, and final screening tests to more completely characterize useful adhesive products. The second step consists of reporting test data and of developing realistic materials and process specifications for the adhesives. The third step consists of tabulating information for use in a data-retrieval system, and the maintenance of the system.

Adhesive materials

When establishing an indexing-retrieval subject descriptor file for adhesive materials, one must contend with a wide range of chemical compositions as well as trade names or proprietary name products, e.g., Proseal 727, Epoxylite 8722 adhesive, etc. These can be readily accommodated by a computer or data processing system. Moreover, for optimum subject retrieval, it is essential that a chemical-family identification of these proprietary or trade-name resins be known, i.e., epoxy adhesives, urethane adhesives, etc., because most adhesive specialists know the general characteristics or behavior of the resin-family system. The computer or data processing system print-out of information and/or literature informs the adhesives specialist or technician of the many and varied successful as well as unsuccessful uses or applications to which any specific adhesive product(s) or chemical family group(s) have been applied or tried. In Table 11.1 are listed numerous adhesive materials descriptors in alphabetical order which can be used for indexing adhesive literature references. Adhesive types can also be classified by function and/or characteristics: ceramic, electrically-conductive, solvent type, composite, hot melt, etc. These subject descriptors and names may also be used for computer and data processing storage and retrieval purposes.

Bonding methods

A number of bonding methods have been described in Chap. 6. Alphabetical listing of terms (bonding) for storing and retrieving purposes can be obtained from the index of Chap. 6 to satisfy the needs of the activity.

Surface preparation

Chapter 5 has discussed a large number of surface-preparation methods but does not serve as a reference list, hence, in Table 11.2 is an alphabetical listing of the various surface-preparation methods useful for storage and retrieval indexing purposes.

Joint and joint configuration

The numerous types of joints enumerated in Chap. 4 of this book can be indexed as specific indexing descriptors or lumped under such a general category. The user will have to exercise his own judgement as well as be governed by the number of literature items which may appear after a time period and the amount of time spent in searching the general category print-out bibliography.

Adhesion theories

As in the above category, specific theories can be indexed individually or lumped under one general indexing term, "Adhesion Theories."

Adhesive formulations

This category can be used for storing and retrieving adhesive formulations prepared by a laboratory group and given laboratory test designations. In some instances, they will appear as proprietary names, alpha-numeric codes, laboratory code or experiment numbers, etc., and provisions must be made to properly cross-index and retrieve them.

Applications

In view of the wide range of applications possible with adhesives, arbitrary or pertinent indexing terms may have to be established: space applications, electrical applications, hydrospace, nuclear applications, cryogenic applications, wings, antennas, radomes, honeycomb structures, etc.

Adherends and substrates

These materials combinations will range broadly and the number of permutations are large:

Metal-to-metal (ferrous or nonferrous)
Metal-to-paper
Metal-to-wood
Metal-to-fibers
Metal-to-plastics
Metal-to-cork
Metal-to-rubber
Metal-to-ceramics

Wood-to-wood
Wood-to-metal
Wood-to-paper
Wood-to-fibers
Wood-to-plastics
Wood-to-cork
Wood-to-rubber
Wood-to-ceramics, etc.

For indexing storage and retrieval purposes, the user will have to select his own combinations or perhaps treat them in specific named material fashion, e.g.:

Ti-6A1-4V alloy to neoprene rubber
Dacron film to kraft paper
2024 Aluminum alloy to polyethylene sheet, etc.

Environmental effects and parameters

Since adhesives are often designed for and used in specialized or diverse environments, some method of indexing, storing, and retrieving such information is imperative. Subject descriptors such as the following are useful for this purpose:

*Cryogenic
*Room temperature
*Medium temperature
*High temperature or pyrotemperature
Vacuum
Air
Inert gas and vapors
Immersion (water, salt water, solvent, chemical solutions, etc.)
Radiation (nuclear, ultraviolet, electron, x-ray, etc.)
Humidity
Salt spray

Testing and evaluation

In this indexing category may be grouped the entire spectrum of tests: mechanical, optical, physical, chemical, thermal, electrical, biological, destructive and nondestructive, ultrasonic, Fokker bond testing, etc. Table 11.3 contains a list of useful testing indexing terms arranged alphabetically. The American Society for Testing and Materials (ASTM) has published test methods covering practically every phase of adhesive bonding:

ASTM D-1002	Lap shear
ASTM D-897	Tensile
ASTM D-1205	Tensile
ASTM D-1344	Tensile
ASTM D-950	Shear-Impact
ASTM D-1781	Peel
ASTM D-1062	Cleavage
ASTM D-1780	Creep

Such indexing descriptors may also be used when greater specificity is desired.

* These temperature ranges are established arbitrarily to serve the purpose of the particular company.

Specifications and standards

Specifications and standards documents form an important part of adhesives technology, because adhesives are procured, tested, and qualified to them more often than not. Most adhesives suppliers indicate, in their vendor data sheets, which specification documents their products can or will meet. Qualified products listing (QPL) information is often a major factor in military contract procurement procedures (see Chap. 12) and this information is easily stored and retrieved in the computer system described above. Irving Katz has prepared "A Guide to Military and Federal Specifications for Rubber-Bonding Adhesives" in the February, 1965 issue of *Adhesives Age*, pages 24 to 31, and more extensive coverage of specifications in his book titled, "Adhesives Materials: Their Properties and Usage," Long Beach, Calif., Foster Publishing Company, 1964. Standard test methods on adhesives are compiled in Part 16 of ASTM Standards: "Structural Sandwich Constructions; Wood; Adhesives." The Department of Defense has developed a "Military Standardization Handbook on Adhesives" (MIL-HDBK-691A, 17 May 1965) to provide basic and fundamental information on adhesives and related bonding processes for the guidance of engineers and designers of military materiel. The handbook is not intended to be referenced in purchase specifications except for informational purposes, nor to supersede any specification requirements.

Patents and miscellaneous literature

The *U.S. Patent Gazette* is the official source for adhesives patent information. Many indexing and abstracting services list adhesives patents (both foreign and U.S.) and such information can be incorporated in the retrieval system; patent number, author, title, adhesive materials, processes, applications can be readily indexed. Robert S. Willard has published a special book report covering the primary patents on adhesives issued by the U.S. Patent Office in the 9-year period from 1955 to 1963. Over 1,000 primary patents are included for liquid adhesives and pressure-sensitive tapes. It also provides data on patents for release coatings for adhesive-coated surfaces and devices for dispensing adhesive tapes. Among the primary categories covered are:

1. Starch and dextrine adhesives
2. Synthetic resin adhesives
3. Rubber adhesives
4. Alkali metal silicate adhesives
5. Cellulose liquor adhesives
6. Protein adhesives

7. Carbohydrate gum adhesives
8. Animal glues
9. Pressure-sensitive tapes.
10. Laminated adhesive tapes
11. Medicated adhesive tapes
12. Core oil binders
13. Adhesive testing procedures
14. Adhesive tape rolls

The book is available from the Palmerton Publishing Company, 101 West 31st St., New York, N. Y., 10001. For those readers interested in filing a patent or searching the adhesives patent literature, Lowell H. McCarter's article, "Patents and Adhesives," *Adhesives Age*, May, 1967, pp. 32–34, gives some guidance. In the U.S. Patent Office, adhesives are classified in several different classes. To give an idea of the areas in which adhesives might be classified, McCarter cites a portion of the index to the classification schedule as follows:

TABLE 11.1 Adhesive Materials Name Index

Subject Descriptors

Adhesives, acrylic
, acrylic acid diester
, acrylonitrile
, acrylonitrile-phenolic (Nitrile Rubber-phenolic)
, alkyd-epoxy
, cellulosic
, ceramic
, composite
, conductive
, cryogenic
, cyanoacrylate (Eastman 910)
, duplex (see Adhesives, composite)
, epoxy
, epoxy-nitrile-phenolic
, epoxy-phenolic
, epoxy-phenolic-rubber
, epoxy-phenolic-vinyl
, epoxy-polyamide (Epoxy-Nylon)
, epoxy-polysulfide
, epoxy-polysulfide-silicone
, halocarbon (includes polychloroprene adhesives)
, melamine
, neoprene rubber
, neoprene rubber-phenolic
, nitrile rubber
, nitrile rubber-phenolic
, phenolic
, phenolic-vinyl
, phenoxy

Subject Descriptors

Adhesives, phthalocyanine
, phthalocyanine
, polyamide (Nylon)
, polyamide-phenolic
, polyamide-neoprene-phenolic
, polybenzimidazole (PBI)
, polychloroprene (see Adhesives, halocarbon)
, polyimide (PI)
, polyester
, polysulfide
, polyurethane
, pressure-sensitive
, rubber
, rubber-phenolic
, silicone
, silicone-rubber
, silicone-epoxy
, silicone-phenolic
, vinyl
Cement, graphite
, optical
Glue, animal
, casein
, wood

TABLE 11.2 Adhesive Processes & Preparation

Bonding, electrical
Cleaning, acid
, chemical
, electrochemical
, emulsion
, solvent and vapor
Descaling
Passivation
Pickling
Polishing, mechanical
Surface preparation

TABLE 11.3 Adhesive Properties Name Index

Acoustic properties and tests
Bearing strength and tests
Bond strength and tests
Chemical properties and tests
Compressive strength and tests
Creep rupture
Creep tests
Dielectric constant
Dielectric strength
Dissipation factor
Electrical properties and tests

TABLE 11.3 Adhesive Properties Name Index *(Continued)*

Fatigue strength and tests
Flexure strength and tests
Impact strength and tests
Mechanical properties
Peel strength and tests
Physical properties
Poisson's ratio
Shear modulus
Shear strength
Tensile strength
Young's modulus

TABLE 11.4 An Index to the Classification Schedule

Adhesive	*Class*	*Subclass*
Applying		
With laminating	156	—
Compositions		
Alkali metal silicate	106	74+
Biocide with	167	42
Carbohydrate gum	106	205+
Cellulose liberation liquor	106	123
Core oils	106	38.2+
Protein	106	124+
Rubber	260	726+
Starch	106	210+
Synthetic resin	260	29.1+
Vermin catching	167	49
Insect catching and destroying	43	136
Insect trap	43	114+
Joining or joint	154	
Moisteners	118	
Envelope sealing combined	120	6
Separator for sheet feeding	271	33
Tape	117	122
Coated on both sides	117	68.5
Holder with edge for tearing	225	6+
Laminated	154	53.5
Medicated	167	84
Rolls	206	59
Testing	73	150

REFERENCES

Milek, John T.: A Novel Retrieval System for Adhesive Materials Technology, *Adhesives Age*, May, 1966, pp. 31–33.

McCarter, Lowell H.: Patents and Adhesives, *Adhesives Age*, May, 1967, pp. 32–34.

Swanson, Frank: How to Develop and Use an Adhesive Evaluation System, *Adhesives Age*, April, 1967, pp. 30–34.

CHAPTER 12

Specifications Applicable to Adhesive Bonding

PART I MILITARY SPECIFICATIONS

MIL-A-928 Adhesive; Metal to Wood, Structural.
Scope: Covers adhesives for structural bonding of aluminum to wood substrates (includes sandwich materials). Covers types and conditions, application, cure cycles, pot-life requirements and qualified products.

MIL-A-1154 Adhesive; Bonding, Vulcanized Synthetic Rubber to Steel.
Scope: Adhesives for bonding vulcanized synthetic rubbers to steel. Covers properties, processes, performance, storage, and qualified products.

MIL-C-1219 Cement, Iron and Steel.
Scope: Intended for repairing minor defects in iron and steel castings; includes classifications, properties, usage, forms, and applications.

MIL-C-3316 Adhesives; Fire Resistant, Thermal Insulation.
Scope: For installation and repair pertinent to fiberglass insulation on piping, machinery, and various metallic surfaces. Covers classifications, properties, processes, performance, and qualified products.

MIL-C-4003 Cement; General Purpose, Synthetic Base.
Scope: Intended for general bonding application; includes classification, performance, storage, application, and qualified products.

MIL-A-5092 Adhesive; Rubber (Synthetic and Reclaimed Rubber Base).
Scope: Non-structural, general purpose rubber cements. Covers classifications, properties, processes, performance, forms, and qualified products.

MIL-A-5534 Adhesive; High Temperature Setting Resin (Phenol, Melamine and Resorcinol Base).
Scope: Covers adhesives for laminating aircraft structural parts (wood). Defines forms, classifications, properties, processes, performance, and mechanical requirements.

MIL-C-5539 Cement; Natural Rubber.
Scope: To be used for bonding fabrics. Covers classifications, properties, performance curing, shelf life, pot life, forms, and qualified products.

MIL-A-5540 Adhesive; Polychloroprene.
Scope: For bonding neoprene coated fabrics. Covers classifications, properties, processes, performance standards, forms, and mechanical requirements.

MIL-A-8576 Adhesive; Acrylic Monomer Base, For Acrylic Plastic.
Scope: For bonding acrylic plastics. Covers classifications, properties, processes, performance, and forms.

MIL-A-8623 Adhesive; Epoxy Resin, Metal-to-Metal Structural Bonding.
Scope: For structural bonding of metal, plastic laminates, wood, glass, or combinations. Covers classifications, properties, processes, performance requirements, forms, and qualified products.

MIL-A-9117 Adhesive; Sealing, For Aromatic Fuel Cells and General Repair.
Scope: Intended for use for bonding rubbers and repairs where fuel resistance is necessary. Covers classifications, properties, performance standards, forms, processes, and qualified products.

MIL-C-10523 Cement, Gasket, For Automobile Applications.
Scope: Intended for bonding cylinder block and head gaskets and fuel lines. Includes classification, properties, processes, performance requirements, and forms.

MIL-S-11030 Sealing Compound, Noncuring Polysulfide Base.
Scope: For sealing metal-to-metal, glass-to-metal, and acrylic-to-metal components (for optical use). Covers classification, properties, performance requirements, forms, and qualified products.

MIL-S-11031 Sealing Compound, Adhesive: Curing, Polysulfide Base.
Scope: For bonding metal-to-metal or glass-to-metal adherends for optical use. Covers classification, properties, processes, performance requirements, forms, and qualified products.

MIL-A-11238 Adhesive, Cellulose Nitrate (For Ordnance Use).
Scope: For assembly of ammunition and general-purpose applications. Covers classification, properties, processes, performance standards, and forms.

MIL-C-12850 Cement, Rubber.
Scope: For bonding natural rubbers. Covers classifications, properties, performance requirements, and forms.

MIL-A-13554 Adhesive for Cellulose Nitrate Film on Metals.
Scope: Intended for bonding cellulose nitrate film to metals, which includes steel, aluminum, brass, and various platings. Covers classifications, properties, processes, performance requirements, and forms.

MIL-C-13792 Cement, Vinyl Acetate Base Solvent Type.
Scope: General-purpose joining of metals. Covers classification, properties, processes, performance requirements, and forms.

MIL-A-13883 Adhesive, Synthetic Rubber (Hot or Cold Bonding).
Scope: For general-purpose bonding. Covers classifications, properties, processes, performance requirements, and forms.

MIL-A-14042 Adhesive, Epoxy.
Scope: For structural bonding of metals, plastics, glass, wood, or combinations. Covers classification, properties, processes, property requirements, form, and qualified products.

MIL-C-14064 Cement: Grinding Disk.
Scope: For bonding abrasive discs to metals. Covers classification, properties, processes, performance requirements, and forms.

MIL-P-14536 Polyisobutylene Binder.
Scope: For fabricating explosive composites. Covers classifications, processes, and form.

MIL-I-15126 Insulation Tape, Electrical, Pressure-Sensitive Adhesive and Pressure-Sensitive Thermosetting Adhesive.
Scope: Intended for use in construction and repair of electrical and electronic equipment. Covers classification, properties, processes, performance requirements, and forms.

MIL-C-18726 Cement, Vinyl Alcohol-Acetate.
Scope: General-purpose bonding; covers classifications, properties, processes, performance, and forms.

MIL-A-22010 Adhesive, Solvent Type, Polyvinyl Chloride.
Scope: Intended for assembly of polyvinyl chloride pipe and fittings. Covers classification, properties, performance standards, and forms.

MIL-A-22397 Adhesive, Phenol and Resorcinol Resin Base.
Scope: For bonding wood for ship and boat use. Covers classifications, properties, processes, performance requirements, forms, and qualified products.

MIL-A-22434 Adhesive, Polyester, Thixotropic.
Scope: Intended for bonding glass fabric reenforced epoxy laminates to metal. Covers classification, properties, processes, performance standards, and forms.

MIL-C-22608 Compound, Insulating, High Temperature.
Scope: For thermal insulation of rocket engines. Covers classification, properties, processes, performance standards, and forms.

MIL-A-22895 Adhesive, Metal Identification Plate.
Scope: Adhesive for bonding metal plates to painted or unpainted surfaces. Covers classifications, properties, processes, performance standards, and forms.

MIL-C-23092 Cement, Natural Rubber.
Scope: For vulcanized and unvulcanized bonding of natural, SBR, and neoprene rubbers. Covers classification, properties, performance requirements, forms, and qualified products.

MIL-A-25055 Adhesive, Acrylic Monomer and Polymer Base, For Acrylic Plastics.
Scope: Intended for use in bonding acrylic plastics. Covers classification, properties, processes, performance standards, and forms.

MIL-A-25457 Adhesive, Air-Drying, Silicone Rubber.
Scope: For bonding silicone rubber to aluminum or to itself. Covers classifications, properties, processes, performance requirements, forms, and qualified products.

MIL-A-25463 Adhesive, Metallic Structural Honeycomb Construction.
Scope: For fabricating metal sandwich structures. Covers classifications, properties, processes, cure conditions, performance requirements, forms, and qualified products.

MIL-A-46028 Adhesive, Flashout, Cold-Setting (Water Cured).
Scope: For bonding cellular polystyrene plastic to steel. Covers classification, properties, performance, and form.

MIL-A-46050 Adhesive, Special; Rapid Room Temperature Curing, Solventless.
Scope: Intended for rapid assembly of porous and nonporous surfaces. Covers classification, properties, processes, performance requirements, and forms.

MIL-A-46051 Adhesive, Room-Temperature and Intermediate-Temperature Setting Resin (Phenol, Resorcinol, and Melamine Base).
Scope: Intended for wood bonding, laminates, or plastic substrates. Covers classification, properties, processes, performance forms, and qualified products.

MIL-A-52194 Adhesive, Epoxy (For Bonding Glass Reinforced Polyester).
Scope: For bonding reinforced polyester laminates. Covers classification, properties, processes, performance, and forms.

MIL-A-9067C Adhesive Bonding, Process and Inspection Requirements For.
Scope: Provides a detailed guidance for the preparation and requirements to be included in the contractor process specification, required for the processing and inspection of adhesive bonded parts, including sandwich constructions bonded with metal-to-plastic and plastic-to-plastic adhesives. Covers the contractor requirements, materials, bonding procedures, prefitting of parts, surface preparation, handling, adhesive application, assembly, curing, pressure requirements, rework and repair, workmanship, sampling and inspection, test methods, and definitions applicable to contractors.

MIL-C-7438 Core Material, Aluminum, For Sandwich Construction.
Scope: Establishes requirements for aluminum core material for structural sandwich construction. Covers bonding process requirements, trade names and code numbers, configurations, sizes, cures, mechanical properties, color coding, workmanship, quality assurance, acceptance tests, test methods, preservation, and packaging.

MIL-C-8073 Core Material, Plastic Honeycomb, Laminated Glass Fabric Base, For Aircraft Structural Applications.
Scope: Covers the requirements for glass fabric base plastic honeycomb core materials for aircraft structural applications, including aircraft exterior parts, such as radio and radar antenna housings. This specification covers requirements, fabrics, resins, core material property requirements, density, workmanship, qualification tests, and quality assurance provisions.

MIL-C-21275 Core Material, Metallic, Heat-Resisting, For Structural Sandwich Construction.
Scope: Establishes the requirements for heat-resisting welded metallic core material for structural sandwich constructions that can be used at elevated temperature. Covers classifications, requirements, materials, quality control, mechanical properties, quality assurance provisions, test methods, and packaging.

MIL-H-9884 Honeycomb Material, Cushioning, Paper.
Scope: Covers paper honeycomb structure for use as an energy-dissipating medium for the landing shock to which air-dropped items are subjected. Covers classifications, adhesives, requirements, materials, dimensions, density, workmanship, inspection, acceptance tests, test methods, and packaging.

MIL-S-9041A Sandwich Construction, Plastic Resin, Glass Fabric Base, Laminated Facings For Aircraft Structural Applications.
Scope: Covers requirements, core material, mechanical properties, electrical properties, dimensions and weight, workmanship, sampling instructions, test procedures, and shipping instructions.

PART II FEDERAL ADHESIVE SPECIFICATIONS

MMM-A-181 Adhesive, Room-Temperature and Intermediate-Temperature Setting Resin (Phenol, Resorcinol and Melamine Base).

Scope: Covers types, forms, quality assurance provisions, and packaging.

MMM-A-00185 Adhesive, Rubber.

Scope: For paper bonding; covers classification, types, properties, performance, and forms.

MMM-A-00187 Adhesive, Synthetic, Epoxy Resin Base Paste Form, General Purpose.

Scope: General-purpose bonding; covers classifications, properties, processes, performance, and forms.

MMM-A-132 Adhesives, Heat Resistant, Airframe Structural, Metal-to-Metal.

Scope: Outlines the requirements for heat-resistant adhesives for use in bonding primary and secondary airframe structures to be exposed to a temperature range of −67 to 500°F. Covers classifications, qualifications, materials, forms, curing agents, solvents, fillers, formulation, identification, processes, sampling, inspection and test procedures. Test procedures defines tensile, peel, fatigue, and creep tests. This specification also defines environmental tests which include elevated-temperature tests, low-temperature tests and conditions, salt-spray fluid immersion, humidity, and fatigue tests under various environmental test conditions.

Federal Test Method Standard 175, Adhesives; Methods of Testing.

Scope: Standard methods for routine sampling, inspection and testing of adhesives and constructions fabricated with adhesives which are purchased under Federal Specifications. Covers requirements, test-specimen fabrication, bonding, testing conditions, and reporting results. Test coverage includes: Applied Weight of Dried Solids (Method 3011), Applied Weight of Liquid Adhesives (Method 3012), Blocking Adhesives (Method 2041), Cleavage (1071-T), Delaminations (Method 2021), Effect of Moisture and Temperature (Method 2052-T), Fatigue (Method 1061), Impact (Method 1051), Peel (Method 1041), Climbing Drum Peel (Method 1042-T), pH of Adhesives (Method 4011), Resistance to Chemicals (Method 2011), Resistance to Water (Method 2031), Cyclic Aging (Method 2051-T), Shear by Compression (Method 1031), Shear by Flexural Loading (Method 1021), Shear by Tension Loading (Method 1033.1-T), Lap Shear Strength (Method 1033), Shear Strength for Plywood Construction (Method

1032), Tensile Properties of Adhesives (Method 1011.1), Tensile Strength for Elastomeric Materials (Method 1012), Solids Content of Adhesives (Method 4021).

MIL-STD-401 Sandwich Constructions and Core Materials; General Test Methods.

Scope: Covers the general requirements and methods for testing sandwich core materials which are used primarily in the aerospace industry. The following areas are defined: design data requirements, quality assurance provisions, test conditions, test apparatus, and reporting. The following detailed test methods are outlined: core density and specific gravity, thermal conductivity, core compression, core shear, core tension, water migration (core), sandwich compression, sandwich tension, sandwich flexure and shear properties, sandwich bonding properties, sandwich peel, sandwich thermal conductivity, plus illustrations of various sandwich tests.

PART III ASTM ADHESIVE SPECIFICATIONS

ASTM C271-61 Density of Core Materials for Structural Sandwich Constructions.

Scope: Establishes a standard method for determination of the density of core materials to be used in structural sandwich construction.

ASTM C272-53 Water Absorption of Core Materials for Structural Sandwich Constructions.

Scope: Procedure for determining the relative rate of water absorption by various types of core materials, including honeycomb structures, wood, cellulated materials, and resin-impregnated reinforcing materials.

ASTM C273-61 Shear Test in Flatwise Plane of Flat Sandwich Constructions or Sandwich Cores.

Scope: Establishes method to determine shear properties of sandwich construction associated with shear distortion of planes parallel to the edge plane of the sandwich. This method covers the determination of shear strength parallel to the plane of the sandwich, and the shearing modulus associated with the strains in a plane normal to the facings.

ASTM C297-61 Tension Test of Flat Sandwich Constructions in Flatwise Plane.

Scope: Covers the procedure for determining the strength in tension, flatwise of the core, or of the bond between the core and face sheets, of an assembled sandwich panel. This is a test of subjecting a sandwich construction to tensile loads normal to the plane of the sandwich, which is accompanied by utilizing thick loading blocks, bonded to the face sheets, through which the load is transmitted.

ASTM C363-57 Delamination Strength of Honeycomb Type Core Material.

Scope: Establishes a method for determining the delamination strength (node) of honeycomb core materials. This test is useful in determining if cores can be handled during cutting and machining without delaminating.

ASTM C364-61 Edgewise Compressive Strength of Flat Sandwich Constructions.

Scope: Covers a procedure for determining compressive strengths of flat structural sandwich constructions in a parallel direction to the plane of the face sheets of the sandwich.

ASTM C365-57 Flatwise Compressive Strength of Sandwich Cores.

Scope: Establishes a method for determining compressive properties of sandwich cores (deformation and crushing). Properties pertinent to the above are normally used for design purposes.

ASTM C366-57 Measurement of Thickness of Sandwich Cores.

Scope: Covers procedures for measuring the thickness of flat sandwich core material using the roller-type thickness tester and the disk-type thickness tester.

ASTM C393-62 Flexure Test of Flat Sandwich Constructions.

Scope: Establishes a procedure for determining properties of flat sandwich constructions subjected to flatwise flexure in such a manner that the applied force produces curvatures of the plane of the face sheet of a sandwich structure.

ASTM C394-62 Shear Fatigue of Sandwich Core Materials.

Scope: Covers a procedure for determining the effect of repeated shear loads on sandwich core materials. If the facings of a sandwich construction are designed so they are elastically stable, the most critical stress to which the core is subjected is shear. The effect of repeated shear stresses on the core material is important.

ASTM C480-62 Flexure Creep of Sandwich Constructions.

Scope: Covers a procedure for determining the creep characteristics and creep rate of sandwich constructions loaded in flexure at any desired temperature. The creep rate provides information pertinent to the behavior of sandwich structures under constant load.

ASTM C481-62 Laboratory Aging of Sandwich Constructions.

Scope: Covers a procedure for determining the resistance of sandwich panels to severe exposure as measured by the change in selected properties of the material after exposure. Tests for selected properties are made on specimens of the material as received and after exposure to prescribed aging conditions and the results are compared. This includes water immersion, heat, steam, etc.

ASTM D553-42 Viscosity and Total Solids Content of Rubber Cements.

Scope: Covers a procedure for determining the viscosity and total solids content of rubber cements containing organic solvents.

ASTM D805-63 Standard Methods of Testing Veneer, Plywood and Other Glued Veneer Constructions.

Scope: Covers procedures for determining the following properties: compression, static bending, tension, panel shear, shear strength, palate shear, toughness, Rockwell hardness, swelling and recovery, moisture absorption, glue block shear test, Plywood glue shear test, moisture content, and specific gravity.

ASTM D816-55 Standard Method for Testing Rubber Cements.

Scope: Covers methods of testing adhesive properties of adhesives that may be applied in a plastic or fluid form and are manufactured from natural rubber, reclaimed rubber, synthetic elastomers, or combinations of these materials. The tests in this standard are divided into two groups. Group one includes those procedures in which the adhesive is applied to specimens of materials to be bonded after which the quality of the glue line is evaluated and includes adhesion strength, bonding range, softening point, and cold flow. Group two is physical property tests which include viscosity, stability, cold brittleness, weight per gallon, and plastic deformation.

ASTM D896-64 Standard Test Method for Resistance of Adhesive Bonds to Chemical Reagents.

Scope: Covers the testing of all types of adhesives for resistance to chemical reagents (strength loss). Includes immersion tests applicable to hydrocarbons, jet fuels, and silicone oils.

ASTM D897-49 Tensile Properties of Adhesives.

Scope: Tensile testing of adhesive bonded specimens; covers tensile testing machine, specimen preparation, testing, conditioning, calculations, and reporting.

ASTM D898-51 Applied Weight Per Unit Area of Dried Adhesive Solids.
Scope: Covers method for determining the quantity of adhesive solids applied in a spreading or coating operation. Covers test specimens, conditioning, procedure, calculations, and reporting.

ASTM D899-51 Applied Weight Per Unit Area of Liquid Adhesives.
Scope: Covers procedure for determining the quantity of liquid adhesives applied in a spreading or coating operation. Includes definition of test apparatus, test specimens, conditioning, procedure, calculations, and reporting.

ASTM D903-49 Peel or Stripping Strength of Adhesives.
Scope: Method for determining the comparative peel or stripping characteristics of adhesives when tested on standard-size specimens and under defined conditioning of pretreatment, temperature, and testing machine. Defines standard testing machine, test specimen, specimen preparation, conditioning, test procedure, calculations, and reporting.

ASTM D904-57 Determining the Effect of Artificial (Carbon-Arc Type) and Natural Light on the Permanence of Adhesives.
Scope: Covers recommended practice intended to define conditions for the exposure of adhesives in the form of transparent or translucent assemblies to artificial or natural light. May be used for films or any other suitable forms. Covers test apparatus, test specimens, method of test, and reporting.

ASTM D905-49 Strength Properties of Adhesives in Shear by Compression Loading.
Scope: Method of test for determining the comparative strengths of adhesive, used for bonding wood and similar materials, tested on a standard specimen under specified conditioning and compression loaded. Covers test apparatus, preparation of specimens, conditioning, calculations, and reporting.

ASTM D950-54 Impact Strength of Adhesives.
Scope: Method for determining comparative impact strength of adhesives in shear and covers test apparatus, holding fixtures for testing, conditioning, test specimen preparation, testing procedure, calculations, and reporting.

ASTM D1002-64 Strength Properties of Adhesives in Shear by Tension Loading (Metal-to-Metal).
Scope: Method for determining the shear strength of adhesives for bonding metals, when using a standard test specimen and specified conditions of preparation and testing. Cov-

ers testing-machine requirements, test specimen preparation, testing procedure, calculations, and reporting.

ASTM D1062-51 Cleavage Strength of Metal-to-Metal Adhesives.

Scope: Method for determining the cleavage properties of adhesives when tested on standard shaped metal specimens, under standard conditions and test procedures. Includes description of test apparatus, test specimen, conditioning, calculations, and reporting.

ASTM D1084-63 Consistency of Adhesives.

Scope: Methods intended for testing consistency of adhesives and includes four methods for various viscosities. Covers test devices, procedures, conditioning, and reporting.

ASTM D1144-57 Determining Strength Development of Adhesive Bonds.

Scope: A recommended practice intended for use in determining the strength development of adhesive bonds when utilizing a standard specimen, under specified conditions of preparation and testing. Applies to all adhesive systems and adherends. Covers test specimens and their preparation, testing procedure, and reporting.

ASTM D1146-53 Blocking Point of Potentially Adhesive Layers.

Scope: Method for determining the blocking point of a thermoplastic or hygroscopic layer or coating of potentially adhesive material. Two varying degrees of blocking are defined: first-degree blocking, second-degree blocking. Two types are covered: cohesive blocking and adhesive blocking. Defines apparatus, conditioning, test specimens, and their preparation, and reporting.

ASTM D1151-61 Effect of Moisture and Temperature on Adhesive Bonds.

Scope: Defines conditions and test methods for determining the performance of adhesive bonds when exposed to specified conditions of moisture and temperature. The performance is expressed as a percentage based on the ratio of strength retained after exposure as compared to the original strength (controls). Covers test apparatus, test specimens, conditioning, test procedure, calculations, and reporting.

ASTM D1174-55 Effect of Bacterial Contamination on Permanence of Adhesive Preparations and Adhesive Bonds.

Scope: Method for evaluating the effect of bacterial contamination on the permanency of adhesive preparations and adhesive bonds. Covers apparatus, test organisms, preparation of cultural media, incubation, preparation of test specimens, inoculation of specimens, viscosity determination, conditioning of specimens, conditioning of test specimens after incubation, determination of bond strength, calculations, and reporting.

D1183-61T Resistance of Adhesives to Cyclic Laboratory Aging Conditions.

Scope: Methods of testing for determining the resistance of adhesives to cyclic conditions by exposing bonded specimens to conditions of high and low temperatures and high and low relative humidities. Covers equipment, test specimens and their preparation, conditioning, procedure, testing, calculations, and reporting.

ASTM D1184-55 Strength of Adhesives on Flexural Loading.

Scope: Test procedure for determining the comparative properties of adhesive-bonded assemblies when subjected to flexural stress with standard specimens under specified conditions. Covers test equipment, test specimens, conditioning, procedures, calculations, and reporting.

ASTM D1205-61 Adhesives for Brake Lining and Other Friction Materials.

Scope: Methods for testing the properties of adhesives used as friction materials, classified in three groups as follows: (1) methods for evaluating the strength and permanence of bonds, (2) methods for measuring the shelf life, (3) method for the use of ultrasonic testing equipment for nondestructive testing for defective areas. Defines equipment, specimens, test procedures, and reporting.

ASTM D1286-57 Effect of Mold Contamination and Permanence of Adhesive Preparations and Adhesive Bonds.

Scope: Methods for determining the effect of mold contamination on the permanency of adhesive bonds and adhesive preparations. This is a comparison method, comparing viscosity and strength of contaminated specimens to control specimens. Covers equipment and materials, test organisms, preparation of culture medium, preparation of inoculae, preservation of cultures, preparation of test specimens, inoculation, determination of viscosity, incubation, bond-strength determination, calculations, and reporting.

ASTM D1304-60 Adhesives Relative to Their Use as Electrical Insulation.

Scope: Methods for testing adhesives in liquid, highly viscous, solid or set states which are intended to be cured by electronic heating, electrical insulation, or for use in electrical assemblies. Covers power factor and dielectric constant of liquid adhesives, direct-current conductivity, extract conductivity, acidity and alkalinity, pH value, power factor and dielectric constant of dried or cured adhesive film, volume and surface resistivity, arc resistance, and specific procedures.

ASTM D1337-56 Storage Life of Adhesives by Consistency and Bond Strength.

Scope: Method applies to all adhesives having relatively short storage life and is determined by utilization of consistency tests and bond strength tests or both. Covers testing equipment, conditioning, procedures, storage of adhesives, and reporting.

ASTM D1344-57 Cross-Lap Specimens for Tensile Properties of Adhesives.

Scope: Method to cover simplified tension test procedure for determining the comparative strength of adhesives by the use of cross-lap assembly under specified conditions. Test especially designed for glass bonded to itself or other adherends. Covers testing equipment, conditioning room, test specimens, bonding, conditioning, procedures, calculations, and reporting.

ASTM D1488-60 Amylaceous Matter in Adhesives.

Scope: Test covers procedures for determining the presence of starch-like material in phenol, resorcinol and melamine resin adhesives. Covers reagents, sampling, procedure, and reporting.

ASTM D1489-60 Nonvolatile Content of Aqueous Adhesives.

Scope: Covers a procedure for determining the nonvolatile content of aqueous adhesives such as dextrin, starch, casein, gelatin, etc. Covers equipment and materials, sampling, procedure, calculations, and reporting.

ASTM 1490-64 Nonvolatile Content of Urea-Formaldehyde Resin Solutions.

Scope: Test procedure for determining the nonvolatile content of the urea-formaldehyde resin solutions for use as wood adhesives. Covers equipment, sampling, procedure, calculations, and reporting.

ASTM D1579-60 Filler Content of Phenol, Resorcinol and Melamine Adhesives.

Scope: Test method suitable for measuring the filler content of phenol, resorcinol, and melamine resin-base adhesives, mixed with hardener or catalyst, that cure at room, intermediate, and high temperatures. Results are based on the nonvolatile content of the mixed liquid adhesive. Covers equipment, sampling, procedure, calculations, and reporting.

ASTM D1582-60 Nonvolatile Content of Phenol, Resorcinol and Melamine Adhesives.

Scope: Test method for determining the nonvolatile content or total solids of phenol, resorcinol, and melamine adhesives with or without hardener. Covers equipment, sampling, procedure, calculations, and reporting.

ASTM D1588-61 Hydrogen-Ion Concentration of Dry Adhesive Films.
Scope: Method to determine the hydrogen-ion concentration (pH) acidity or alkalinity of organic adhesives in the cured dry film form. Covers equipment, sampling, procedures, and reporting.

ASTM D1759-64 Conducting Shear-Block Test for Quality Control of Glue Bonds in Scarf Joints.
Scope: Method utilizes shear-block test to evaluate the quality of wood failure. Covers test specimens and their preparations, procedure, and reporting.

ASTM D1780-62 Conducting Creep Tests of Metal-to-Metal Adhesives.
Scope: Covers test for the determination of the amount of creep of metal-to-metal adhesive bonds due to the combined effects of temperature, tensile shear stress, and time. Covers test equipment, temperature controls and measurements, vibration control, test specimens and their fabrication, procedure, measurements, plotting results, and reporting.

ASTM D1781-62 Climbing Drum Peel Tests for Adhesives.
Scope: Test to determine the relative peel resistance of adhesive bonds between a relatively flexible adherend and a rigid one. May also be used for determining the peel resistance of adhesives in sandwich structures where the face sheets are flexible. Covers equipment, test specimens and their preparation, test procedures, equipment calibrations, conditioning, procedures, calculations, and reporting.

ASTM D1828-61T Atmospheric Exposure of Adhesive Bonded Joints and Structures.
Scope: Method for direct exposure of adhesive bonded joints and structures to natural atmospheric environments and sheltered atmospheric environments. Covers exposure sites, test apparatus, test specimens, procedures, and reporting.

ASTM D1875-61T Density of Adhesives in Fluid Form.
Scope: Covers procedure for the measurements of density (weight per gallon) of adhesives, and components, when in liquid form. This test is particularly applicable where the fluid has too high a viscosity or where a component is too volatile for a specific-gravity balance determination. Covers equipment, sampling, procedures, calculations, and reporting.

ASTM D1876-61T Peel Resistance of Adhesives (T-Peel Test).
Scope: Method for determining peel resistance of adhesive bonds between two flexible adherends. Covers testing equipment, test specimens and their preparation, conditioning, procedure, calculations, and reporting.

ASTM D1878-61T Pressure-Sensitive Tack of Adhesives.
Scope: Method covers a test procedure for measuring the pressure sensitive tack or "quick stick" of pressure-sensitive adhesives. (More applicable to pressure-sensitive adhesives whose backing is reasonably rigid). Covers equipment, methods, test specimens, procedures, and reporting.

ASTM D1879-61T Exposure of Adhesive Specimens to High Energy Radiation.
Scope: Method defining conditions for the exposure of polymeric adhesives in bonded form to high energy radiation to determine changes in physical or chemical properties. Covers x-ray radiation, electron or beta radiation, neutrons, and mixtures of these, such as reactor radiation. Defines significance, effects, test specimens, exposure at various temperatures, determination of exposure and reporting.

ASTM D1916-61T Penetration of Adhesives.
Scope: Method for determination of penetration of adhesives under pressure where at least one of the adherends is porous (particularly starch base). Covers apparatus, reagents, specimens, procedure, calculations, and reporting.

ASTM D2093-62T Preparation of Surfaces of Plastic Prior to Adhesive Bonding.
Scope: Describes surface preparation for plastic adherends prior to adhesive bonding of test specimens. Covers cleaning procedure for cellulose acetate, cellulose acetate butyrate, cellulose nitrate, methylstyrene, polycarbonate, polystyrene, cellulose propionate, vinyl chloride, polymethylmethacylate, ethylcellulose, epoxy, polyester, phenolic, urea-formaldehyde, diallyl phthalate, melamine, nylon, polyurethane, polyolefin, polypropylene, and Teflons. Covers solution preparation and use.

ASTM D2094-62T Preparation of Bar and Rod Specimens for Adhesion Tests.
Scope: Describes bar and rod type butt-joined adhesive test specimens and procedures for preparing and bonding them. Covers test specimens, surface preparation, procedures, and reporting.

ASTM D2095-62T Tensile Strength of Adhesives by Means of Bar and Rod Specimens.

Scope: Test for determination of tensile strength of adhesives by the bar or rod type butt-joined specimens under defined conditions. Covers equipment, test specimens, conditioning, procedures, calculations, and reporting.

ASTM D2181-64T Vinyl Acetate Resin Emulsion Adhesive.
Scope: Covers vinyl acetate resin emulsion adhesives generally found in home or office. Covers detail requirement, sampling, methods of test, rejection, packaging, and marketing.

ASTM D2182-63T Strength Properties of Metal-to-Metal Adhesives by Compression Loading (Disc Shear).
Scope: Covers test method for the determination of shear strength of adhesives under specified conditions of preparation and loading in compression—and intended primarily for metal-to-metal adhesives. Covers testing equipment, test specimens, conditioning, procedures, calculations, and reporting.

ASTM D2183-63T Flow Properties of Adhesives.
Scope: Test for determining flow properties of adhesives under prescribed heating or curing conditions. Intended for film form or liquid which is dried to tack-free state before mating. Covers equipment, preparation, and selection of film for testing, determination of weight ratio, calculations, and reporting.

ASTM D2293-64T Creep Properties of Adhesives in Shear by Compression Loading (Metal-to-Metal).
Scope: Test covers determination of creep properties of adhesives for bonding metals when tested on a standard specimen under specified conditions. Covers equipment, test specimens, procedure, and reporting.

ASTM D2294-64T Creep Properties of Adhesives in Shear by Tension Loading (Metal-to-Metal).
Scope: Method covers determination of creep properties of adhesives for bonding metal when tested on standard specimens under specified conditions. Covers test equipment, test specimens, procedures, and reporting.

ASTM D2295-64T Strength Properties of Adhesives in Shear by Tension Loading at Elevated Temperatures (Metal-to-Metal).
Scope: Test method for determination of shear strength when tested at elevated temperatures. Covers equipment, test specimens and their preparation, procedures, and reporting.

A List of Manufacturers and Adhesive Suppliers for the Aerospace Industry

Name and Address	*Trade Names*
Adhesive Engineering Division of Hiller Aircraft Corp. 1411 Industrial Road San Carlos, Calif.	AEROBOND, ADSEAL, CONCRESIVE, ELECTROBOND, METLHESIVE, GLASSHESIVE
Adhesive Products Corp. 1660 Boone Avenue New York, N.Y., 10460	RESGRIP, ADOPOX, METALSTIX, LUTEX, ADRUB, APCO, DRI-TAC, GRIPSTIX, GRIFTEX, GRIPWELD, KWICK, PLASTIX, PLASTIGRIP, NU-RUB
Adhesive Products, Inc. 520 Cleveland Avenue Albany, Calif., 94710	API, ADIOK, ADLOK
Allaco Products 130 Wood Street Braintree, Mass.	ALL-BOND, TWENTY-TWENTY, MONOBOND, MINIT-CURE, FLEXOBOND
Allied Chemical Corporation Plastics Division 40 Rector Street New York, N.Y.	PLASKON

Name and Address	*Trade Names*
American Cyanamid Company Bloomingdale Department Havre de Grace, Md., 21078	FM --- HT --- BR --- CORFIL, PREBOND PRESS-TO-FLO
Armstrong Products Company Argonne Road Warsaw, Ind.	ARMSTRONG
B. B. Chemical Company Division of United Shoe Machinery Corp. 784 Memorial Drive Cambridge, Mass.	BOSTIK
Carl H. Biggs Company, Inc. 1547 14th Street Santa Monica, Calif.	HELIX
The Borden Chemical Company Division of the Borden Company 350 Madison Avenue New York, N.Y., 10017	CASCAMITE, CASCO, CASCALA CASCOPHEN, CASCO RESIN, CASCOREZ, EPIPHEN, RESLAC GLUE-ALL, PLACCO, ELMER'S
Catalin Corporation Division of Ashland Oil and Refining One Park Avenue New York, N.Y., 10016	CATALIN
Chemical Development Company Box 2 Danvers, Mass.	DEVCON, CEPOX CHEM-O-SOL
Chromerics 380 South Street Plainville, Mass.	CHROMERIC
Coast Pro-Seal & Manufacturing Co. 2235 Beverly Boulevard Los Angeles, Calif., 90057	COAST
Commercial Chemical Company 1021 Sumner Street Cincinnati, Ohio, 45204	NEW EPO EPO-COAT EPO-SEALANT EPO-PATCH
Peter Cooper Corporation Gowanda, N.Y., 14070	INDUSTRIAL ADHESIVES

Name and Address	*Trade Names*
Cycleweld Chemical Products Division of The Chrysler Corporation 5437 West Jefferson St. Trenton, Mich.	CYCLASTIC, CYCLEBOND, CYCLEFUSE, CYCLEWELD
Daubert Chemical Company 2000 Spring Road Oak Brook, Ill., 60523	DAUBOND
Dennis Chemical Company 2701 Papin Street St. Louis, Mo., 63103	DENNIS ———
Dow Chemical Company Midland, Mich., 48640	DOW, ETHOCEL
DRR, Inc. 600 Cartlandt Street Belleville, N.J., 07109	DPR
E. I. DuPont de Nemours & Co., Inc. 1007 Market Street Wilmington, Del., 19898	DUPONT, LUCITE, DUCO, ELVACET, ELVANOL
Eastman Chemical Products, Inc. Subsidiary of Eastman Kodak Company Kingsport, Tenn., 37662	EASTMAN 910
Emerson & Cumings, Inc. 869 Washington Street Canton, Mass.	ECCOBOND, ECCOSORB, STYCAST, ECCOFOAM, ECCOCAST, ECCOSPHERES, ECCOSEAL, ECCOGEL, ECCOBILD
Epoxylite Corporation 1428 North Tyler Avenue P. O. Box 3397 South El Monte, Calif.	EPOXYLITE RESIWELD, RESIWOOD
Epoxylite Corp. of New York 42 Breckenridge St. Buffalo, N.Y., 14213	EPOXYLITE
H. B. Fuller Company 1150 Eustis Street St. Paul, Minn.	
Furane Plastics, Inc. 4516 Brazil Street Los Angeles, Calif., 90039	UROLANE, EPOCAST, EPIBOND, RESIN-X

Name and Address	*Trade Names*
Gates Engineering Company Division of the Glidden Company 100 South West Street Wilmington, Del., 19899	GACO
General Electric Company One Campbell Road Schenectady, N.Y., 12306	GE
Gilbreth Company 212 E. Courtland Street Philadelphia, Pa., 19120	GILBRETH
Girder Process, Inc. P. O. Box 96 Carlton Hill, N.J., 07073	GIRDER, GP
Glidden Company Gates Engineering Division 100 South West Street Wilmington, Del., 19899	GLIDDEN
B. F. Goodrich Company 500 Main Street Akron, Ohio	GOODRICH, GP PRE/SET, PLASTIKON, PLASTILOCK, VULCALOK, METAFIL, ANODEX
Goodyear Tire & Rubber Co. 1411 East Market Street Akron, Ohio	BONDOLITE, PLIOBOND, PLIOGRIP, PLIOTAC
Hepol Corporation 322 Houghton Avenue Olean, N.Y.	HEPOL
C. B. Hewitt & Brothers, Inc. 23-25 Greene Street New York, N.Y., 10013	GOOD BOND, NEWHOLD, HOLDBOND
Honeycomb Corporation of America 1225 Connecticut Avenue Bridgeport, Connecticut	HONEY-CO
Hughson Chemical Corporation Division of Lord Corporation 1635 West 12th Street Erie, Pa.	CHEMLOK
Isochem Cook Street Lincoln, R.I., 02865	ISOCHEMBOND, ISOCHEMREZ ISOSILIPOXY

Name and Address	*Trade Names*
Johns-Manville 22 East 40th Street New York, N.Y., 10016	DUTCH BRAND
Leffingwell Chemical Company P. O. Box 1187, Perry Annex Whittier, Calif., 90603	LEFKOWELD
Loctite Corporation North Mountain Road Newington, Conn., 06111	LOCTITE
Marblette Corporation 3731 30th Street Long Island City 1, N.Y.	MARAFOAM, MARAGLAS, MARASET, MARAWELD, MARAWOOD, MARBLETTE
Mereco Products Metachem Resins Corporation 530 Wellington Avenue Cranston, R.I., 02910	META-GRIP
Minnesota Mining & Mfg. Company 2501 Hudson Road St. Paul, Minn., 55119	SCOTCHWELD, AF ———, WEATHERBAN
Mobay Chemical Company Penn-Lincoln Parkway West Pittsburgh, Pa.	MULTRON, MONDUR, MONDUR MULTRON
Goodloe E. Moore, Inc. 2811 Vermilion Street Danville, Ill., 61832	TUFF-BOND
Pittsburgh Plate Glass Company Adhesive Products Division 225 Belleville Avenue Bloomfield, N.J., 07003	BONDMASTER
Plastic Associates Laguna Beach, Calif.	PA ———
Polymer Industries, Inc. Springdale, Conn., 06879	
Products Research & Chemical Corp. 2919 Empire Avenue Burbank, Calif.	PR ———
Raybestos Manhattan Company 75 East Main Street Stratford, Conn.	RAY-BOND, RAYCO

Name and Address	*Trade Names*
Reichold Chemicals, Inc. 525 North Broadway White Plains, N.Y.	PLYACIEN, PLYAMINE, PLYAMUL, LITHGOW, PLYOPHEN, NOBELAC
Ren Plastics 5656 South Cedar Street Lansing, Mich., 48909	REN
Reyolin, Inc. 1651 18th Street Santa Monica, Calif.	EPOLITE
Rohm and Haas Resinous Products Washington Square Philadelphia, Pa.	TEGO, UFORMITE, AMBERLITE, RHOPLEX
Rubber and Asbestos Corporation 225 Belleville Avenue Bloomfield, N.J.	PLYMASTER TREADMASTER
Savereisen Cements Company Pittsburgh, Pa., 15215	SAVEREISEN
Shell Chemical Company Plastics and Resins Division P. O. Box 831 Pittsburg, Calif.	SHELL EPON
Shur-Lok Bonded Structures Adhesives & Test Division 1300 East Normandy Place Santa Ana, Calif.	REDUX HI-DUX
Synco Resins Bethel, Conn., 06801	SYNCO
Thermo-Resist Corporation Fullerton, Calif.	THERMO GLOM-ON
Thiokol Chemical Corporation 780 North Clinton Avenue Trenton, N.J., 08607	THIOKOL TIPOX
Union Carbide 270 Park Avenue New York, N.Y., 10017	CELLOSIZE ERL
U. S. Rubber Company 1230 Avenue of the Americas New York, N.Y., 10020	LOTOL, KOTOL, KRALAC, KRALASTIC, INDU-SEALZ, NITREX

Name and Address	*Trade Names*
Whittaker Corporation Narmco Materials Division 600 Victoria Street Costa Mesa, Calif., 92627	NARMTAPE, METLBOND, NARMCO
Will Barbeau Associates 36 Kennedy Plaza Providence, R.I.	LOCTITE
X-Pando Corporation 43-15 36th Street Long Island City, N.Y., 10001	EXPENDOTITE
Xylos Rubber Company Division of Firestone Tire & Rubber Akron, Ohio, 44301	

Glossary of Terms Associated With Adhesive Bonding

A-stage: An early stage in the reaction of certain thermosetting resins during which the material is fusible and still soluble in certain liquids.

Abhesive: A material that resists adhesion. Abhesive coatings are applied to surfaces to prevent sticking, heat sealing, etc.

Ablative adhesives: This description applies to a material which absorbs heat (while part of it is being consumed by heat) through a decomposition process known as pyrolysis that takes place in the near-surface layer exposed to heat.

Abrasion cycle: The number of revolutions to which a specimen is subjected in evaluation of the abrasion resistance.

Abrasion resistance: Ability to resist surface wear; term usually applied to adhesive coatings, ablative adhesives, paint, etc.

Absolute viscosity: Of a fluid adhesive, the tangential force on unit area of either of two parallel planes at unit distance apart when the space between the planes is filled with the fluid in question and one of the planes moves with the unit differential velocity in its own plane. The unit of measurement is the centipoise.

Absorption: (1) The capillary or cellular attraction of adherend surfaces to draw off the liquid adhesive film into the substrate. (2) Term applied to the removal of oxygen, or other surface undesirables from the adherend surface during the curing process by the absorption properties of the adhesive system.

Accelerator: A substance that hastens a reaction or the solidification of an adhesive caused by a catalyst.

Accelerated test: A test in which conditions are intensified to reduce the time required to obtain the necessary or desired data; to reproduce the results expected during long-time service in a few hours.

Acceptance tests: All tests performed in connection with the fabrication of adhesive-bonded aerospace components which, directly or indirectly, contribute to the accepting for use of a material, detail, or assembly.

Acetal resin: High-molecular-weight, stable linear polymers of formaldehyde; an oxygen atom which joins the repeating units in an ether rather than ester type link.

Acetate: A salt or ester or acetic acid; cellulose acetates.

Acetone resin: A synthetic resin formed by the reaction of acetone with another compound such as phenol or formaldehyde.

Acid value: A determination of the free-acid content of an adhesive, normally expressed as the number of milligrams of potassium hydroxide required to neutralize 1 g of the adhesive using phenolphthaline as the indicator.

Acrylic resin: A synthetic resin prepared from acrylic acid or a derivative of acrylic acid.

Acrylonitrile: Butadiene-Styrene resins (ABS), thermoplastic resins with uniform molecular structure, high impact strength, high heat-distortion strength, good electrical properties, and low temperature properties; resistant to the actions of most solvents, oils, and chemicals.

Activation: The process of inducing radioactivity in a specimen by bombardment.

Activator: See Accelerator.

Additive: The addition of a material to the basic resin to change its physical properties.

Adherend: A body or material held to another body or material by an adhesive.

Adherend failure: Rupture of an adhesive bond, such that the separation appears to be within an adherend; sometimes erroneously called "cohesive failure."

Adhesion: The state in which two surfaces are held together by interfacial forces which may consist of interlocking action (mechanical means), or valence forces, or both.

Adhesion failure: Adhesion failure occurs when no bond is formed between an adhesive and an adherend or when a very poor bond is formed.

Adhesive: A substance capable of holding materials together by surface attachment.

Adhesive: (1) *Cold setting*, an adhesive system which sets at temperatures below 65°F; (2) *hot setting*, usually refers to an adhesive that sets above 260°F; (3) *intermediate setting*, cures in the temperature range of 100 to 260°F; (4) *pressure sensitive*, an adhesive designed to adhere to a surface at room temperature or is very soft and flexible; (5) *room-temperature setting*, an adhesive that sets between 65 and 100°F.

Adhesive age: The length of time in days from the date of shipment from the manufacturer's plant to the present date.

Adhesive batch: A homogenous quality of finished adhesive manufactured under controlled conditions at one time or representing a blend of several manufactured units of finished adhesives of the same formulation and processing.

Adhesive blocking: The adhesion between a potentially adhesive face and a standard test paper.

Adhesive dispersion: A two-phase adhesive system in which one phase is suspended in a liquid.

Adhesive evaluation tests: Those tests conducted to assure that an adhesive meets the requirements of the specification governing its use in the production of aerospace components.

Adhesive failure: A failure in the bond occurring at the adhesive-adherend interface.

Adhesive lot: All of the adhesive from one adhesive batch received in one shipment.

Adsorption: The adhesion of the molecules of gases, dissolved substances, or liquids in more or less concentrated form to the surfaces of solids or liquids with which they are in contact.

Affinity: An attraction or polar similarity between adhesive and adherend.

After bake: See Post Cure.

Aggregate: A hard fragmented material used with an epoxy binder (or other resin), common in plastic tooling.

Aggressive tack: See Dry Tack.

Aging: The change in properties of a material with time under stated conditions.

Air bubble: Air entrapment within the bond line or between the substrates causing an inclusion or void.

Air dry: Drying freshly applied wet adhesive by exposure to air at about 65 to 95°F.

Aliphatic: Organic compounds whose molecules do not have their carbon atoms arranged in a ring structure. This category therefore includes all the paraffin hydrocarbons and their saturated and unsaturated derivatives of all types.

Aliphatic hydrocarbons: Saturated hydrocarbons having an open chain structure. Examples are naphtha, gasoline, propane.

Alkyd resins: (1) Made by the union of diabasic acids or anhydrides, usually phthalic anhydride, with a polybasic alcohol such as glycerine. Hard resin types are produced by using rosin or similar resins as modifying agents. They have good weathering properties, heat resistance, good adhesion, and are noted for ease of application; (2) polyester resins made with fatty acids as a modifier. (See Polyester and Fatty Acids.)

Alloy: Composite adhesive made by blending polymers or copolymers with other polymers or elastomers under controlled conditions.

Alpha cellulose: A very pure cellulose prepared by special chemical treatment.

Amino: Indicates the presence of an $-NH_2$ or $-NH$ group.

Amino resin: A large class of thermosetting resins made by the reaction of an amine with an aldehyde. The aldehyde is usually formaldehyde and the most important amines are urea and melamine. In general, these resins are hard and brittle and are utilized primarily with fillers and modifiers.

Amorphous: Noncrystalline; without descriptive physical form or selective structure.

Amylaceous: Pertaining to starch; starchy.

Anaerobic: An adhesive that cures in the absence of oxygen.

Anisotropic: Exhibiting different properties when tested along axes in different directions.

Annealing: A process of holding a material at a temperature near, but below, its melting point, the objective being to permit stress relaxation without distortion of shape.

Anodizing: The application of a protective oxide film on aluminum, magnesium, or other light metals by passing a high-voltage electric current through a bath in which the metal is suspended. The metal serves as the anode. Three common ingredients used in anodized parts are sulfuric, chromic, and oxalic acids.

Antioxidant: Substance which prevents or slows down oxidation. For example, arsenic pentoxide is sometimes used as an anti-oxidant in the polyaromatic adhesives.

Antiozonant: A substance that retards or prevents the action of ozone on elastomeric systems, when exposed statically or dynamically to an atmosphere containing ozone.

Applicator mechanism: Devices such as wheels, stencils, brushes, etc., used to transfer adhesive from adhesive reservoirs onto adherend surfaces.

Aqueous: Water-containing or water-based. Refers to adhesive systems such as starch, dextrine, natural gums, animal glue, etc., which use water as the carrier system for the adhesive.

Arc resistance: The time required (in seconds) for a standard arc adjacent to the surface of a material to establish a conductive path in the material.

Aromatic hydrocarbons: Derived from or characterized by presence of unsaturated resonant ring structures.

Arsenic pentoxide: Derived by the action of an oxidizing agent such as nitric acid on arsenious oxide and is used in adhesives as an antioxidant agent.

Aspect ratio: The ratio of length to diameter of a fiber, used as a filler for adhesives.

Asphalt: A dark-colored viscous-to-solid hydrocarbon complex including (1) the easily fusible bitumens often associated with a mineral matrix, not having a waxy luster or unctuous feel; (2) fusible residuums obtained from the distillation of bitumens.

Assembly: A group of materials or parts, including adhesive, which has been placed together for bonding or which has been bonded together.

Assembly adhesive: An adhesive which can be used for bonding parts together, as in the manufacture of a boat, an airplane, furniture, or the like.

Assembly series: Groups of not less than five nor more than nine parts with successive sequence numbers, together with the correspondingly numbered process control assemblies, which have been cleaned, processed, assembled, and bonded with validated adhesive of the same type within the required time limits.

Assembly time: The time interval between the spreading of the adhesive on the adherend and the application of pressure or heat, or both, to the assembly.

ASTM: Abbreviation for American Society for Testing Materials, a national technical society which developed and standardized numerous tests pertinent to adhesives.

Atactic: A molecular chain in which the positions of the methyl groups are more or less in random order.

Atomic weight: Relative weight of an atom of any element as compared with that of one atom of oxygen taken at 16.

Autoclave: A large container in which controlled heat and/or pressure can be applied to adhesive-bonded assemblies for curing of the adhesive.

Autoclave bonding: Modification of the pressure-bag method for adhesive bonding. After final fit-up, an entire assembly is put into a steam or electrically heated autoclave at elevated pressure. Additional pressure achieves higher reinforcement loadings and improved removal of air.

Average (arithmetic mean): The sum of a group of test values divided by the number of test values summed.

Axial: Pertaining to, or forming an axis.

B-stage: An intermediate stage in the curing of a thermosetting adhesive during which the adhesive softens when heated and swells in contact with certain liquids but does not entirely fuse or dissolve. The resin in an uncured thermosetting adhesive is usually in this stage. Sometimes referred to as "Resitol."

Backing plate: A plate used with bonding tools as a support for cavity blocks, guide pins, etc.

Bactericide: Material used in small percentages to kill bacteria which may occur in liquid adhesive forms, or which may attack cast carbohydrate or proteinaceous adhesive films.

Bag bonding: A method of bonding involving the application of fluid pressure to a flexible cover which, usually in connection with a rigid die, completely encloses and exerts pressure on the assembly being bonded.

Barcol hardness: A hardness value obtained by measuring the resistance to penetration of a sharp steel point under load. The apparatus used to obtain this reading is called a barcol impressor and the readings are sometimes associated with the degree of cure but more often with aging characteristics of adhesive coated fabrics.

Bare compressive strength: The compressive strength of honeycomb materials loaded without stabilization of the plane surfaces. Bare compressive strength data are subject to a considerable variation in test values due to the difficulty of the honeycomb. Careful surface preparation and the use of self-aligning loading heads help to reduce the test scatter. Because the test setup and specimen preparation are simple, the bare compressive test has the advantage of giving a quickly obtainable indication of the relative strength of a honeycomb material.

Barium titanate: A ferroelectric ceramic used as a crystal for nondestructive testing devices, especially the resonant frequency ultrasonic technique.

Bathtub: The area of a sandwich panel slug material that has been machined out in the shape of a bathtub for weight reduction of the slug.

Bearing stress: The applied load in pounds divided by the bearing area of a specimen.

Bentonite: A clay containing large portions of the clay mineral montmorillonite with aluminum and silicates, sometimes used with magnesium and iron. Bentonite is widely used as an adhesive filler.

Benzene ring: The basic structure of benzene, the most important aromatic chemical. It is an unsaturated, resonant six-carbon ring, having three double bonds. One or more of the six hydrogen atoms of benzene may be replaced by other atoms or groups.

Binder: An adhesive substance, usually of liquid or molten form, used to create adhesion between aggregates, globules, etc. Distinguished from an adhesive in that it performs an internal adhesive function rather than a surface adhesive function.

Bisphenol A: A condensation product formed by the reaction of two molecules of phenol with acetone. This polyhydric phenol is the standard intermediate resin that is reacted with epichlorohydrin in the production of epoxy resins.

Bite: The ability of the adhesive to penetrate or dissolve the uppermost portions of the adherends.

Bleed: (1) To give up color when in contact with water or solvents. (2) An undesired movement of certain materials in adhesive (sometimes plasticizers) to the surface of the bonded article (especially plastics) or into adjacent material; (3) to evacuate air or gases from an assembly during the cure cycle.

Bleeder cloth: A cloth (usually glass or cotton) used to provide an avenue of escape for gases during the bonding cycle.

Blister: An elevation of the surface of an adherend, somewhat resembling in shape a blister on the human skin; its boundaries may be indefinitely outlined and it may have burst and become flattened. It may be caused by insufficient adhesive; inadequate curing time, temperature, or pressure; or trapped air, water, or solvent vapor.

Block polymer: A polymer whose molecule is made up of comparatively long sections that are of one chemical composition, those sections being separated from one another by segments of different chemical character, for example, blocks of polyvinyl chloride interspersed with blocks of polyvinyl acetate.

Blocking: (1) An undesired adhesion between touching layers of materials, such as occurs under moderate pressure during storage or use; (2) preventing adhesives from wetting in certain areas by the use of a substrate to which the adhesive will not adhere (Teflon, polyethylene).

Bloom: A visible surface exudation usually caused by a plasticizer and common in elastomeric adhesives.

Body: The consistency of an adhesive; thickness; viscosity.

Body putty: A paste-like mixture of an adhesive and filler used to repair small damaged areas or scratches on an assembly.

Bond: The union of materials by adhesives. To unite materials by means of an adhesive.

Bond strength: The unit load applied in tension, compression, flexure, peel impact, cleavage, or shear, required to break an adhesive assembly with failure occurring in or near the plane of the bond.

Bonded fabric: A web of fibers held together with an adhesive and does not form a continuous sheet of adhesive material, i.e., not impregnated completely but more a form of joining.

Bonded structure: The structure resulting when a combination of alternating dissimilar simple or composite materials are assembled and intimately attached to each other by applying a structural adhesive to the faying surfaces followed by curing of the adhesive by subjecting the structure to pressure, heat, or both.

Bonding operation: The curing of adhesives applied to details to effect the joining of parts into a single component.

Bonding procedure: All operations connected with bonding, from the prefit of the formed details through their assembly for final bonding.

Bonding process: All operations connected with the manufacture of assemblies whose parts are joined by adhesives.

Bonding range: See Tack Range.

Borate: To add borax to a starch adhesive to improve adhesive tack and viscosity.

Boss: Protuberance on a bonded part designed to add strength, to facilitate alignment during assembly, to provide for fastenings, etc.

Branched: In molecular structure of an adhesive polymer, this refers to side chains attached to the main chain, which may be long or short.

Brashness: A property of dry animal glues referring to the relative flexibility of a cast glue film. A brittleness resulting from drying, plasticizer migration, etc.

Breathing: When referring to plastic sheeting that is used for bagging, "breathing" indicates permeability to air.

Brittlepoint: The highest temperature at which an elastomer fractures in a prescribed impact test procedure.

Buckle: Sometimes used to describe the wrinkle or collapse of a bonded face sheet.

Built-up laminated wood: An assembly made by joining layers of lumber with mechanical fastenings so that the grain of all laminations is essentially parallel.

Bulk density: The mass per unit volume of adhesive as determined in a large volume.

Burr: A projection of metal beyond the normal plane of the surfaces establishing an edge.

Butyl acetate: Derived from esterification and distillation, after contact of butyl alcohol with acetic acid in the presence of a catalyst such as sulfuric acid. This material used as a dope for aircraft structures.

C-stage: The final stage in the reactions of a thermosetting adhesive in which the material is relatively insoluble and infusible. Thermosetting resins in a fully cured adhesive are in this stage. Sometimes referred to as "Resite."

Cab-O-Sil: Trade name for anhydrous silica used as a thixotropic agent for adhesives, usually as a filler.

Calendar: (1) To prepare adhesive film material by pressure between two or more counterrotating rolls; (2) the machine performing this operation.

Caprolactam: A cyclic amide-type compound, containing six carbon atoms. When the ring is opened, caprolactam is polymerizable into nylon resin known as "type-6 nylon" or "polycaprolactam."

Carving: Shaping, cutting honeycomb to the desired configuration.

Casein: A protein material precipitated from milk by the action of rennet or dilute acid. The acid casein is primarily used for the production of adhesives.

Cast: (1) To form a plastic object by pouring a fluid resin into an open mold where it hardens. This method is used to produce "dog bone" type specimens for testing. (2) Casting; the finished product.

Casting resin: A resin which can be cast and hardened in a mold to form a desired configuration; used in plastic tooling for adhesive bonding.

Catalyst: A chemical substance added to thermosetting resin adhesives to speed up the cure time of these adhesives and to increase the crosslinking of the synthetic polymer; a reactant material added to accelerate adhesive drying.

Catastropic failures: Gross failures of an unpredictable nature.

Caul: A sheet of material employed singly or in pairs in hot or cold pressing of assemblies being bonded. Cauls are employed usually to protect either of the faces of the press platen or both against marring and straining, to prevent sticking, and to facilitate press loading.

Caulk: To fill voids with plastic or semiplastic materials; to fill crevices in the adherend surface with adhesive materials; to provide a seal against moisture or solvent intrusion.

Cavity tool: A tool that is machined, cast, etc. where the desired contour of the part is represented by the cavity, may be used with press, vacuum bag, autoclave, etc.

Cellulose: A natural high-polymeric carbohydrate found in most plants; the main constituent of dried woods, flax, and cotton.

Cellulose acetate: An acetic acid ester of cellulose. It is obtained by the action under rigidly controlled conditions of acetic acid and acetic anhydride on purified cellulose. When compounded with suitable plasticizers, it produces a tough thermoplastic material.

Cement: To bond together or to adhere with a liquid adhesive; a liquid adhesive employing a solvent base of the synthetic elastomer or resin variety; an inorganic paste.

Cement, rubber: General term for solution of rubber in a hydrocarbon or other solvent solution.

Centipoise: One hundredth of a poise; a unit for measuring viscosity; a unit of the Bureau of Standards for viscosity measurement.

Chip: (1) To break away an adhesive flash utilizing a chisel or blade of some type; (2) a small test sample of adhesive.

Chromic acid: Crystallized out of a solution of sulfuric acid and sodium dichromate, used for surface preparation of metals for adhesive bonding.

Chucking: Applied to holding honeycomb core in place during machining.

Cleanliness: Refers to the condition of the structural components from the time they leave the cleaning tank until they are removed from the autoclave or oven as bonded structures.

Cleavage strength: The tensile load in terms of pounds per inch of width required to cause separation of a test specimen 1 in. in length as described in Standard Method 1071-T of Federal Test Method Standard 175.

Closed assembly time: The time interval between completion of assembly of the parts for bonding and the start of the cure cycle.

Cobwebbing: A phenomenon observed during the spray application of an adhesive characterized by the formulation of web-like threads along with the usual droplets as the adhesive leaves the nozzle of a spray gun.

Coefficient of expansion: The fractional change in dimensions of a material pertinent to a change in temperature.

Cohesion: The propensity to adhere to itself; the internal attraction of molecular particles toward each other; the ability to resist partition from the mass; internal adhesion.

Cohesive blocking: The blocking of two similar, potentially adhesive faces.

Cohesive failure: A failure which occurs within the adhesive itself. That is, if the adhesive has adhered to the metal involved and no voids are visible, and failure is completely in the adhesive layer, then 100 percent cohesive failure is obtained.

Cold flow: The movement of the adhesive interface due to exposure of the assembly to conditions of moisture, heat, etc., exceeding the resistance limits of the adhesive interface. (See Creep.)

Cold pressing: An assembly method in which the bonded structures are held in place without the application of heat or drying air until the adhesive interface has solidified and reached proper shear proportions.

Colloidal: A state of suspension in a liquid medium in which extremely small particles are suspended and dispersed but not dissolved in the liquid medium. The protection offered by proteins, such as casein and animal glue, to synthetic lattices is an example of colloidal protection.

Compressive modulus: The modulus of elasticity of honeycomb material measured parallel to the cell axes and denoted as E_c.

Compressive strength: The yield strength in compression of a material, expressed in pounds per square inch. For honeycomb materials, the load is applied normal to the plane surface and parallel to the axis of the cells.

Condensation: A chemical reaction in which two or more molecules combine with the separation of water or other substance. If a polymer is formed, it is called "polycondensation."

Condensation resin: A resin formed by polycondensation, e.g., the alkyd, phenol-aldehyde and urea-formaldehyde resins.

Consistency: The resistance of a liquid adhesive to deformation comprised of viscosity, plasticity, and other fundamental properties.

Contact failure: A failure which occurs when no adhesive layer is present over an area intended to be bonded. This may be caused by uneven surfaces, lack of uniform pressure distribution, or insufficient adhesive.

Contact-pressure adhesives: Liquid adhesives which thicken or resinify when bonding laminates, and require only enough pressure to assure intimate contact.

Contaminant: An impurity or foreign matter present in a bonded assembly which affects its usefulness.

Control: The authority (as given the quality control department) to require conformance to the applicable document and specifications establishing the quality of the materials and the processing requirements to be used in producing adhesive-bonded assemblies.

Controlled material: Any material which is subject to periodic testing after its acceptance to ensure that the material continues to meet the quality requirements of its specifications.

Cooling channels: Channels or passageways located within the body of a bonding fixture through which a cooling medium can be circulated to control temperature on the fixture surface.

Copolymer: A polymer formed by the reaction of two or more different monomers.

Copolymerization: Polymerization involving two or more different monomers.

Core: A lightweight natural, synthetic, or fabricated material bonded between the facings of sandwich construction in order to separate the facings and to support them against buckling under stress.

Core splicing: The joining of honeycomb core details into a single component by bonding the details together with structural adhesives.

Corrosion: Attack by the adhesive of one or both of the adherend surfaces due to chemically reactive materials in the adhesive film.

Cottoning: A phenomenon observed during machine application of an adhesive characterized by the formation of weblike filaments of adhesive between machine parts or between machine parts and the receiving surface during transfer of the liquid adhesive onto the réceiving surface.

Coverage: A measure of the ability of the adhesive to be spread over adherend surfaces; the total amount of adhesive required per 1000 sq ft of bonded assembly.

Crack: (1) An actual separation of adhesive. (2) Surface cracking exists only on the surface of the adhesive. These terms are usually associated with a nondestructive testing of adhesive joints.

Crater: Small, shallow, crater-like surface imperfection.

Crazing: (1) Fine cracks at the surface of the adhesive; (2) common in the acrylics, causes distorted optical properties; (3) fine cracks which may extend in a network on or under the surface or through a layer of a cured adhesive material.

Creep: The dimensional change with time of a material under load, following the initial instantaneous elastic deformation. Creep at room temperature is sometimes called "cold flow."

Crimp: To fold over and fasten under pressure; to indent the adherend surface in order to obtain more positive contact at the adhesive interface.

Cross laminated: A laminate in which some of the layers of material are oriented at right angles to the remaining layers with respect to the grain or strongest direction in tension.

Cross linking: Applied to polymer molecules, the setting up of chemical links between the molecular chains. When extensive, as in most thermosetting resins, crosslinking makes one infusible supermolecule of all the chains.

Crush strength: The load required to continue deformation of a honeycomb material after the initial compressive failure. Crush strength is used to determine the efficiency of a honeycomb material as an energy-absorbing material.

Cryogenics: The science of low-temperature environments, especially very low temperatures, i.e., testing with liquid oxygen (−298°F) or liquid helium (−452°F).

Cure: To further change the physical properties of an adhesive (which has set) by chemical reaction, which may be condensation, polymerization, or vulcanization. (See also Dry and Set.)

Cure cycle: The interval elapsing between the time the adhesive bond reaches a temperature of 120°F during the initial temperature rise and the time the adhesive bond has been cooled to a temperature of 180°F following the cure period.

Cure load: All assemblies and/or subassemblies cured on the same cure platen or in the same oven during the same cure cycle.

Cure period: The elapsed time the adhesive bond is required to remain at a specified cure temperature to produce a cure.

Curing agent: That part of a two-part adhesive which combines with the resin (binder) to produce a cured adhesive film.

Curing temperature: The temperature to which an adhesive or an assembly is subjected in order to cure the adhesive.

Curing time: The period of time during which an assembly is subjected to heat or pressure, or both, to cure the adhesive.

Curling: Warping or distortion of the bonded assembly due to the introduction of moisture or solvents into the adhered surfaces and due to the unequal contraction and expansion properties of the adherend surfaces involved.

Cut-back: Asphalt adhesive which has been liquified or diluted with solvents such as petroleum distillates; for use as low-strength adhesives, waterproofing, and protective coating.

Cycle: The complete, repeating sequence of operations in a process or part of a process. In adhesive bonding, the cycle time is the period or elapsed time between a certain point in one cycle and the same point in the next.

D-glass: A high-boron-content glass made especially for laminates requiring controlled dielectric constant; is also used, finely chopped, as a filler for liquid-type adhesives.

Daylight opening: Clearance between two platens of a bonding press in the open position.

Damping: The decay with time of the amplitude of free vibrations of a specimen.

Decompose: A chemical and physical material breakdown due to excess exposure to heat and oxidation or due to the effects of bacterial contamination of the adhesive.

Deflashing: Covers the range of finishing techniques used to remove the flash (excess, unwanted material) on a bonded assembly.

Deflection temperature under load: The temperature at which a bonded beam or honeycomb sandwich deflects a given amount under a specific load.

Deformation: (1) Any change of form or shape in an assembly or specimen; (2) the linear change of dimension of a body in a given direction produced by the action of external forces. (See Creep.)

Degassing: Removal of air or gases from adhesives, usually accomplished by subjecting the materials to a vacuum.

Degradation: A change in an adhesive material due to deteriorating influences.

Degree of polymerization (DP): The number of structural units, or mers, in the "average" polymer molecule in a particular sample. In most adhesives, the DP must reach several thousand if worthwhile physical properties are to be had.

Delamination: The separation of layers in a laminate because of failure of the adhesive, either in the adhesive itself or at the interface between the adhesive and the adherend, or because of cohesive failure of the adherend.

Denier: A filament or yarn numbering system in which the yarn number is equal numerically to the weight in grams of 9000 m. The lower the denier, the finer the strands of the material.

Density: Weight per unit volume of a material, expressed in grams per cubic centimeter, pounds per cubic foot, etc. The material involved may be core, adhesives, facings, or inserts.

Desiccant: Substance which can be used for drying purposes because of its affinity for water.

Destructive test: Actual destruction of a bonded assembly for the purpose of evaluating the bond properties. Destructive tests are usually made initially for qualification of materials and tools and periodically for tool and process control.

Detail: A single piece of material having no attachments and frequently referred to as a "detail part."

Deterioration: A permanent change in the physical properties of an adhesive evidenced by impairment or reduction of its functional adequacy.

Dielectric: A nonconductor of electricity; the ability of an adhesive to resist the flow of electrical current.

Dielectric constant: Normally, the relative dielectric constant. For practical purposes, the ratio of the capacitance of an assembly of two electrodes separated solely by an adhesive insulating material to its capacitance when the electrodes are separated by air.

Dielectric curing: The curing of a synthetic thermosetting resin by the passage of an electric charge produced from a high-frequency generator through the adhesive joint. Dielectric curing is usually employed with nonconductive materials such as wood. High-frequency current passing through the liquid adhesive interface creates rapid molecular motion and accelerates setting of the adhesive film.

Dielectric strength: The electric voltage gradient at which an insulating material (adhesive) is broken down or "arced through" in volts per mil of thickness.

Diluent: An ingredient added to an adhesive to reduce the concentration of bonding materials.

Dimensional stability: Ability of an adhesive-bonded part to retain the precise shape in which it was originally bonded.

Dip coating: Applying an adhesive coat (primer) by immersing the substrate into a tank of liquid resin.

Discoloration: Any change from an initial color possessed by an adhesive after or during cure.

Dispersion: (1) A two-phase system in which one phase, which may be solid or a liquid, is suspended in the other, which is a liquid; (2) finely divided particles of a material in suspension in a resin.

Doctor-bar: A scraper mechanism which regulates the amount of adhesive applied to the spreader roll or to the surface being coated.

Doctor-blade: See Doctor-bar.

Doctor-roll: A roller revolving at a different surface speed, or in an opposite direction, from the spreader roll, resulting in a wiping action for regulating the amount of adhesive supplied to the spreader roll.

Doping: (1) Coating a mold or bonding tool with a mold release; (2) application of a dope (adhesive) to a structure.

DPA: Diphenolic acid; basically a bisphenol-A molecule with a carboxylic acid group added.

Draft: The degree of taper of a side wall or the angle of clearance designed to facilitate removal of parts from a fixture or tool.

Drawing: The process of stretching a thermoplastic sheet or rod to reduce its cross-sectional area.

Driflex: A granular mixture of animal glue, sugars, and magnesium chloride used as a nonwarp paper adhesive after being reconstituted with water and heated to application temperature.

Dry: To change the physical state of an adhesive or an adherend by the loss of solvent constituents by evaporation, absorption, or both.

Dry bond: See Contact bonding.

Dry box: An enclosure used for mixing or handling hazardous materials; usually utilizes a pair of gloves that can be utilized from outside. An inert gas blanket is used inside the box.

Dry spot: Area of incomplete surface film on laminated plastics; an area over which the interlayer and the glass have not become bonded. A term common to nondestructive testing.

Dry strength: The strength of an adhesive joint determined immediately after drying under specified conditions or after a period of conditioning in the standard laboratory atmosphere.

Dry tack: The property of certain adhesives, particularly nonvulcanizing rubber adhesives, to adhere on contact to themselves at a stage in the evaporation of volatile constituents, even though they seem dry to the touch. Sometimes called "agressive tack."

Drying oil: One of many natural, usually vegetable oils—the glyceryl esters of unsaturated fatty acids—that harden in air by oxidation to a resinous skin. Typical drying oils are linseed and tung oils. Drying oils are the binding agents of oil paints and varnishes and some adhesive coatings.

Drying temperature: The temperature to which an adhesive on an adherend or in an assembly, or the assembly itself, is subjected to dry the adhesive.

Drying time: The period of time during which an adhesive, an adherend, or an assembly is allowed to dry with or without the application of heat, pressure, or both.

Durometer hardness: The hardness of a material (usually applied to elastomerics) as measured by a Shore durometer.

Dyes: Synthetic or natural organic chemicals that are usually soluble in solvents characterized by good transparency, high tinctorial strength, and low specific gravity.

Dynamic fatigue: Failure of a part under cyclic loading similar to fatigue in testing of metals.

Dynamic stress: Stress which originates or is induced into an adhesive joint by shocking (mechanical, usually rapid).

E-glass: A borosilicate glass; the type used for glass fibers for adhesive fillers and laminates. This type is also called "electric glass" because of its high electrical resistivity.

Edgemember: The structural member around the perimeter of a sandwich panel used to secure the panel to adjacent structure.

Edgewise: The application of forces in directions parallel to and actually in the plane of a sheet of sandwich.

Elastic limit: The greatest stress which an adhesive material is capable of developing without any permanent strain remaining upon complete release of the stress.

Elasticity: The extensible property of adhesive films or adhesive interfaces to contract and expand in such a manner as to overcome the differential contraction and expansion rates that the bonded adherends may exhibit.

Elastomer: A material which, at room temperature, can be stretched repeatedly to at least twice its original length and, upon release of the stress, will immediately return with force to approximately its original length.

Elastomeric: Capable of distension or deformation under stress; adhesives based on natural or synthetic rubbers.

Elongation: The fractional increase in length of a material stressed in tension.

Elongation at break: Elongation recorded at the moment of rupture of a bonded specimen, expressed as a percentage of the original length.

Embrittlement: The drying and solidification of the adhesive interface to the point where it exhibits fissures and stress cracks under low impact conditions; a condition resulting from the migration of the adhesive plasticizer into the adherend substrate; the solidification and stratification of a cast adhesive coating due to exposure to atmospheric oxidation.

Emulsion: A suspension of fine droplets of one liquid in another.

Encapsulating: Enclosing an article (usually an electronic component or the like) in a closed envelope of plastic by immersing the object in a casting resin and allowing the resin to polymerize it, or if not, to cool it.

Encapsulization: The enclosure of adhesive particles with a protective film which prevents adhesive-particle coalescence until such time as proper pressure or solvation is applied.

Endothermic: Requiring heat to produce reaction.

Entrained air: Globules of air forced into liquid adhesive systems by the action of applicator mechanisms working in the adhesive mass. Differentiated from foam particles in that the entrained air is not readily dissipated and tends to produce false viscosity properties.

Environment: The aggregate of surrounding conditions which influence the performance of the adhesive system. This may include temperature, humidity, radiation, electrical fields, salt spray, etc.

Epichlorohydrin: A basic chemical used in the production of epoxies. It contains an epoxy group and is highly reactive with polyhydric phenols such as bisphenol A.

Epoxy resins: An important class of structural adhesives; based on ethylene oxide, its derivatives or homologs, epoxy resins; form straight-chain thermoplastics and thermosetting resins, e.g., by the condensation of bisphenol and epichlorohydrin.

Ester: The reaction product of an alcohol and an acid.

Etching: The act or process of chemically cleaning material surfaces by immersing the material in an acid or a basic solution capable of dissolving the material being cleaned and allowing a desired amount of surface material to be removed by dissolution; may also involve the application of an electric current to cause or promote the chemical reaction.

Ether: One of a class of organic compounds in which any two organic radicals are attached directly to a single oxygen atom.

Ethyl cellulose: Thermoplastic material obtained by ethylation of cellulose, which is realized by a treatment with diethylic sulfate or ethylic halogenures and a base.

Exothermic: Evolving heat during reaction (cure).

Expiration time: The date and hour the usable life of a lot of adhesives ends and beyond which the adhesive should not be used, unless it is retested and a new usable life established.

Extender: A low-cost substance, generally having some adhesive action, added to an adhesive to reduce the concentration of the primary binder required per unit area. (See also Binder, Diluent, and Filler.)

Extensions: Pieces of bonded assemblies, which may be cut off after the structure has been bonded, used to supply specimens for destructive tests of adhesive bond properties.

Extrude: To expel or force through a measured orifice, to pump adhesive onto the adherend interface; to apply a molten thermoplastic at the adhesive interface.

Facing: One of the two outer layers which have been bonded to the core of a sandwich.

Fade-ometer: An apparatus for determining the resistance of resins and other materials to fading. This apparatus accelerates the fading by subjecting the article to high-intensity ultraviolet rays of approximately the same wavelength as those found in sunlight.

Failure: An event which occurs when a bonded joint does not meet the design standards.

Fatigue: (1) A condition of stress created by repeated flexing or impact force upon the adhesive-adherend interface; (2) the failure or decay of mechanical properties after repeated stress applications. Fatigue tests provide information on the ability of an adhesive to resist the development of cracks which will bring about failure as a result of continued cyclic bonds.

Fatigue ratio: The ratio of fatigue strength to tensile strength. Mean stress and alternate stress must be tabulated.

Fatigue strength: The maximum cyclic load a material can withstand for a specified number of cycles before rupture occurs; the residual strength after being subjected to fatigue loading.

Fatty acid: An organic acid obtained by the hydrolysis of natural fats and oils, e.g., stearic and palmitic acids. These acids are monobasic, may or may not have double bonds, and contain 16 or more carbon atoms.

Faying surface: The surface of an object which comes in contact with another surface to which it is bonded; the bonding surface.

Feathering: See Cottoning.

Fiber: This term usually refers to relatively short lengths of very small cross-sections of various materials. Fibers can be made by chopping filaments (converting). Staple fibers may be ½ to a few inches in length and usually 1 to 5 denier.

Fiber tear: The dislocation and rupture of paper or cloth fibers during the separation of the adhesive-adherend interface. Used to evaluate the effectiveness of adhesive bonding by rating the total percentage of area exhibiting fiber rupture after cleavage of the bonded assembly.

Filament: A variety of fiber characterized by extreme length, which permits its use in yarn with little or no twist and usually without the spinning operation required for fibers.

Filler: A relatively nonadhesive substance added to an adhesive to make it less costly, or to improve physical properties, particularly hardness, stiffness, impact strength, workability, permanence, color, and electrical properties.

Filler sheet: A sheet of deformable or resilient material which, when placed between the assembly to be bonded and the pressure applicator, or when distributed within a stack of assemblies, aids in providing uniform application of pressure over the area to be bonded.

Fillet: A rounded filling of the internal angle between two surfaces of an adhesive-bonded assembly.

Film: In the plastics industry, an optional term for a thermoplastic material having a thickness not greater than 0.010 in.

Film adhesive: An adhesive that has been placed on a carrier or calendered into a thin film (0.002 to 0.016 in.).

Film forming: The property or ability of an adhesive substance to cast a dimensionally stable continuous film. Also refers to the relative strength of a cast adhesive film. Adhesives with good film-forming characteristics are those which tend to deposit more uniform films of high structural strengths.

First-degree blocking: An adherence between the surfaces under test of such degree that when the upper specimen is lifted, the lower specimen will cling thereto but may be parted with no evidence of damage to either surface.

Fish eye: Small globular mass which has not blended completely into the surrounding material and is particularly evident in a transparent or translucent material.

Flake: Used to denote the dry, unplasticized base of cured cellulosic adhesives.

Flammability: Measure of the extent to which an adhesive will support combustion.

Flash: Excess adhesive exuded from a bonded assembly along the edges of faying surfaces. It must be removed before the part can be considered finished.

Flatwise: Describes the application of forces in a direction normal to the plane of sandwich. Thus, flatwise compression and flatwise tension designate forces applied to compress the sandwich core and to pull the facings from the core, respectively. Flatwise flexure designates bending so as to produce curvature of the plane of a sheet or sandwich.

Flexible resin: A resin that is, in comparison with others under consideration, less stiff. Loosely, any material in which the modulus of elasticity is so low that it is of no practical importance.

Flexibilizer: An additive that makes a resin or rubber more flexible, i.e., less stiff; also a plasticizer.

Flexural shear test: A type of honeycomb shear strength test in which the honeycomb is bonded between two facings and tested by loading the composite as a beam in bending. Flexure tests are of dubious value for obtaining the true core shear values of the honeycomb alone, since the action of the sandwich in bending and the strength and modulus of the facings tend to modify the values obtained. Shear strengths obtained by flexure can rarely be reliably duplicated from one test facility to another. The flexural shear test can only be considered a time test of the facings, adhesives, and core acting as a composite sandwich structure and of the geometry tested. Shear values obtained by the flexure test are often higher than by the plate shear test.

Flexural strength: The strength of a bonded assembly in bending, expressed as the tensile stress of the outermost fibers of a bent test sample at the instant of failure.

Floating platen: A platen located between the main head and the press table in a multidaylight press and capable of being moved independently of them.

Flow: A qualitative description of the fluidity of an adhesive material during the process of bonding, before the adhesive is set.

Foamed adhesive: An adhesive, the apparent density of which has been decreased substantially by the presence of numerous gaseous cells dispersed throughout its mass.

Foaming agents: Chemicals added to plastics and rubbers that generate inert gases on heating, causing the resin to assume a cellular structure.

Force dry: Drying of an adhesive by exposing it to circulating air at a temperature of 220° to 235°F for a specified period of time, usually 1 hr.

Foreign object (metallic): Metallic particles included in an adhesive which are foreign to its composition.

Foreign object (nonmetallic): Particles of a substance included in an adhesive which seem foreign to its composition.

Fracture: Rupture of surface without complete separation of laminate.

Frit: Calcined flint, sand, or glass, ground finely after fusing and used in a body paste or glazes so as to reduce by chemical composition any tendency of the ceramic materials to dissolve in water.

FRP: Fiberglass-reinforced plastics, used commonly in core materials; thus, such materials are labeled "FRP core."

Fully automatic press: A hydraulic press which operates continuously, the timing of the operations being controlled mechanically, electrically, hydraulically, or by a combination of any one of these methods.

Furan resins: Dark colored, thermosetting resins available primarily as liquids ranging from low-viscosity polymers to thick, heavy sirups.

Furnish: Synonymous with composition or makeup. Refers to ingredient composition for cellulosic sheets such as paper, particle board, paperboard, etc.

Fusible: Capable of being melted and formed into a continuous adhesive film; the property of adhesive melting in combination with substrate melting to form a homogeneous mass at the interface.

Gel: A semisolid mass, capable of deformation by heat or pressure; a system of solid aggregates dispersed through a liquid carrier medium; a nonflowing adhesive mass exhibiting strong cohesive forces with low shear resistance.

Gelation: Formation of a gel.

Gelation time: That interval of time, in connection with the use of synthetic thermosetting resins, extending from the introduction of a catalyst into a liquid adhesive system until the interval of gel formation.

Glue: An adhesive prepared from hides, tendons, cartilage, bones, etc., of animals by heating with water. Through general usage, this term is now synonymous with the term "adhesive." (See Adhesive, Cement, Mucilage, Paste, and Sizing.)

Glue joint: The area of a bonded assembly where the adhesive and adherend are in contact.

Glue-laminated wood: An assembly made by bonding layers of veneer or lumber with an adhesive so that the grain of all laminations is essentially parallel.

Glue line: The adhesive layer between two adherends.

GPD: Grams per denier.

Gum: Any of a class of colloidal substances, exuded by or prepared from plants, sticky when moist, composed of complex carbohydrates and organic acids, and that are soluble or swell in water. (See Adhesive and Resin.)

Hand layup: The process of placing successive plies of reinforcing material and application of the resin (or prepreg) manually.

Hard fibers: Fibers produced from leaves.

Hardener: A substance or mixture of substances added to an adhesive to promote or control the curing reaction by taking part in it. The term is also used to designate a substance added to control the degree of hardness of the cured film. (See Accelerator and Catalyst.)

Hardness: The resistance of surface indentation usually measured by a prescribed hardness tester.

Head block: A large, thick piece of lumber used for bottom and top of a bale of plywood during pressing and clamping.

Heat-activated adhesive: A dry adhesive film which is rendered tacky or fluid by application of heat or heat and pressure to the assembly.

Heat-convertible resin: A thermosetting resin that can be converted by heat to a solid.

Heat-distortion point: The temperature at which a standard test bar (ASTM D-648) deflects 0.010 in. under a stated load of either 66 or 264 psi.

Heat endurance: The time of heat aging that an adhesive can withstand before failing at a specified load (physical test).

Heat resistance: The ability of an adhesive to resist the deteriorating effects of elevated temperatures.

Heat seal: An adhesive film tended to be reactivated by the application of heat to one or both of the adherend surfaces; the process of bonding plastic substances by bringing the adherend surfaces to their melt points under complete contact and continuing pressure. Used frequently for sealing vacuum bags in bonding operation.

Heat sink: A contrivance for the absorption or transfer of heat away from a critical detail.

Heat treat: Term used to cover annealing, hardening, tempering, etc.

Heteropolymer: See Copolymer.

Heteropolymerization: See Copolymerization.

Hexa: Short for hexamethylenetetramine; source of reactive methylene for curing novolaks.

High elasticity: The property that a material presents when elongated (stretched) to rebound quickly without inherent damage.

High-load melt index: The rate of flow of a molten resin through a 0.0825-in. orifice when subjected to a force of 21,600 g at 190°C.

High-pressure laminates: Laminates molded and cured at pressures not lower than 1000 psi and more common in the range of 1200 to 2000 psi.

Homogeneity: Consisting of similar parts or elements.

Honeycomb core: A material produced primarily by applying continuous adhesive strips on metal foil and then stacking in such a way that after cure of the adhesive strips, the stack, when expanded, will result in a matrix of hexagonal cells. It is used primarily as core material for sandwich construction.

Hot melt: A general term referring to thermoplastic synthetic resins used as 100 percent solid adhesives at temperatures between 250° and 400°F.

Hot pickup gum: A rosin-based material, used for affixing paper labels to metallic cans on automatic can-labeling equipment.

Hot pressing: The accelerated curing of thermosetting resin adhesives by the application of heat and pressure to the bonded assembly. Primarily restricted to plywood manufacture.

Hot-setting adhesive: An adhesive which requires a temperature at or above 100°C (212°F) to set it.

Hydraulic press: A press in which the platens are pressured by pressure exerted on a fluid.

Hydrogen peroxide: A colorless, unstable, oily liquid, H_2O_2, the aqueous solution of which is used as an antiseptic and a bleaching agent.

Hydrolysis: Chemical decomposition of a substance involving the addition of water.

Hydrophilic: Capable of absorbing water.

Hydrophobic: Capable of repelling water.

Hygroscopic: Tending to absorb moisture.

Hysteresis: The noncoincidence of the elastic loading and unloading curves under cyclic stressing; the percent energy loss per cycle of deformation or the ratio of energy absorbed to the total energy output.

Identical bonding operations: Bonding operations, in the same cure load, in which all the assemblies have received the same processing, have been assembled with adhesive from the same lot and with the same expiration time.

Impact tests: Measures the energy necessary to fracture or break an adhesive joint as a result of a swinging pendulum or other impact device.

Impact resistance: Relative susceptibility of adhesives to fracture by shock, e.g., as indicated by the energy expended by a standard pendulum-type impact machine in breaking a standard specimen in one blow.

Impact shock: A stress transmitted to the adhesive interface resulting from the sudden jarring or vibration of the bonded assembly.

Impact strength: (1) The ability of a material to withstand shock loading; (2) the work done in fracturing, under shock loading, a specified test specimen in a specified manner.

Impregnate: To provide liquid penetration into a porous or fibrous material; the dipping or immersion of a fibrous substrate into an adhesive liquid.

Inert: Not capable of reaction; a surface, such as polyethylene, which is not readily affected by adhesive solvent.

Inert filler: A material added to an adhesive to alter properties by physical rather than chemical means.

Infrared: Part of the electromagnetic spectrum lying outside the visible light range at its red end. Radiant heat is in this range, and infrared heaters are used as a source of heat in adhesive bonding.

Infusible: Does not melt when heated.

Inhibitor: A substance which slows down chemical reaction. Inhibitors are sometimes used in certain types of adhesives to prolong storage or working life. (See Retarder.)

Inorganic: Applies to the chemistry of all elements and compounds not classified as organic; matter other than vegetable, such as earthy or mineral matter.

Inorganic pigments: Natural or synthetic metallic oxides, sulfides, etc., calcined during processing at 1200 to 2200°F. They impart heat stability, weathering resistance, color, etc., to adhesives.

Instron: An instrument utilized to determine the tensile and compressive properties of material.

Insulation: A coating of a dielectric or essentially nonconducting material whose purpose it is to prevent the transmission of heat or electricity.

Insulator: (1) An adhesive material of low electrical conductivity; (2) an adhesive with low thermal conductivity.

Interface: The adherend surfaces immediately adjacent to and in contact with the adhesive layer; the area of contact between the adhesive and adherend surfaces.

Interlaminar shear strength: The maximum shear stress existing between layers of laminated substrates.

Intermesh: The positioning of adjacent pieces of honeycomb so that the outermost edge of one piece falls within the outermost edge of the adjacent piece.

Internal stress: Stress created within the adhesive layer by the movement of the adherends at differential rates or by contraction or expansion of the adhesive layer.

Irreversible: Not capable of redissolving or remelting. Refers to thermosetting synthetic resins; refers to chemical reactions which proceed in a single direction and are not capable of reversal.

Isocyanate resins: Resins containing organic isocyanate radicals. They are generally reacted with polyols such as polyester or polyether and the reactants are joined through the formation of the urethane linkage. Synonym for "polyurethanes."

Isomeric: Composed of the same elements united in the same proportion by weight, but differing in one or more properties because of difference in structure.

Isomers: Two or more compounds having the same kind and number of atoms but with different molecular structures or properties.

Isotropic: Material, either facings or cores, having the same properties in all directions.

Izod impact test: A destructive test designed to determine the resistance of an adhesive to the impact of a suddenly applied force.

Jacket: An enveloping hollow metal cover for holding circulating steam or water used to heat or cool the mechanism it covers (such as a mixing vessel, mold, or platen).

Jig: A clamping device used to immobilize the bonded assembly until complete solidification of the adhesive film has taken place.

Joint: The location at which two adherends are held together with a layer of adhesive.

Joint aging time: See Joint conditioning time.

Joint conditioning time: The time interval between the removal of the joint from the conditions of heat, pressure, or both, used to accomplish bonding, and the attainment of approximately maximum bond strength. Sometimes called "joint aging time."

Joint glue: See Assembly adhesive.

Kinetic viscosity: The viscosity/density ratio.

Lack of fill-out: An area of reinforcement that has not been wet with resin; mostly seen at the edge of a laminate.

Lacquer: Solution of natural or synthetic resins, etc., in readily evaporating solvents, which is used as a protective coating.

Laminate (noun): A product made by bonding together two or more layers of material with an adhesive.

Laminate (verb): To unite layers of material with adhesive.

Lamination: The process of preparing a laminate; also any layer in a laminate.

Lap joint: A joint made by placing one adherend partly over another and bonding together the overlapped portion.

Latex: An emulsion of rubber or resin particles dispersed in an aqueous medium; a natural or synthetic elastomeric dispersion in an aqueous system.

Lay-flat: An adhesive material with good noncurling and nondistension characteristics; the property of nonwarping in laminating adhesives.

Layup: A term used to indicate the placing of materials in a tool in preparation for bonding. Used frequently pertinent to sandwich structure or reinforced plastic laminates.

Let-go: An area in laminated glass oven which the initial adhesion between interlay and glass has been lost.

Limiting viscosity number: The limiting value at infinite dilution of the ratio of the specific viscosity of the polymer solution to its concentration.

Linear molecule: A long chain molecule as contrasted to one having many side chains or branches.

Liquifier: A material—such as urea, ammonium thiocyanate, etc.—used to reduce the gel point and viscosity of carbohydrate or proteinaceous systems.

Long time exposure: A relative term, found in most military specifications for adhesives, used normally to designate an exposure period of 192 hours' duration to any desired medium or condition.

Loss factor: The product of the power factor and the dielectric constant.

Low-pressure laminates: In general, laminates molded and cured in the pressure range of 400 psi to contact pressure.

M-glass: A high beryllia glass designed especially for high modulus of elasticity.

Mallet-tapping: A light tapping with a mallet or other object to determine if voids exist in adhesive joints, based on the sound transmission theory.

Mandrel: The core around which paper, fabric, or resin-impregnated fibrous glass is wound to form pipes or tubes.

Manufactured unit: A quantity of finished adhesive or finished adhesive components, processed at one time. It may be a batch or a part thereof.

Mark-off or printing: An indentation or imprinting of the skin surface.

Mastic: A highly viscous, organic adhesive or sealant, usually of putty-like consistency; also, any of various pasty cements, especially those made by boiling bituminous matter such as tar or asphalt with an extender such as fine sand. Mastic often implies usage for sealing or caulking, and lack of high-bond strength. (See Adhesive, Cement, Gum, and Resin.)

Mat: A randomly distributed felt of glass fibers used in reinforced plastics layup molding.

Matrix: The part of an adhesive which surrounds or engulfs embedded filler or reinforcing particles and filaments.

Maturing temperature: The temperature as a function of time and bonding condition, which produces desired characteristics in bonded components.

Mechanical adhesion: Adhesion between surfaces in which the adhesive holds the parts together by interlocking action.

Melamine formaldehyde resin: Classified as a synthetic resin derived from the reaction of melamine (2,4,6-triamino-1,3,5 triazine) with formaldehyde or its polymers.

Melting point: The temperature at which a resin changes from a solid to a liquid.

Mer: The repeating structural unit of any high polymer.

Metal bond: The process of joining metals by use of adhesives, heat and pressure generally being used to produce a final bond.

Metallographic inspection: As used in this manual, metallographic inspection refers to microscopic examination of metallic materials for the determination of microstructure.

Micron: One $\mu = 0.001$ mm $= 0.00003937$ in.

Migration: A condition of extraction whereby a solvent, be it water or an organic solvent, selectively dissolves a portion of the adhesive film; the transferral of the plasticizer portion of an adhesive into the adherend resulting from the attraction of the adherend composition to the plasticizer used in the adhesive material composition.

Migration of plasticizer: Loss of plasticizer from an elastomeric plastic compound with subsequent absorption by an adjacent medium of lower plasticizer concentration.

Mileage: See Coverage.

Miscible: Capable of being mixed; mutually soluble in each other.

Moca: Trademark for methylene-bisortho-chloroaniline, used as a curing agent for polyurethanes and epoxy resins.

Modifier: Any chemically inert ingredient added to an adhesive formulation that changes its properties but does not react chemically with the binder.

Modulus in flexure: The ratio of the flexure stress to the strain in the adhesive over a range for which this value is constant.

Modulus, initial: Young's modulus.

Modulus in shear: The ratio of the shear stress to the strain in the adhesive, over the range for which this value is constant.

Modulus in tension: The ratio of the tension stress to the strain in the adhesive over the range for which this value is constant.

Modulus of elasticity: The ratio of stress to strain in a material that is elastically deformed.

Modulus of resilience: The energy that can be absorbed per unit volume without creating a permanent distortion, calculated by integrating the stress/strain curve from zero to the elastic limit, and dividing the original volume of the specimen.

Moisture absorption: The pickup of water vapor from the air by an adhesive. This relates only to vapor taken from the air by a material and must be distinguished from water absorption which results in a weight gain due to water pickup by immersion.

Mold release: A liquid, powder, or other "slip agent" used to prevent adhesion to tooling during a bonding operation.

Monomer: A molecule or compound usually containing carbon, and of relatively low molecular weight and simple structure, which is capable of conversion to polymers, synthetic resins, or elastomers by combination with itself or other similar molecules or compounds.

Mooning: A permanent deflection in the material usually caused by supporting the weight of the material by its end.

Movable platen: The large upper platen of an adhesive-bonding press. This platen is moved either by a hydraulic ram or a toggle mechanism.

Mucilage: An adhesive prepared from a gum and water. Also, in a more general sense, a liquid adhesive which has a low order of bonding strength.

Multiple-layer adhesive: A film adhesive, usually supported with a different adhesive composition on each side; designed to bond dissimilar materials such as the core-to-face bond of a sandwich composite.

MVT: Moisture vapor transmission. A measure of the adhesive film's ability to prevent the transmission of moisture through the adhesive film into the opposing substrates.

Necking: The localized reduction in cross-section which may occur in a material under tensile stress.

Node bonds: The areas of honeycomb core where the sheets of foil are bonded together to form the honeycomb cells.

Noncontrolled material: Any material which is not subject to periodic testing to insure that the material continues to meet the quality requirements of its specifications.

Nondestructive testing: Any method of testing (visual, radiographic, ultrasonic, microscopic, magnetic particle, penetrant, gauging, audio, etc.) used to determine conformance of adhesive bonded structures to applicable specifications or engineering drawings without detrimentally affecting the mechanical properties of the structure.

Nonhygroscopic: Will not absorb or retain an appreciable quantity of moisture.

Nonpolar: Having no concentrations of electrical charge on a molecular scale, thus incapable of significant dielectric loss. Examples among resins are polystyrene and polyethylene.

Nonrigid plastic: A nonrigid plastic is one which has a stiffness or apparent modulus of elasticity of not over 50,000 psi at 25°C when determined according to ASTM Test Procedure D747-43-T.

Nonwarp: Refers to an adhesive system which does not distend, curl, shrink, wrinkle, etc., the laminated structure; a plasticized animal glue used for paper-box manufacture; a plasticized dextrine adhesive used for paper-mounting operations.

Notch sensitivity: The extent to which the sensitivity of a material to fracture is increased by the presence of a surface inhomogeneity such as a notch, a sudden change in section, a crack, or a scratch. Low notch sensitivity usually is associated with ductile materials, and high notch sensitivity with brittle materials.

Novolak: A phenolic resin which, unless a source of methylene groups is added, remains permanently thermoplastic.

Nylon: A generic name for all synthetic polyamides.

Offset yield strength: The stress at which the strain exceeds by a specific amount (the offset) on extension of the initial proportional portion of the stress-strain curve. It is expressed in force per unit area, usually pounds per square inch.

Oil-soluble resin: Resin which at moderate temperatures will dissolve in, disperse in, or react with drying oils to give a homogeneous film of modified characteristics.

Opalescence: The limited clarity of vision through a sheet of cured adhesive at any angle, because of diffusion within or on the surface of the material, which becomes a consideration in bonding optics.

Open-assembly time: The time interval between the application of the adhesive on the adherend and the completion of assembly of parts for bonding.

Orange peel: Surface roughness somewhat resembling the surface of an orange.

Organic: Designating or composed of matter originating in plant or animal life or composed of chemicals of hydrocarbon origin, either natural or synthetic.

Orthotropic: Material, either facings or cores, having different strength and elastic properties in different directions.

Out-of-process failure: Failure resulting from errors made by the personnel performing the operation (incorrect timing, sequence, application and curing of the adhesive, tooling trouble due to incorrect layup, etc.).

Out-of-station bonding: A bonding operation not performed in a bonding shop or normal bonding shop facilities; an example is an in-field repair or installation.

Overlap: A simple adhesive joint in which the surface of one adherend extends past the leading edge of another.

Overlay sheet (surfacing mat): A nonwoven fibrous mat (either in glass, synthetic fiber, etc.) used as the top layer in a cloth or mat layup to provide a smoother finish or to minimize the appearance of the fibrous pattern.

Over-the-hill: Synonymous with "kicked." Both expressions used to indicate the adhesive is no longer in workable state relative to "pot-" or "work" life.

Oxidation: The chemical reaction involving the process of combining with oxygen to form an oxide; the deterioration of an adhesive film due to atmospheric exposure; the breakdown of a hot-melt adhesive due to prolonged heating and oxide formation.

Parallel laminate: A laminate in which all the layers of material are oriented approximately parallel with respect to the grain or strongest direction in tension.

Parameter: An arbitrary constant, as distinguished from a fixed or absolute constant. Any desired numerical value may be given as a parameter.

Part: Synonymous with assembly, subassembly, or detail part.

Parting agent: A material used to prevent the adhesive from bonding other materials together or adhering to other materials.

Paste: An adhesive composition having a characteristic plastic-type consistency, a high order of yield value, and low bond strength. A typical paste is prepared by heating a mixture of starch and water and subsequently cooling the hydrolyzed product. (See Adhesive, Glue, Mucilage, and Sizing.)

Paste-back: An increase in viscosity; a retrogradation of a vegetable adhesive form.

Peel strength: Bond strength in pounds per inch width, obtained from peel test.

Peel test: A test made on bonded strips of metals by peeling the metal strips back and recording the adhesive strength values.

Penetration: The entering of an adhesive into an adherend; penetration implies wetting of an adherend by an adhesive.

Permanence: The resistance of an adhesive bond to deteriorating influences.

Permanent set: The increase in length, expressed as a percentage of the original length, by which an elastic material fails to return to original length after being stressed for a standard period of time.

Permeability: (1) The passage or diffusion of a gas, vapor, liquid, or solid through a barrier without physically or chemically affecting it; (2) the rate of such passage.

pH: The log of the reciprocal of the hydrogen-ion concentration; a measure of the acidity or alkalinity in an empirical scale ranging from 1 through 14, with those values below 7 being recorded as acid and those values above 7 being recorded as alkaline.

Phenolic resin: A synthetic resin produced by the condensation of an aromatic alcohol with an aldehyde, particularly of phenol with formaldehyde. Phenolic resins form the basis of thermosetting molding materials, laminated sheet, and stoving varnishes. They are also used as impregnating agents and as components of paints, varnishes, lacquers, and adhesives.

Pick: To experience tack; to transfer unevenly from an adhesive applicator mechanism due to high surface tack; to offset onto opposing surfaces; the relative integral strength of a cellulosic substrate relating to its ability to resist fiber distension when applied to a tacky surface and removed.

Pickup roll: An adhesive spreading device where the roll for picking up the adhesive and supplying it to a spreader roll runs in a reservoir of adhesive.

Piezoelectricity: Electricity produced by pressure, as in a crystal subject to pressure along a certain axis.

Pimple: Small, sharp or conical elevation, (usually resin-rich) on the surface of a plastic, whose form resembles a pimple in the common meaning.

Pit (pinhole): Small regular or irregular crater in the surface of a plastic or adhesive usually with width approximately of the same order of magnitude as its depth.

Plastic deformation: Change in dimensions of an object under load that is not recovered when the load is removed.

Plastic tooling: Tools constructed of plastics, generally laminates or casting resins.

Platen press: A unit with flat or formed plates which applies the necessary pressure and temperature for curing adhesive-bonded assemblies.

Platens: The mounting plates of a press used to apply heat and pressure during the bonding cycle of an adhesive-bonded assembly.

Plasticity: A property of adhesives which allows the material to be deformed continuously and permanently without rupture, upon the application of a force that exceeds the yield value of the materials.

Plasticizer: A material incorporated in an adhesive to increase its flexibility, workability, or ductibility. The addition of the plasticizer may cause a reduction in melt viscosity, lower the temperature of the second-order transition, or lower the elastic modulus of the solidified adhesive.

Plastometer: An instrument for determining the flow properties of a thermoplastic resin by forcing the molten resin through an orifice of specific size at a specified temperature and pressure.

Plate shear test: A type of honeycomb shear strength test in which the honeycomb is bonded between two thick steel plates which are displaced relative to each other to place the specimen in shear. Displacement is accomplished by loading either in tension or in compression. The plate shear test represents the best currently known method for obtaining true shear data on honeycomb material.

Plywood: A cross-laminated assembly made of (1) layers of veneer joined with adhesive, or (2) veneer in combination with a lumber core or plies joined with an adhesive. Two types of plywood are recognized: veneer plywood and lumber core plywood.

Plywood adhesive: An adhesive suitable for bonding wood veneers together in the manufacture of plywood.

Poise: A unit of viscosity in which the shearing stress is expressed in degrees per square centimeter to produce a velocity gradient of 1 cm/sec/cm.

Poisson's ratio: When a material is stretched, the cross-section area changes as well as the length. Poisson's ratio is the constant relating these changes in dimensions.

Polarity: Refers to the relative surface change of a material resulting from the molecular structure of the adherend surface.

Polyacrylate: A polymer of an ester of acrylic acid or its derivatives.

Polyamide: A polymer in which the structural units are linked by amide or thioamide groupings. Many polyamides are fiber-forming.

Polybutylene: A material composed of polymers of butylene ranging in consistency from a viscous liquid to a rubbery solid.

Polycarbonate resins: Polymers derived from the direct reaction between aromatic and aliphatic dihydroxy compounds with phosgene, or by the ester-exchange reaction with appropriate phosgene-derived precursors.

Polycondensation: A chemical reaction in which the molecules of a monomer link together to form a polymer, releasing water or a similar simple substance. (See Polymerization.)

Polyurethanes: Synthetic polymers that may be either thermosetting or thermoplastic and range from soft and rubber-like to hard and brittle. They are usually made by the action of toluene diisocyanate or another diisocyanate with polyols, polyethers, polyesters, amines, or other materials containing active hydrogens.

Polyester: A resin formed by the reaction between a dibasic acid and a dihydroxy alcohol, both organic. Modification with multifunctional acids and/or bases and some unsaturated reactants permits crosslinking to thermosetting resins. Polyesters modified with fatty acids are called alkyds.

Polymer: A large molecule of high molecular weight, formed by the reaction of simple molecules (monomers) having functional groups which permit their combination to proceed to high molecular weights under suitable conditions. Polymers may be formed by polymerization (addition polymer) or polycondensation (condensation polymer). When two or more monomers are involved, the product is called a "copolymer."

Polymerization: A chemical reaction in which the molecules of a monomer link together to form large molecules (polymer) whose molecular weight is a multiple of that of the original material without releasing any other substance. When two or more monomers are involved, the process is called "copolymerization" or "heteropolymerization."

Porosity: Presence of numerous visible pits (pinholes).

Post-cure bonding: Post curing at elevated temperatures after the parts have been removed from the autoclave or press to obtain higher heat-resistant properties of the adhesive bond.

Pot life: See Working life.

Potting: See Encapsulization.

Pregel: An extra layer of cured resin on part of the surface of the laminate (not including gel coats).

Preimpregnation: See Prepreg.

Preloaded: Term used to denote preloading (stress) due to thermal expansion or thermal contraction during a mechanical test before the testing mechanism is set into motion.

Premix: An adhesive that has all the ingredients mixed to produce a joint and is then sealed or frozen for preservation until ready for use.

Prepreg: Ready-to-use adhesive film or impregnated carrier; usually in a gel state or B-stage condition.

Press bonding: Using a platen press or unit tool to apply pressure to the structures being bonded.

Pressure bags: Bags made of rubber, rubberized fabric, plastic, or other impermeable materials, and used to provide a flexible, impermeable barrier between the pressure medium and the part being bonded.

Pressure-sensitive adhesive: An adhesive formulated so as to adhere to a surface at room temperature by briefly applied pressure alone.

Primer: A coating applied to a surface, prior to the application of an adhesive, to improve the performance of the bond.

Process control: The continual surveillance of all the steps associated with adhesive bonding to ensure adherence to accepted and approved procedures.

Process-control chart: A combined tabulated and graphical arrangement of test results and allied data for each production assembly arranged in chronological order of assembly.

Process test failure: Failure due to elements beyond the control of the personnel performing the operation (inferior adhesives, contaminations, or out-of-balance cleaning or treating solutions, etc.).

Progressive bonding: A method of curing a resin adhesive in successive steps or stages by application of heat and pressure.

Promoter: A chemical, itself a feeble catalyst, that greatly increases the activity of a given catalyst.

Proof resilience: The tensile strength required to elongate an elastomer from zero elongation to the breaking point, which is expressed in foot pounds per cubic inch of original dimension.

Proportional limit: The greatest stress which a material is capable of sustaining without deviation from proportionality of stress and strain (Hook's law). It is expressed in force per unit area, usually in pounds per square inch.

Proteinaceous: Of protein base; refers to adhesive materials such as animal glue, casein, soya, etc., which are protein materials.

Prototype: A model or unit of an assembly to be manufactured for testing purposes to validate the design, materials, and processes.
psi: Pounds per square inch.
Pull strength: Bond strength in pounds per square inch; results of a tensile test.
Purchaser inspection: The examination and/or testing of any material for conformance to specification requirements prior to acceptance from the vendor.
Pyrolysis: Material decomposition by heat.

Qualification tests: An investigation, independent of a procurement action, performed on an adhesive product to determine whether or not the product conforms to all requirements of the proposed application.
Quality assurance: The function of evaluating product quality and the procedures taken to ensure that the final product conforms to the specification requirements. It has for its purpose the continuing assurance of the customer that the product he receives is of, or better than, the quality level he expects.
Quality control: Monitoring and controlling the usefulness of the end product.

Range: The difference between the extreme high and low test values obtained from specimens cut from one test assembly.
Refractory: A difficult-to-fuse material such as ceramic adhesives which requires extremely high temperatures for fusion.
Reinforcement: A strong inert material bound into an adhesive to improve its strength, stiffness, and impact resistance. Reinforcements are usually long fibers of glass, sisal, cotton, etc., in woven or nonwoven form. To be effective, the reinforcing material must form a strong adhesive bond with the resin.
Relaxation: Decrease in stress under sustained constant strain or creep under constant load.
Release agent: A material which is placed on tooling or other objects to prevent the adhesive from adhering to them during a bonding operation. The agent may be a mineral spray release or a plastic sheeting such as Teflon.
Release for production use: An adhesive shall be considered released for production use when, after initial quality-assurance testing, it is issued for production use.
Reliability: The probability that an assembly or part will function properly in the operating environment for the expected service life.
Reliability, initial: Degree level of reliability designed into the product.
Reliability, plateau: The ultimate performance which might be obtained under maximum operating production levels in all phases of a particular manufacturing venture.
Remoistening adhesive: An adhesive system, such as dextrine, animal glue, gum arabic, etc., which is reactivated by the deposit of water upon the adhesive film.
Requalification: Repetition of the qualification requirements, in whole or in part, as specified.

Resilience: The ratio of energy returned on recovery from deformation to the work input to produce the deformation. Usually expressed as a percentage; the ability to quickly regain an original shape after being stretched or distorted.

Resiliency: Ability to quickly regain an original shape after being strained or distorted.

Resin: Any of a class of solid, semisolid, or pseudosolid organic materials of natural or synthetic origin, generally of high molecular weight, insoluble in water, having no tendency to crystallize, exhibiting a tendency to flow when subjected to stress, usually having a softening or melting range, and usually fracturing conchoidally. (See Gum.)

Resin applicator: A device for applying or spreading an adhesive resin system.

Resin content: The amount of resin in a laminate expressed as a percent of total weight or total volume.

Resin pocket: An apparent accumulation of excess resin in a small, localized area.

Resin-rich edge: Insufficient reinforcing material at the edge of the molded laminate.

Resin-starved: Describing an area having an insufficient amount of resin.

Resinoid: Any of the class of thermosetting synthetic resins, either in their initial temporary fusible state or in their final infusible state.

Resite: An alternate term for C-stage.

Resitol: An alternate term for B-stage.

Resol: An alternate term for A-stage.

Restricted adhesive: An adhesive which for any reason cannot be validated and therefore cannot be assigned a usable life. Such an adhesive cannot be used for structural bonding.

Retarder: A substance added to an adhesive to slow down chemical reaction. (See Inhibitor.)

Retrogradation: A process of deterioration; a reversal to a simpler physical form; a chemical reaction involving vegetable adhesives, characterized by a reversion to a simpler molecular structure.

Rheology: Pertaining to viscosity behavior; the viscosity performance of an adhesive material under conditions of shear; the plastic flow properties of an adhesive interface.

Rib: A reinforcing member of a fabricated or bonded assembly.

Rigid resin: One having a modulus high enough to be of practical importance, e.g., 10,000 psi or greater.

Room temperature: As used in most specifications for adhesives and adhesive bonding, this term applies to temperatures within the range of approximately 75 ± 5°F.

Room-temperature-setting adhesive: An adhesive which sets in the temperature range of 20 to 30°C (68 to 86°F).

Room-temperature vulcanization (RTV): Vulcanization or curing by chemical reaction at room temperature; usually applies to elastomeric systems.

Rosin: A resin obtained as a residue in the distillation of crude turpentine from the sap of the pine trees (gum rosin) or from an extract of the stumps and other parts of the tree (wood rosin).

Rupture: A cleavage or break in the adhesive film, resulting from physical stress.

S-glass: A magnesia-alumina-silicate glass, especially designed to provide high tensile strength.

Safety hardener: A curing agent which causes only a minimum amount of toxic effect on the human body, either on contact with the skin or as vapor in the air.

Sandwich panel: A laminar construction, consisting of thin facings bonded to a relatively thick lightweight core, resulting in a rigid and lightweight panel.

Saturated compounds: Organic compounds which do not contain double or triple bonds and thus cannot add on elements or compounds.

Scarf joint: A colinear joint made by cutting away similar angular segments of two adherends, then bonding the adherends with the two cut faces fitted together.

Scratch: Shallow mark, groove, furrow, or channel normally caused by improper handling or storage.

Secant modulus: The ratio of total stress to corresponding strain at any specific point on the stress/strain curve. It is expressed in force per unit area, usually in pounds per square inch, and reported together with the specified stress or strain.

Second-degree blocking: An adherence of such degree that when surfaces under test are parted, one surface will be found to be damaged.

Secondary bonding: The joining of nonstructural members which are not primary load-carrying joints.

Self-curing: A term pertaining to an adhesive which cures without the application of heat; a term used in the rubber industry.

Self-extinguishing resin: A resin formulation which will burn in the presence of a flame but which will extinguish itself within a specified time after the flame is removed.

Self-ignition temperature: The temperature of a material at which spontaneous combustion takes place, when the temperature rises slowly.

Self-seal: An adhesive joint which is accomplished by coating both adherend surfaces and bringing them together under pressure; an elastomeric adhesive used on envelope flaps, box closures, etc.; an adhesive film which will bond only to itself.

Self-vulcanizing: A term pertaining to an adhesive which undergoes vulcanization without the application of heat.

Separate-application adhesive: An adhesive consisting of two parts, one part of which is applied to one adherend and the other part to the other adherend with the two being brought together to form a joint.

Separator sheet: A parting agent in film (sheet) form. Usually refers to the protective material covering one or both surfaces of adhesive in tape form.

Sequence number: Daily, consecutive numbers assigned to parts (and corresponding test assemblies) in the order of their release from the tape-assembly room, prior to crushing or bonding, for the purpose of ensuring chronological processing control.

Service conditions: The heat, cold, flexation, shock, impact, vibration, etc., that an adhesive will be subjected to in service.

Set: To convert an adhesive into a fixed or hardened state by chemical action (such as condensation, polymerization, oxidation, vulcanization, gelation, hydration) or by physical action (such as evaporation of volatile constituents). (See Dry and Cure.)

Setting temperature: The temperature to which an adhesive or an assembly is subjected to set the adhesive.

Setting time: The period of time required for an assembly to be subjected to heat or pressure, or both, in order to set the adhesive.

Shear modulus: The modulus of rigidity of honeycomb material, denoted as G_c. It is the ratio of the shearing stress to the shearing strain at low loads, or simply the initial slope of the stress/strain diagram for shear. Values are reported for both the L and W directions.

Shear strength: The strength properties, in pounds per square inch, of an adhesive or a honeycomb material when loaded to produce shear distortion of planes parallel to the plane of the adhesive bond or to the edge planes of the honeycomb. Shear strengths are strongest in the L or "ribbon" direction of the honeycomb where the continuous length of ribbon is parallel to the span of a beam. Shear strengths are weakest in the W or "transverse" direction of honeycomb, where the continuous length of ribbon is parallel to the width of a beam.

Shelf life: The period during which the manufacturer guarantees that an adhesive stored at some specified temperature will produce specified mechanical properties when used.

Shore hardness: A measure of the resistance of material to indentation by a spring-loaded plunger; used primarily with cure control of the elastomeric compounds.

Shortness: A qualitative term that describes an adhesive that does not string, cotton, or otherwise form filaments or threads during application.

Short-time exposure: A relative term, found in most military specifications for adhesives, used normally to designate an exposure period of 10 minutes' duration to any desired medium or condition.

Shot: One complete cure cycle.

Shot bag: A fabric bag filled with small lead beads; sometimes used for dead weight during adhesive cure.

Silicone: One of the family of polymeric materials in which the recurring chemical group contains silicone and oxygen atoms as links in the main chain. At present these compounds are derived from silica (sand) and methyl chloride. The present forms obtainable are characterized by their resistance to heat. Silicones are used in the following applications: (1) Greases for lubrication; (2) rubber-like sheeting for gaskets, etc.; (3) heat-stable fluids and compounds for waterproofing, insulating, etc.; (4) thermosetting insulating varnishes and resins for both coating and laminating.

Size: A chemical substance, such as rosin or a synthetic polymer, coated on an adherend surface to reduce water absorption, scuffing, oil penetration, etc.

Sizing: The process of applying a material to a surface in order to fill pores and thus reduce the absorption of the subsequently applied adhesive or coating; also, to otherwise modify the surface properties of the substrate to improve adhesion; also, the material used for this purpose. The latter is sometimes called Size or Sizing.

Slip: The physical property of an adhesive, referring to the ability to move or position the adherends after an adhesive has been applied to their surfaces.

Slippage: Undesired movement of adherends with respect to each other during the bonding process.

Slug: The structural member used in core areas and between faces of a sandwich panel for splicing and/or attaching the panel to support structure.

Softener: A material added to an adhesive film to prevent embrittlement of the adhesive layer; an additive to elastomeric adhesive films to increase their long-term flexibility properties.

Solids content: The percentage by weight of the nonvolatile matter in an adhesive.

Solvent: A chemical substance capable of dissolving another material; a liquid used to clean adhesive contamination from machine parts and for adhesive cleanup.

Solvent cement: A liquid adhesive utilizing an organic solvent, such as benzene, MEK, toluol, etc., as the vehicle for the deposit of the adhesive polymer.

Solvent reactivated: Describing an adhesive bonding method in which a resinous adhesive is deposited on one or both adherend surfaces, dried, and then reactivated just prior to bonding by the wiping or deposit of an adhesive solvent on the coated adherend surface.

Specific adhesion: Adhesion between surfaces which are held together by valence forces (of the same type as those which cause cohesion).

Specific gravity: The density (mass per unit volume) of any material divided by that of water at a standard temperature, usually 4°C; since water's density is nearly 1.00 g/cc, density in g/cc and specific gravity are numerically nearly equal.

Splice line: The boundary between core details that are bonded together with a structural adhesive.

Spread: The quantity of adhesive per unit joint applied to an adherend, usually expressed in pounds of adhesive per thousand square feet of joint area. (1) Single spread refers to application of adhesive to only one adherend of a joint; (2) double spread refers to application of adhesive to both adherends of a joint.

Stabilized core: (1) Core in which the cells have been filled with a specified reinforcing material for the purpose of supporting the cell walls during machining; (2) core in which the cell walls have been reinforced with a specified reinforcing material.

Stabilized honeycomb compressive strength: The compressive strength of honeycomb materials for which the plane surface of the test specimens has been stabilized with either a plastic resin or by the attachment of facings. The scatter of test results is reduced, and the compressive strength of a honeycomb tested in the stabilized condition is generally higher than the same honeycomb tested in the bare condition. The stabilized test more closely approximates the honeycomb compressive strength expected in a sandwich application.

Starved joint: A joint which has an insufficient amount of adhesive to produce a satisfactory bond.

Static fatigue: Failure of a part under continued static load; analogous to creep-rupture failure in metals testing, but often the result of aging accelerated by stress.

Static modulus: The ratio of stress to strain under static conditions. It is calculated from static-strain tests in shear, compression, or tension; expressed in psi unit strain.

Static-stress: A stress in which the force is constant or slowly increasing with time.

Stencil: An adhesive application method, involving the deposit of adhesive in predetermined patterns; applying adhesive through stencil cutouts; a machine component which prints the adhesive layer onto the adherend surface in a preselected pattern.

Storage life: The period of time during which a packaged adhesive can be stored under specified temperature conditions and remain suitable for use. Storage life is sometimes called "shelf life." (See Shelf life.)

Strain: The ratio of the extension to the original length of the measured elongating section of the test specimen, i.e., the change in length per unit original length.

Strain relaxation: Referred to as "Creep." (See Creep.)

Stress: The internal force per unit area which resists a change in size or shape of a body.

Stress concentration: The magnification of the level of an applied stress in the region of a void or inclusion in the bond line.

Stress-strain curve: Simultaneous readings of load and deformation, converted to stress and strain, are plotted as ordinates and abscissae to obtain a stress-strain diagram.

Stress wrinkles: Distortions in the face of the laminate caused by uneven web tensions, by insufficient adhesive setting speed, by selective absorption of the adherends, or by reaction of the adherends to materials in the adhesive.

Stringiness: The property of an adhesive that results in the formation of filaments or threads when adhesive transfer surfaces are (such as rolls or stencil) separated. (See Webbing.)

Stringing: A condition occurring during an adhesive transfer process, characterized by webbing or incomplete breakoff when an adhesive film is split between rolls or between a stencil applicator mechanism; the uneven transfer of an adhesive to an adherend surface.

Structural adhesive: An adhesive having sufficiently high mechanical properties, when cured, that it may safely be used for bonding parts in assemblies where human lives, valuable equipment, or both, are involved.

Structural-adhesive bonding: The art of using an adhesive to join two or more details by controlled processes in order to provide a high strength-to-weight ratio, excellent fatigue characteristics, and good aerodynamic smoothness in aerospace components.

Structural sandwich construction: A laminar construction comprising a combination of alternating dissimilar simple or composite materials assembled and intimately fixed in relation to each other so as to use the properties of each to attain specific structural advantages for the whole assembly.

Substrate: A material upon the surface of which an adhesive-containing substance is spread for any purpose, such as bonding or coating. This term is used in a broader sense than the term adherend.

Surface tension: A property of liquid or solid matter due to unbalanced molecular forces near a surface and the measure thereof; an apparent tension in an actual nonexistent surface film associated with the capillary phenomena of adhesion and cohesion.

Supported adhesive film: An adhesive supplied in a sheet or in a film form with an incorporated carrier that remains in the bond when the adhesive is applied and used.

Syneresis: The exudation of small amounts of liquid by gels on standing.

Tack: The property of an adhesive that enables it to form a bond of measurable strength immediately after adhesive and adherend are brought into contact under low pressure.

Tack-free: An intermediate stage where a catalyzed resin has gelled, but remains workable. Surface contact will indent the resin surface but will not adhere to the contacting media.

Tack range: The period of time in which an adhesive will remain in the tacky-dry condition after application to an adherend, under specified conditions of temperature and humidity.

Tack stage: The interval of time during which a deposited adhesive film exhibits stickiness or tack, or resists removal or deformation of the cast adhesive.

Tackifier: A material, such as a rosin ester, added to synthetic resins or to elastomeric adhesives to improve the initial and extended range of the deposited adhesive film.

Tacky-dry: Pertaining to the condition of an adhesive when the volatile constituents have evaporated or been absorbed sufficiently to leave it in a desired tacky state for wet assembly; not related to "dry tack."

Tangent modulus: The slope of the line at any point on a static stress-strain curve expressed in psi per unit strain. This is the tangent modulus at that point in shear, extension, or compression.

Telegraphing: A condition in a laminate or other type of composite construction in which irregularities, imperfections, or patterns of an inner layer are visibly transmitted to the surface. Occasionally referred to as photographing.

Tensile strength: The pulling stress, in pounds per square inch, required to break a given specimen. Area used in computing strength is usually the original, rather than the necked-down area.

Tensile test: A test in which specimens are subjected to an increasing pull until they fracture.

Thermal conductivity: Ability of a material to conduct heat; physical constant for quantity of heat that passes through unit cube of a substance in unit of time when difference in temperature of two faces is 1°.

Thermal decomposition: Breakdown of an adhesive caused by heat.

Thermal expansion (Coefficient of): The fractional change in length (sometimes volume, specified) of a material for a unit change in temperature. Values for plastics range from 0.01 to 0.2 mils/in., °C.

Thermal stress cracking: Crazing or cracking of adhesives as a result of exposure to elevated temperatures.

Thermoplastic (adjective): Capable of being repeatedly softened by heat and hardened by cooling.

Thermoplastic (noun): A material which will repeatedly soften when heated and harden when cooled.

Thermosetting: Having the capability of undergoing a chemical reaction by the action of heat, catalysts, ultraviolet light, etc., leading to a relatively infusible state.

Thinner: A volatile liquid added to an adhesive to modify the consistency or other working properties. (See Diluent and Extender.)

Thixotropic: Materials that are gel-like (paste), but may become fluid when agitated or heated.

Throwing: A characteristic behavior of some adhesives occurring when they are transferred from rollers or rotary stencil mechanism wherein, due to peripheral speed, small droplets of adhesive are thrown from the roller or stencil.

Tooling: Molds, jigs, or fixtures designed for assembly or curing.

Toughness: The energy required to break an adhesive material (cohesively).

Translucent: Adhesives that allow the passage of light but are not transparent.

Trapped-air pocket: An internal pocket of air or gas in a cured adhesive; but not considered "porosity."

Trowel: To spread a high-viscosity adhesive material by spatula, knife, spreader bar, etc.

Tunneling: A condition occurring in incompletely bonded laminates, characterized by release of longitudinal portions of the substrate and deformation of these portions to form tunnel-like structures.

Ultimate elongation: The elongation at rupture.

Ultimate tensile strength: The ultimate stress of an adhesive joint at rupture when destructively tested in tension.

Ultrasonics: The study, effects, and application of sound vibrations at and beyond the limit of the audible range of frequencies.

Ultra-short time exposure: A relative term found in many military specifications for adhesives, used normally to designate an exposure period of 2 minutes' duration to any desired medium or condition.

Undercure: Results of a shortened cure cycle or the applied temperature is not sufficient to polymerize the adhesive.

Uniaxial load: A condition whereby an adhesive is stressed in only one direction.

Unsaturated compounds: Any compound having more than one bond between two adjacent atoms, usually carbon, and capable of adding other atoms at that point to reduce it to a single bond.

Unsupported adhesive film: An adhesive supplied in sheet or film form without an incorporated carrier.

Urea formaldehyde resin: A synthetic resin derived from the condensation of urea (carbamide) with formaldehyde or its polymers.

Usable life: The maximum number of hours allowed to elapse between the time an adhesive is sampled for quality-assurance evaluation testing and the time the adhesive is cured.

Vacuum-bag bonding: A process for bonding with adhesives that utilizes a sealed plastic bag that applies the desired pressure to the adherends by vacuum pressure.

Validated adhesive: An adhesive which has satisfactorily passed its evaluation tests and has a current usable life. Each unit of validated adhesive must have a sticker attached to it showing the expiration time of the adhesive.

Vapor degreasing: A cleaning process that employs the hot vapors of a chlorinated solvent to remove soils—particularly oils, grease and waxes.

Veneer glue: An adhesive used to make plywood, laminates, or other articles that are end-products in themselves; used in the wood industry in contrast with "joint glues." (See Assembly adhesive.)

Viscosity: The ratio of the shear stress existing between laminae of moving fluids and the rate of shear between these laminae; internal frictional resistance of an adhesive to flow (directly proportional to the applied force).

Viscosity coefficient: The shearing stress tangentially applied that will induce a velocity gradient. A material has a viscosity of one poise when a shearing stress of 1 dyne/cm^2 produces a velocity gradient of 1 cm/sec^2.

Void: An air pocket or discontinuity in the bond line.

Volatiles: Gaseous materials in an adhesive formulation that are driven off or liberated during the curing reaction.

Vulcanization: A chemical reaction in which the physical properties of a rubber are changed in the direction of decreased plastic flow, less surface tackiness, and increased tensile strength by reacting it with sulfur or other suitable agents.

Warm-setting adhesives: Term that is used synonymously with "intermediate-temperature-curing adhesives."

Warp: (1) Distortion in an adhesive-bonded assembly after bonding; (2) the yarn running lengthwise in a woven plastic.

Water absorption: Ratio of the weight of water absorbed by an adhesive to the weight of the same material in a dry condition.

Weathering: The exposure of adhesives to an outdoor environment; accelerated weathering (simulated) tests under controlled laboratory conditions.

Webbing: Filaments or threads that may form when adhesive transfer surfaces (such as rolls or stencils) are separated. (See Stringiness.)

Wetting: The relative ability of a liquid adhesive to display interfacial affinity for an adherend and to flow uniformly over the adherend surface.

Wet strength: The strength of an adhesive joint determined immediately after removal from a liquid (most commonly water) in which it has been immersed under specified conditions of time, temperature, and pressure. Also, in reference to elastomeric adhesives, the joint strength, immediately after the joint has been assembled, while the adhesive is still wet.

Wood failure: The rupturing of wood fibers in strength tests on bonded specimens, usually expressed as the percentage of the total area involved which shows such failure.

Wood veneer: A thin sheet of wood, generally within the thickness range of 0.01 to 0.25 in., to be used as an outer surface of a laminate.

Working life: The period of time during which a two- or multipart adhesive—after mixing with catalyst, solvent, or other compounding ingredients—remains usable.

Working time: A measure of the interval of time during which an adhesive may be effectively applied to the adherend surface before adhesive setting retards the flow and application properties of the adhesive.

Yield point: The first stress in an adhesive (or adherend) less than the maximum attainable stress, at which an increase in strain occurs without an increase in stress.

Yield value (yield strength): The lowest stress (either normal or shear) at which a marked increase in deformation occurs without an increase in load.

Young's modulus: The ratio of the tensile stress to tensile strain below the proportional limit.

Index

Index